SPRINGER SERIES IN
MATERIALS PROCESSING

Springer
Berlin
Heidelberg
New York
Barcelona
Budapest
Hong Kong
London
Milan
Paris
Singapore
Tokyo

Bernhard Dischler   Christoph Wild (Eds.)

Low-Pressure Synthetic Diamond

Manufacturing and Applications

With 193 Figures and 26 Tables

Springer

Dr. Bernhard Dischler
Dr. Christoph Wild
Fraunhofer-Institut für Angewandte Festkörperphysik, Tullastrasse 72
D-79108 Freiburg, Germany
E-mail: Bernhard.Dischler@t-online.de
E-mail: wild@iaf.fhg.de

Series Editors:

Professor Dr. H. Warlimont
Institut für Festkörper- und Werkstofforschung e.V., Helmholtzstrasse 20
D-01069 Dresden, Germany

Professor Dr. E. Weber
University of California, Materials Science and Mineral Engineering
587 Evans Hall, Berkeley, CA 94720-1760, USA

ISSN 1434-9795
ISBN 3-540-63619-6 Springer-Verlag Berlin Heidelberg New York

Library of Congress Cataloging-in-Publication Data
Low-pressure synthetic diamond: manufacturing and applications/
[edited by] Bernhard Dischler, Christoph Wild.
p. cm. -- (Springer series in materials processing, ISSN 1434-9795)
Includes bibliographical references.
ISBN 3-540-63619-6 (acid-free paper)
1. Diamonds, Artificial. 2. Chemical vapor deposition.
I. Dischler, Bernhard, 1935- . II. Wild, Christoph, 1961- . III. Series.
TP873.5.D5L68   1998   666'.88--dc21   98-2582 CIP

Typesetting: Camera-ready from the editors
Cover concept: eStudio Calamar Steinen
Cover production: *design & production* GmbH, Heidelberg

SPIN: 10575641   57/3144 - 5 4 3 2 1 0 – Printed on acid-free paper

Preface

The synthesis of diamond by low-pressure methods has developed into a rapidly expanding field during the past 15 years. Therefore an up-to-date reference is needed, which provides practical information on the important aspects of CVD-diamond manufacturing and applications. The term "CVD diamond" refers to the Chemical Vapor Deposition (CVD) growth process used in low-pressure synthesis, in contrast to synthetic HPHT diamond, which is obtained at very High Pressure and High Temperature (HPHT) by a transformation of graphite into diamond.

The great variety of methods for the manufacturing of CVD diamond is impressive, especially in view of their late discovery. The present applications of low-pressure synthetic diamond exceed by far those of natural and HPHT synthetic diamond. For this monograph, a hopefully adequate selection had to be made and the reader may use the great number of references for further reading.

After a short introduction in Part I (Chap. 1), manufacturing methods are described in Part II. At the beginning are the four growth methods that have the potential for commercial CVD-diamond fabrication; that is, microwave-enhanced plasma deposition (Chap. 2), combustion flame deposition (Chap. 3), plasma jet deposition (Chap. 4), and hot-filament deposition (Chap. 5). The substrate temperatures normally used are in the range 700–1000°C, but for some applications much lower substrate temperatures (100–200°C) are required and have been realized (Chap. 6). Other deposition methods that use, for example, halogens or lasers are described in Chap. 7. The important issue of how heteroepitaxial or highly oriented growth of CVD diamond can be obtained is treated in Chap. 8.

In Part III, an overview of relevant applications of CVD diamond is given. Some of these have already entered the market; others are still in a prototype state. CVD diamond is used for thermal management (Chap. 9) or for optical (Chap. 10) and X-ray (Chap. 11) windows. The excellent mechanical properties of diamond are exploited for cutting tools (Chap. 12) and surface acoustic wave filters (Chap. 14). The great potential of diamond for electronic and electrical applications is described in Chap. 13 (Temperature and pressure sensors), in Chap. 15 (Electron emission from cold cathodes), in Chap. 16 (Ultra-violet and particle sensors), and in Chap. 17 (Electronic devices).

The difficult question of the direction in which the manufacturing and applications of CVD diamond will go in the future is addressed in Part IV, Chap. 18.

The editors acknowledge the valuable support of Professor P. Koidl and his CVD diamond group at the Fraunhofer-Institut für Angewandte Festkörperphysik in Freiburg.

Freiburg, February 1998
B. Dischler
C. Wild

Table of Contents

PART III Applications of CVD Diamond

PART IV Outlook

18. CVD Diamond in the 21st Century 363
P. Chalker and S. Lande

Part I

Introduction

1. CVD Diamond:
A New and Promising Material

Bernhard Dischler*

Fraunhofer-Institut für Angewandte Festkörperphysik,
Tullastrasse 72, D-79108 Freiburg, Germany

* Present address:
  P.O. Box 364, D-79003 Freiburg, Germany
  e-mail: Bernhard.Dischler@t-online.de

Springer Series in Materials Processing
Low-Pressure Synthetic Diamond Eds.: B. Dischler and C. Wild
© Springer-Verlag Berlin Heidelberg 1998

1.1 Introduction

The outstanding properties of diamond, including its hardness, chemical inertness, good optical transparency and high thermal conductivity, are well known. The revolutionary finding that diamond can be synthesized at low pressure by the Chemical Vapor Deposition (CVD) method has opened up new fields of technical applications for this interesting material. There exist considerable differences, both with respect to manufacturing and to applications, between low-pressure (CVD) synthetic diamond and High-Pressure/High-Temperature (HPHT) synthetic diamond. The synthesis of CVD diamond can be performed by a variety of methods using conventional laboratory equipment (Chaps. 2–8), while the synthesis of HPHT diamond requires special industrial equipment which is used by very few companies. Many quite different applications for CVD diamond are the object of intensive research and some applications (e.g. heat spreaders) have entered the market (Chaps. 9–17); in contrast, the diamond from HPHT synthesis is almost exclusively used as grit for mechanical applications. Of special interest are market niches, where CVD diamond offers "enabling technologies"; that is the manufacturing of new products such as optical and X-ray windows, special coatings and electronic devices. Only low-pressure synthesis can provide diamond in the shape of wafers and films. The unique possibility of growing layered diamond structures using CVD technology is exploited for adhesion layers on metal tools (Chap. 12) or for δ-doping in electronic devices (Chap. 17).

1.2 New Horizons in Diamond Synthesis

1.2.1 Outside the Phase Diagram

According to the carbon phase diagram, diamond is unstable with respect to graphite at temperatures below 1300°C and pressures below 40 kbar. The conditions for CVD growth of diamond are clearly outside the diamond region in the carbon phase diagram, and the possibility of growing diamond under these conditions was not anticipated. It is still hard to believe that, for example, diamond crystals can be grown using a welding torch (Chap. 3). The deposition of CVD diamond does not depend on the phase diagram; rather, it is governed by the laws of crystal growth from the gas phase, including nonthermal equilibrium and complex surface reactions (e.g. co-deposition of diamond and graphite, and selective etching of graphite by atomic hydrogen or oxygen).

1.2.2 Chronology of CVD-Diamond Deposition

Some important steps towards the present CVD-diamond technology should be mentioned. First, there are the early tube oven experiments by W.G. Eversole from 1953 to 1962 [1.1] and by J.C. Angus et al. from 1966 to 1968 [1.2]. In both cases, diamond overcoats on diamond seeds were obtained, and the co-deposited graphite was removed using a cyclic growth etching procedure.

The research group of B.V. Deryagin, D.V. Fedoseev et al. investigated low-pressure diamond synthesis from 1956 on, and reported successful growth of diamond from the gas phase in 1969 by methods which they later described [1.3].

N. Setaka et al. initiated a research project on CVD-diamond synthesis at the National Institute for Research in Inorganic Materials (NIRIM) in 1974. The experimental details of CVD-diamond deposition with practical growth rates of the order of 1 μm h^{-1} were disclosed in 1981 for the hot-filament method [1.4] (see Chap. 5) and in 1982 for the microwave or radio-frequency activated plasma process [1.5] (see Chap. 2).

Additional methods of CVD-diamond deposition were found later. The use of an oxy-acetylene combustion flame was described in 1988 by Y. Hirose et al. [1.6] (see Chap. 3). In the same year, K. Kurihara demonstrated diamond growth in a direct current (DC) arc plasma jet [1.7] (see Chap. 4).

In the meantime, the above methods for CVD-diamond synthesis have been further developed (see Chaps. 2–5) and others have emerged (see Chaps. 6–8).

1.2.3 High Deposition Rates

In the early experiments [1.1, 2], the growth rates were much too low for practical purposes. Deposition rates of 1 μm h^{-1} [1.3, 4] already marked considerable progress. The limitations inherent in the hot-filament and microwave-plasma methods prevent faster growth than ca. 5–10 μm h^{-1} with these systems. Growth rates an order of magnitude higher (> 40 μm h^{-1}) are possible with the combustion flame process (see Chap. 3). Impressively high deposition rates (> 900 μm h^{-1}) are reported for the plasma jet method (see Chap. 4). However, those high deposition rates are limited to small areas (some mm^2). On large area substrates high-quality diamond films are usually deposited at growth rates below ~10 μm h^{-1} irrespective of the deposition technique.

1.2.4 Large-Area Deposition

An important achievement in recent years is the increase in substrate dimensions. It is now routinely possible to grow uniform diamond films at rates of 5 μm h^{-1} on 6 inch diameter wafers using the microwave-plasma method (see Chap. 2). Using a hot-filament array, homogeneous diamond deposition on rectangular substrates with dimensions of 70×30 cm² and with growth rates of up to 3 μm h^{-1} has been obtained (see Chap. 5). The flat flame burner (see Chap. 3) has been designed for deposition areas greater than 20 cm² (see Sect. 3.3.1).

1.3 Economics: Low Price
via Low-Pressure Diamond Deposition?

Considering the great variety of applications, it is evident that quite different qualities of CVD diamond are needed. The high-quality products are termed "electronic grade" (see Chap. 17) or "optical grade" (for high-power laser windows; see Chap. 10) while lower quality is sufficient for heat spreaders (see Chap. 9) and for mechanical applications (see Chap. 12). Therefore only some general trends will be addressed below.

1.3.1 Cost of Equipment

While inexpensive laboratory equipment is sufficient for CVD-diamond growth on small substrates, the cost of the factory equipment rises steeply with up-scaling to large-area deposition, because severe physical and technical problems must be solved. Since up-scaling is a prerequisite for economic production, a careful balance between high output and low investment has to be found.

1.3.2 Cost of Materials and Energy

An important factor in CVD-diamond production is the cost for materials and energy. These costs can be estimated for the next ten years [1.8]. For the methods described in Chaps. 2–5, these amount to 15–80% of the total cost. It is evident that, with the present technology, all cost-reduction schemes find their ultimate limit at these relatively high percentages. This is also part of the reason why synthetic HPHT diamond grit, with its cheap input material (graphite), has a price (< 1 \$/carat) that is considerably lower than that of CVD diamond (> 3 \$/carat).

1.3.3 Rapid Progress in Cost Reduction

The worldwide interest in CVD-diamond research, and more recently in production, has resulted in amazing innovations for the CVD process in a relatively short period (~15 years; see Sect. 1.2.2). The rapid progress is illustrated in Table 1.1 [1.9]. Within four years, the output (of one reactor) increased by a

Table 1.1: Example of cost reduction in CVD-diamond production [1.9].

Year	Growth rate (g/h)	Deposition cost reduction (\$/carat)
1994	0.3	100
1995	1.4	50
1996	4.0	12
1997	> 6.0	3

factor of 20 while the cost decreased by a factor of 33. However, the production price of 3 \$/carat is now approaching the ultimate limit described in Sect. 1.3.2. This table also demonstrates the importance of a high production output for cost reduction. A similar trend is observed for synthetic HPHT diamond grit, where from 1960 to 1990 the production increased by a factor of 40 and the price decreased by a factor of 20 [1.8].

1.4 Conclusions and Outlook

The further development of low-pressure synthesis of CVD diamond is a great challenge. The transition from the research laboratory to the manufacturing level is not straightforward, but it has been successfully undertaken by several companies. Considering the enormous progress during the past 15 years, further steps forward – or even a major breakthrough – might be expected. The final result of such a substantial progress would be the economic mass production of CVD diamond.

References

1.1   W.G. Eversole, US Patents 3 030 287 and 3 030 188 (1962)
1.2   J.C. Angus, H.A. Will, and W.S. Stanko, J. Appl. Phys. **39**, 2915 (1968)
1.3   B.V. Deryagin and D.V. Fedoseev, Sci. Am. **233**, 102 (1975)
1.4   S. Matsumoto and N. Setaka, 8th Japan Carbon Soc. Fall Meeting (1981)
1.5   M. Kamo, M. Tsutsumi, Y. Sato, and N. Setaka, 43rd Japan Appl. Phys. Soc. Fall Meeting (1982)
1.6   Y. Hirose and N. Kondo, Japan Appl. Phys. Soc. Spring Meeting (1988)
1.7   K. Kurihara, K. Sasaki, M. Kawarada, and N. Koshino, Appl. Phys. Lett. **52**, 437 (1988)
1.8   J.V. Busch and J.P. Dismukes; in Synthetic Diamond: Emerging CVD Science and Technology, ed. K.E. Spear and J.P. Dismukes (Wiley, New York, 1994), pp. 581–624
1.9   S. Lande, Paper 12.1 at Diamond '96 Conference, Tours (France), 8–13 Sept. (1996)

Part II

Methods
of CVD-Diamond Production

2. Microwave-Plasma Deposition of Diamond

Evelio Sevillano

Applied Science and Technology, Inc. (ASTeX),
35 Cabot Road, Woburn, MA 01801, USA
e-mail: Evelio_Sevillano/ASTEX@astex.com

Springer Series in Materials Processing
Low-Pressure Synthetic Diamond Eds.: B. Dischler and C. Wild
© Springer-Verlag Berlin Heidelberg 1998

2.1 Introduction

Microwave-plasma enhanced chemical vapor deposition (MPECVD) is among the most widely used techniques for diamond growth from the gas phase. From its inception in the early 1980s [2.1], the technique has found a lot of success because of its simplicity, flexibility, and the early commercial availability of reactors from New Japan Radio Corporation based on the NIRIM (National Institute for Research in Inorganic Materials) work. The reactor developed at NIRIM consisted of an evacuated quartz tube which was inserted through a waveguide. A hydrogen–methane plasma was formed inside the tube and was used to create the proper gas chemistry for diamond growth. Research programs that used this technique flourished in Japan, and the first attempt at a production unit using NIRIM type reactors was made [2.2].

Although diamond has many unique properties with a potentially broad number of applications, much of the early interest in the growth of CVD diamond centered around electronics applications. Diamond's high band gap, high charged particle mobilities, and the potential for large area that the CVD technique would allow ignited this research field. In the late 1980s, new commercial microwave-plasma reactors were developed by ASTeX, which allowed large-area deposition [2.3] while removing the quartz walls away from the vicinity of the plasma and substrate to reduce contamination. Bias-enhanced nucleation techniques were developed using microwave reactors [2.4], which allowed diamond growth on semiconductor quality Si wafers without any wafer pretreatment. "Heteroepitaxial" diamond growth in near epitaxy with the underlying Si wafer was reported [2.5, 6] and relatively high mobilities in homoepitaxial films have now been achieved [2.7].

Impressive progress in MPECVD capabilities over the past few years has kept the technique at the forefront of commercial applications. In the early 1990s, progress in the power handling capabilities of microwave reactors led to the development of high power density processes. A new plasma chemistry regime, characterized by intense light emission from C_2 species now present in the plasma discharge, was achieved. These new processes led to a significant increase in the deposition rate, with a significant improvement in the material quality. Thick films (several hundreds of microns thick) could now be produced in a matter of hours [2.8, 9] using the microwave technique and a significant reduction in the production costs of the material became a reality. This process has now been used to produce some of the highest-quality optical CVD diamond reported [2.10–12], with transmission properties that closely match those of type IIa diamond. With the same basic process, diamond grown in a microwave-plasma using isotopically pure methane also holds the record for the highest recorded room-temperature thermal conductivity in CVD diamond (26 W/cm K along the crystal growth direction and 21.8 W/cm K across) [2.13].

In this chapter, we review the fundamental physical processes in microwave-plasma discharges. A summary of the state of the art in microwave reactors and some of the evolution to this point is presented. We briefly review some of the capabilities and applications of the technique in achieving specific material properties. Reference is made to other chapters that cover many of these subjects in more detail. We conclude with the fundamentals of an economic analysis, which will prove useful when a commercial evaluation of any deposition method is necessary.

2.2 Fundamentals of Microwave-Plasma Discharges

In a microwave-plasma diamond deposition reactor, process gases are introduced into a reactor chamber, which contains the substrate to be coated. Microwave power is then coupled into the chamber through a dielectric window in order to create a discharge. The chamber is an integral part of an electromagnetic cavity in which the microwave electric field profiles are such that the discharge location can be reproducibly controlled. Typically, the substrate to be coated with diamond is immersed into the plasma within this cavity. The microwaves couple energy into the electrons, which in turn transfer their energy to the gas through collisions [2.14–17]. As the process gas is heated, chemical reactions in the gas phase lead to the formation of diamond precursors which impinge on the substrate surface. If the surface conditions are right, diamond is grown on the substrate.

The steady state that is achieved is the result of the coupling between the global cavity electromagnetic fields and the bulk fluid flow (diffusive and convective) through a complicated set of local processes. These local processes include electron heating by the microwave electric fields and the subsequent electron energy transfer to the neutral gas, which leads to heating, dissociation of the molecules, and the formation of active species. As a result of all these processes, microwave energy is transferred to the fluid. However, the cavity electromagnetic fields are in turn modified by the plasma profile, which is dependent on the fluid flow. A complete solution of this problem requires the coupling of these pieces in a self-consistent manner [2.14,15].

2.2.1 The Boltzmann Equation and the Cold Plasma Dielectric

Plasmas typically used for MPECVD of diamond are weakly ionized and highly collisional. This combination of properties makes these plasmas very different from the widely studied fusion or interstellar plasmas. To understand the relevant physical processes that take place in these plasmas, it is useful to establish some of their basic parameters.

We consider the case of a weakly ionized cold plasma in a background of neutral gas. The plasma is created by coupling microwaves into a cavity. At microwave frequencies, the ions are too massive to be able to respond to the rapid time variation of the electric fields. Only the electrons can gain energy through their interaction with the local electric field. This energy coupling is the most important role of the electrons in the plasmas used for CVD diamond.

Electrons are characterized by their energy distribution function, $f(\mathbf{r},\mathbf{v},t)$, which satisfies the collisional Boltzmann equation,

$$\frac{\partial f}{\partial t} + \mathbf{v} \cdot \nabla_{\mathbf{r}} f + \frac{\mathbf{F}}{m_e} \cdot \nabla_{\mathbf{v}} f = \frac{\partial f}{\partial t}\bigg|_c$$

where $\mathbf{v}$ is the electron velocity vector, $\mathbf{F}$ is the Lorentz force acting on the electrons, m_e is the electron mass and the term on the right-hand side represents the effects of collisions. Macroscopic variables such as density, mean velocity, energy density, and so on, are obtained from the velocity moments of the distribution function averaged over velocity space. Conservation equations are obtained in the same manner. A self-consistent solution of the Boltzmann equation for the electron and Maxwell's equations for the electromagnetic fields in the cavity is needed to properly model the plasma discharge.

The response of the electrons to the microwave fields can be obtained from the momentum conservation equation. This equation is derived from multiplying the Boltzmann equation by $\mathbf{v}$ and integrating over velocity space. The details of how this is done can be found in standard plasma physics textbooks [2.18]. The result for an unmagnetized plasma for which $\mathbf{B} = 0$ and $\mathbf{F} = -e\mathbf{E}$ (e is the charge of the electron) is as follows:

$$m_e \frac{d n_e \mathbf{u}_e}{dt} = -e n_e \mathbf{E} - \nabla p - v_{e,0} n_e m_e \mathbf{u}_e \tag{2.1}$$

where n_e is the electron density, $\mathbf{u}_e$ is the mean electron velocity, and we have assumed isotropic pressure so that only a ∇p term is present. The collision term is expressed in terms of an effective electron-neutral momentum transfer frequency $v_{e,0}$.

When the time variation of the fields is of the form $e^{-i\omega t}$ (e.g. microwave excitation at frequency ω) then, using Maxwell's equations, the current density $\mathbf{J}$ can be expressed as a constant times the electric field $\mathbf{E}$:

$$\mathbf{J} = i\omega(1 - \varepsilon)\mathbf{E} \tag{2.2}$$

where ε is the dielectric constant.

For a cold plasma where $p = 0$, (2.1) can be written as

$$-i\omega m_e n_e \mathbf{u}_e = i\omega e n_e \mathbf{E} - v_{e,0} n_e m_e \mathbf{u}_e$$

Since the current is only carried by the electrons,

$$\mathbf{J} = -e n_e \mathbf{u}_e$$

and, therefore, the frequency-dependent dielectric constant is given by

$$\varepsilon = 1 - \frac{\omega_p^2}{\omega(\omega + i v_{e,0})} \tag{2.3}$$

$$\varepsilon = 1 - \frac{\left(\dfrac{\omega_p}{\omega}\right)^2}{1 + i\dfrac{v_{e,0}}{\omega}}$$

where $\omega_p^2 = 4\pi n_e^2/m_e$ is the square of the plasma frequency. Equation (2.3) is the well known cold plasma dielectric response. The density at which the plasma frequency is equal to the frequency of the microwaves is known as the cutoff density, n_c. For waves at 2.45 GHz, this density is 7.4×10^{10} cm^{-3}, and for 915 MHz it is 1.0×10^{10} cm^{-3}.

Equations (2.2) and (2.3) have an important physical meaning. For collisionless plasmas, where $\nu_{e,0} = 0$, the dielectric constant is real. The current induced in the plasma given by (2.2) is therefore exactly out of phase with the propagating electric field. When the plasma density is small, the dielectric constant is near one and the waves are largely unaffected by the plasma. However, as the plasma density increases and approaches the cutoff density, the current induced in the plasma effectively shields the waves, and therefore the propagating fields are strongly refracted away from the plasma. Also note that, in such a collisionless plasma, there is no power dissipation. Of course, no such plasmas exist in nature.

In collisional plasmas, the ratio of the collision frequency to the microwave frequency determines the size of the imaginary part of the dielectric constant. This corresponds to a resistive response, in which case there is power dissipated in the plasma. When the collision frequency is high compared with the microwave frequency, the plasma acts like a resistive load. The electric fields can penetrate in the plasma and are strongly attenuated. This highly collisional regime is not common but we shall see that this is the regime that best describes the plasmas typically used for diamond CVD, especially at high power densities.

2.2.2 Collisions and Ionization Fraction

We now proceed to calculate the ratio of the collision frequency to that of the microwave field for microwave discharges of interest in CVD diamond. The collision frequency of electrons with neutrals is given by

$$\nu_{e,0} = n_0 \sigma_0 \sqrt{kT_e / m_e}$$

where n_0 is the neutral density, and σ_0 is the cross-section, typically 1×10^{-15} cm^2 and weakly dependent on the electron temperature T_e [2.19]. In practical units (for $T_e = 1$ eV), this equation can be written as

$$\nu_{e,0} = 4 \times 10^{11} \, p(\text{Torr}) / T_g(\text{K}) \tag{2.4}$$

where T_g is the neutral gas temperature.

Throughout this chapter, we will be considering typical low and high power density discharges. For the low power density discharge, we assume that the microwave power is such that, at 30 Torr pressure, the gas temperature is of the order of 2000 K. In a typical ASTeX reactor operating at 2.45 GHz (see Sect. 2.3.2) these conditions are typically attained at 1000 W of microwave power.

The high power density discharge is assumed to be at 180 Torr and 4000 K. These conditions are typical of discharges in large reactors operating at 915 MHz (see Sect. 2.3.7), where the power can be as high as 90 kW.

Using (2.4), the collision frequency for the low power density discharge is calculated to be 6×10^9 sec^{-1}. For the high power density discharge, the collision frequency is 1.8×10^{10} sec^{-1}. These rates are to be compared with the angular microwave frequencies which are 1.5×10^{10} sec^{-1} and 5.7×10^9 sec^{-1} for 2.45 GHz and 915 MHz respectively. The collisionality ratio in (2.3) therefore changes significantly between these two types of discharge. In the low power density discharge the ratio $v_{e,0}/\omega = 0.4$, and in the high power density case $v_{e,0}/\omega = 3.2$, almost an order of magnitude larger. This relatively large ratio of the collision frequency to the microwave frequency indicates how plasmas at high power densities can be highly dissipative and very efficient absorbers of the microwave power. As the frequency of microwaves is lowered in order to scale up reactor size, this ratio is even larger, which leads to a regime that is not typically studied or well understood in the standard plasma physics literature.

Even in the case in which the plasma effectively shields the electric fields, modeling indicates that the electron density can reach a few times the cutoff frequency [2.15]. The ionization fraction, given by the ratio of n_e/n_0 is therefore a few times the ratio n_c/n_0. This ratio is given in practical units by:

$$\frac{n_c}{n_0} = 1.3 \times 10^{-27} \frac{v^2(\sec^{-2})T_g(\mathrm{K})}{p(\mathrm{Torr})} .$$

At 2.45 GHz microwave frequency, the ionization fraction for the typical low energy density plasma conditions above is given by $n_c/n_0 = 5.2 \times 10^{-7}$. For high power density plasmas, the corresponding value is $n_c/n_0 = 2.4 \times 10^{-8}$. Note that the ionization fraction in all cases is much less than one, even when the plasma density is ten times the cutoff frequency. Indeed, typical CVD-diamond plasmas are very weakly ionized, and the role of the electrons is merely to provide a very efficient method to transfer microwave energy to the gas.

2.2.3 Modeling Results

So far, we have given the most elementary details on the physics of microwave absorption in plasmas relevant to CVD diamond. More detailed numerical calculations have been carried out which include self-consistent calculations of the electromagnetic fields in real reactors and their coupling to the fluid flow. In these models, the electron distribution function is calculated from the local value of the electric fields. However, the details of these calculations and the simplifications needed to keep the calculation viable are beyond the scope of the present chapter. The reader is referred to references [2.14–17] for more details. However, such calculations provide very valuable insights into the understanding of the energy transfer in the hydrogen-dominated plasmas most commonly used in CVD diamond. The results show that for low-energy electrons ($E < 0.5$ eV), the dominant energy loss mechanism is through the excitation of rotational levels of the hydrogen molecule. For medium-energy electrons (0.5 eV $< E < 8.5$ eV), the dominant loss is through vibrational excitation of the hydrogen molecule. Because of the shape of the distribution function, this is the dominant energy-loss channel for the electrons. Higher-energy electrons ($E > 8.5$ eV) lead to molecular dissociation of the hydrogen molecule, for which the energy for the transition from the ground state to the triplet state that is repulsive is 8.5 eV. The dissociation leads to the production of two hydrogen atoms with an energy of about 2 eV each (Franck–Condon atoms). Finally, ionization of the molecule and atom have thresholds of 15.4 eV and 13.6 eV respectively.

As soon as the microwave power is turned on and the electric fields are high enough for plasma breakdown, a small number of electrons with energies high enough for ionization of the gas is created. As the plasma is formed, the electron density increases rapidly and their energy distribution quickly relaxes as a result of collisions with the neutral gas. At these early times during the discharge, electronic dissociation dominates. However, after just a few milliseconds into the discharge, the gas temperature has increased significantly because of the efficient energy coupling into the vibrational levels. Once gas temperatures of 2000–3000 K are reached, thermal dissociation of the hydrogen molecule becomes more important than dissociation. Thermal dissociation leads to clamping of the average electron temperature at a relatively low value (1–2 eV).

The situation is very different in atomic gas discharges (e.g. argon) where higher values of the electron temperature are typically attained. Such high electron temperatures lead to the formation of streamers at high neutral pressures. As a consequence, an argon plasma in a cavity designed to operate with hydrogen can have a completely different behavior as compared to hydrogen [2.20]. Small amounts of hydrogen added to the gas mixture is an effective method of avoiding undesirable plasma behavior. On the other hand, discharges with other molecular gases behave like those with hydrogen [2.21,22].

2.2.4 Plasma Chemistry

Undoubtedly, MPECVD is the most flexible technique among those used for CVD-diamond growth in terms of the variety of gas precursors that can be used for diamond growth. In addition to the standard H_2–CH_4 chemistry, almost every conceivable gas mixture that may or may not include H_2 as an inlet gas [2.22] can be used. The use of liquid precursors is also common [2.21]. However, clearly not every possible gas combination will lead to successful diamond growth. To identify and numerically follow all possible plasma chemistries is nearly impossible, because most of the required rate coefficients are not known as a function of temperature. As a result, only relatively simple chemistries can be numerically modeled.

Fortunately, the development of the triangular C–H–O diagram for CVD diamond by Bachmann [2.21] has proven to be a very useful tool in process development when new and unusual gas mixtures are used. Each inlet gas flow combination that contains C, H, and O can be represented by a point within the triangle (for details on how this is done, see [2.21]). From a plot of a large number of CVD-diamond deposition experiments, it became clear that diamond growth occurs only within a relatively narrow range of chemistries, most of which are located near the CO tie line. In the C–H–O triangle, this line extends from the H corner to the middle of the opposite side of the triangle (midway between the C-corner and the O-corner). Below the CO tie line, diamond does not grow (there is too much etching), while too far above the CO tie line only soot grows. Although much of the original region of growth has remained unchanged over time, high power density discharges (Sect. 2.3.5) expanded the available growth space along the H–C line.

It is important to realize that having the correct gas composition is a necessary but not sufficient condition for diamond growth. The correct surface temperature, surface material, surface pretreatment, and so on are also required for successful growth.

2.3 Types of Microwave-Plasma Reactors

The power, size, and deposition rate that can be achieved in microwave-plasma reactors have increased significantly over the past few years. This evolution will likely continue as new commercial applications of diamond appear in the marketplace. Because of limitations of space, we can only cover selected reactor concepts. These have been chosen based on their wide use or because they offer unique capabilities.

2.3.1 The NIRIM Reactor

The microwave-plasma reactor developed at NIRIM for the early work in MPECVD was very popular in the early 1980s among researchers in the field. A schematic of the reactor is given in Fig. 2.1. This reactor was especially attractive to researchers in this rapidly expanding field because of its simplicity of installation and operation, and its commercial availability from New Japan Radio in Japan. Given the broad interest in industry and academia which CVD diamond generated in Japan and elsewhere during the early 1980s, we estimate that several hundred such reactors must have been sold over a period of just a few years.

The reactor is very simple. A quartz discharge tube is inserted through the broad side of a fundamental mode rectangular waveguide appropriate for the propagation of microwaves at 2.45 GHz. The TE_{10} mode that propagates in this waveguide is parallel to the short dimension of the guide (y-direction) and varies sinusoidally along the broad length of the guide (x-direction), as follows:

$$E_y = E_0 \sin(\pi x \, / \, a) \qquad (2.5)$$

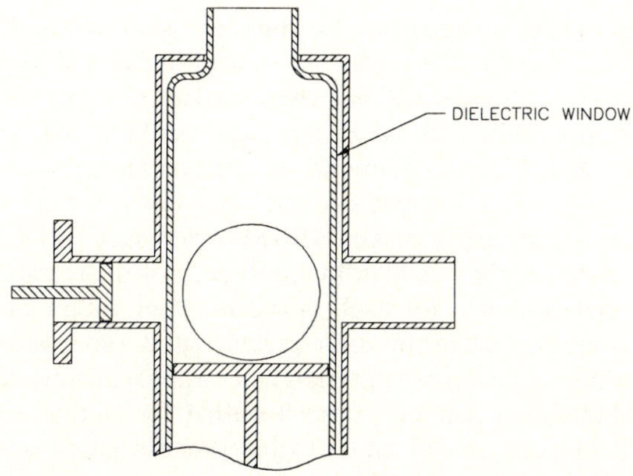

Fig. 2.1: A schematic diagram of a tube reactor. The dielectric window in this case is a quartz tube in which the plasma discharge is maintained.

where a is the broad dimension of the guide. This electric field is independent of y. The field is peaked on the center of the guide ($a/2$) but it is rapidly changing along this dimension.

Along the propagation direction and beyond the discharge tube location, the waveguide is terminated by either a fixed short (located at a location $\lambda_g/4$, where λ_g is the microwave wavelength inside the guide)[1] or a sliding short which could be adjusted to maintain the plasma location in the middle of the discharge tube. The choice of the position of the short is such that a maximum of the microwave electric field is obtained at the plasma location. A tuner is placed between the microwave source and the discharge tube to ensure that all the power is absorbed in the plasma load.

The substrate to be coated is introduced from the bottom of the discharge tube using a dielectric rod to prevent microwave leakage to the outside. Although this reactor design has enjoyed tremendous popularity, it has several disadvantages, as follows. (1) The deposition area is small (of the order of 1–2 cm^2) since the plasma size is limited by the rapid field variation (see (2.5)), the presence of the nearby walls, and the microwave wavelength. (2) Plasma etching of the nearby reactor walls often leads to contamination of the diamond deposited. (3) The substrate temperature is determined by plasma conditions and the position of the sample with respect to the plasma. Temperature control is therefore poor, since the temperature depends on the power and pressure used as well as on the exact position of the sample in the chamber. (4) The power that can be coupled into this

---

[1] The wavelength of the propagating electromagnetic fields within a waveguide is different than the vacuum wavelength. This is a result of the boundary conditions for the electromagnetic fields in the guide. For the TE$_{10}$ mode this relation is

$$\lambda_g = \lambda_0 / \sqrt{(1-(\lambda_0/2a)^2)} ,$$ where λ_0 is the vacuum wavelength and a is the long length of the guide.

configuration and the plasma operating pressure is limited by the possible destruction of the discharge tube. As a consequence, both the linear and mass deposition rates possible with this reactor are small and of the order of 0.5 mg/hr.

Even with all of these limitations, the NIRIM design was used for the first attempt at the commercialization of MPECVD diamond for the coating of cutting tools. Idemitsu Petrochemical of Japan commissioned the construction of production units which consisted of an array of six NIRIM-style reactors [2.2]. The microwave power from a single supply was distributed to each of the reactors. Gas flow and pressure control was provided for each of the reactors within each unit. A single cutting tool could be coated in each reactor at a time and a complicated loading and unloading mechanism was designed. Approximately ten such devices, or a total of 60 individual reactors, were installed for cutting tool production. Idemitsu pioneered the use of CO and CO_2 as process gases, in a unique departure from the standard methane and hydrogen mixtures of the time. Unfortunately, such an approach to commercialization was likely to fail. Vacuum leaks were common with frequently required discharge tube exchanges. Achieving temperature and process reproducibility reactor to reactor was a challenge, especially in a design in which substrate position determines its temperature. Manpower requirements to sustain production were significant and only a few tools could be coated simultaneously. As a result, production costs per tool were too high, which prevented commercialization. In the early 1990s Idemitsu discontinued all efforts in CVD diamond.

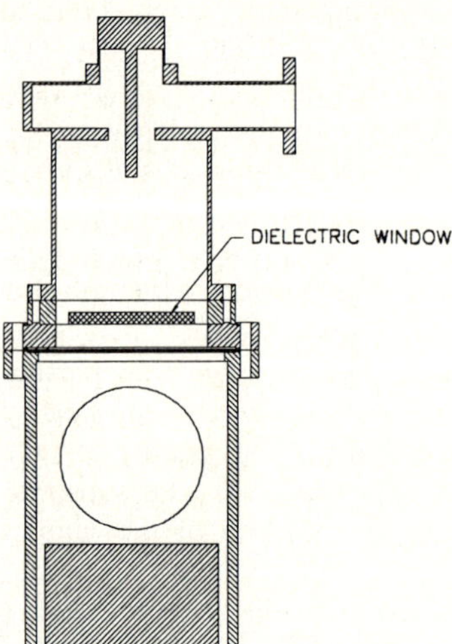

Fig. 2.2: A schematic diagram of a high-pressure plasma source. Microwaves are mode converted from the TE_{10} to the TM_{01} mode in the waveguide structure shown above the reactor chamber. A stable plasma discharge is generated above the substrate stage, capable of holding substrates up to 10 cm in diameter.

2.3.2 The High-Pressure Microwave Source

In the late 1980s, Bachmann and Smith [2.3] set out to develop a microwave reactor that would soon overcome many of the limitations of the NIRIM design. This basic design, shown in Fig. 2.2, has been commercialized by Applied Science and Technology, Inc. (ASTeX). In this reactor design, microwaves are coupled into a water-cooled metal cavity through a dielectric window (quartz). The inner chamber diameter is chosen so that only one microwave radial mode can be sustained inside the cavity at 2.45 GHz. The substrate is located on top of an inductively heated substrate stage, which allows substrate temperature control independently of plasma conditions. Substrates as large as 10 cm in diameter can be coated in such reactors and microwave powers of up to 1.5 kW can be coupled into the plasma.

To achieve azimuthally symmetric plasma deposition over the large area possible with this design, the fundamental propagation mode in rectangular guide, the TE_{10} mode, is converted using a patented design to the TM_{01} cylindrical mode. The TM_{01} mode was chosen because it is an azimuthally symmetric mode in which the z-component of the cavity electric field peaks on-axis (see, for example, [2.23]).

Since the substrate holder is a conductor, the boundary conditions for the vacuum electric fields require them to be normal to the surface at this location. Therefore, only the z-component of the electric field is large near the holder. As a result, a stable and reproducible plasma ball is formed on top of the substrate stage almost independently of the location of the stage. The plasma ball moves with the substrate stage, and therefore attempts to change the distance of the plasma to the substrate by moving the stage along the reactor length usually do not work. In addition to the inherent axial plasma stability of the design, there are other factors that enhance plasma formation on top of the stage. When the heated stage is operated, the local neutral density is reduced as the gas in the vicinity of the stage is heated. As the local particle density decreases it becomes easier to break down the plasma in this region.[2] In addition, depending on the shape of the substrate, there can be a local increase of the electric field near conductors placed on the stage that can also cause preferential breakdown at this location.

As a consequence of the axial periodicity of the electric fields, it is possible to obtain discharges near the dielectric window in these reactors. These occur when the pressure in the chamber is too low for a given microwave power. Figure 2.3 gives the operating boundary for a typical ASTeX chamber. The stability boundary is not sharp and depends on how the boundary is approached. This plasma stability behavior can now be understood given the minimum in the breakdown curve.[2]

---

[2] In hydrogen and many other gases, the breakdown electric field as a function of pressure displays a broad minimum that occurs at a pressure lower than the deposition pressures typically used in most CVD-diamond deposition. Therefore, as the pressure is reduced, the plasma is easier to break down. However, if the pressure is too low, the electric fields for breakdown increase again and the addition of a magnetic field is needed to break down, as in the case of magnetized plasmas (see Sect. 2.3.3).

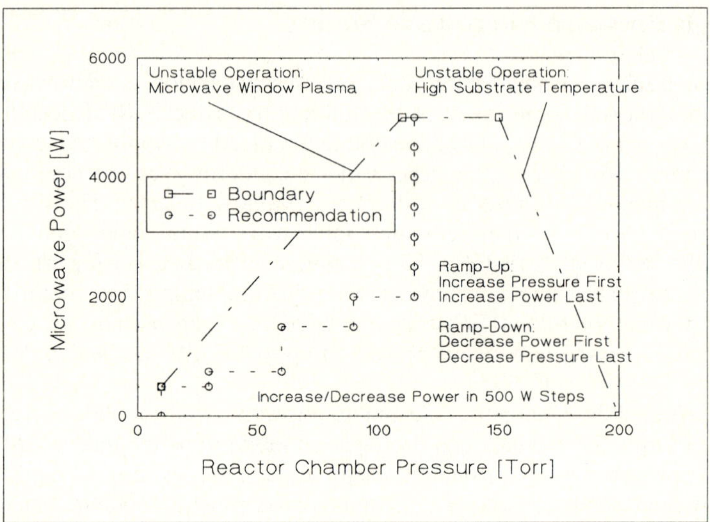

Fig. 2.3: The stable operating boundary for the reactor shown in Fig. 2.2. An operational path always within the reactor stable region avoids window discharges. The operating boundary is for the case of a reactor that can be operated at up to 5 kW further discussed in Sect. 2.3.5.

Standard plasma operation at high pressures occurs on the positive slope of this curve. This implies that as the pressure drops, the electric field required to sustain the discharge decreases. If the pressure drops too much for a given power, that is for a given electric field in the cavity, the field near the window can become high enough to cause a plasma to form there. The plasma is observed to jump to the window in this case. However, with proper operation of the reactor within the stability boundary, long-term (many hundreds of hours) and stable discharges are obtained with this type of reactor.

The choice of the TM_{01} mode for the cylindrical cavity reactor successfully allows the deposition of diamond films over large area with excellent azimuthal uniformity. However, good radial uniformity is more difficult to achieve. Measurement of the diamond deposition profile on a 10 cm diameter silicon wafer in this cylindrical reactor yields a Gaussian profile. This profile is consistent with that that is obtained from a point source of activated species located above the substrate. It should be obvious that a uniform deposition profile that extends much beyond the size of the plasma is not physically possible. When the stage is much larger than the plasma, the need to satisfy the electromagnetic boundary conditions at 0the stage makes it impossible to improve the radial deposition uniformity. Instead, the plasma stays a fixed distance from the substrate and moves with it. However, substrates that are smaller than the plasma can be used to greatly improve the radial deposition uniformity. In this case, the plasma coupling to the edge of the substrate changes with substrate height. For example, when substrates of the order of 5 cm in diameter are used, the deposition profile can be adjusted such that profiles which are peaked on-axis have a minimum on-axis, or are nearly flat can be obtained. The proper height choice results in a radial deposition uniformity much better than $\pm 10\%$.

2.3.3 Magnetized Plasma and the ECR Deposition Reactor

At neutral pressures below approximately 1 Torr, the addition of a magnetic field near the electron cyclotron resonance (ECR) at the microwave frequency greatly reduces the electric field required for plasma breakdown and sustainment [2.24]. Although the magnetic field effects on breakdown and sustainment are present at these pressures, the ratio $v_{e,0}/\omega$ determines whether the plasma is magnetized (there is an effect of the magnetic field in the breakdown and sustainment characteristics) or true ECR conditions are achieved. True ECR conditions are achieved when $v_{e,0}/\omega \ll 1$. In this case, the electron executes several orbits in the confining magnetic field before collisions that change its direction take place. Resonant absorption of the microwave energy takes place in the region in which the magnetic field is near the cyclotron resonance. At 2.45 GHz, the pressure required for true ECR conditions is near 1 mTorr. ECR plasmas are used in the etching and deposition of a wide variety of materials (e.g. silicon carbide, silicon nitride, diamond-like carbon, and many others). However, at pressures near 1 mTorr, diamond deposition has not been reported. Instead, magnetized plasmas at one or two orders of magnitude higher pressures have been used to obtain diamond growth.

Hiraki and co-workers have led the efforts in magnetized plasma diamond deposition [2.25]. Most experiments have been carried out near 100 mTorr, with an emphasis on the achievement of low-temperature deposition using oxygen-containing (CO and CO_2) gas mixtures with hydrogen. Plasmas at these pressures are more diffused and the deposition uniformity tends to be somewhat better than for higher-pressure plasmas. Parabolic diamond deposition profiles are observed instead of the Gaussian profile at high pressures. The major limitation of this technique in commercial applications is that the deposition rates are small and the diamond that is deposited consists of small grains. This limits the range of products for which this material can be used. Attempts at diamond deposition at pressures near 10 mTorr and temperature near 500 °C have yielded diamond films with deposition rates of the order of 0.04 µm/hr [2.26]. Gas mixtures of CO and H_2 were used. A more complete review of magnetized plasmas for CVD-diamond deposition at low temperature is given in Chap. 6 of the present volume.

2.3.4 The Microwave Torch Reactor

In the early 1990s there was a need to expand the diamond deposition area that could be coated uniformly. The main application was for infrared optical coatings of interest to the military. A remote plasma microwave system was developed by

ASTeX and named the Large Area Deposition System (LADS) [2.27]. There were several unique features in this device: (1) a new type of plasma torch applicator was developed capable of operation at 5 kW of microwave power, (2) the device relied on convection of the excited species to the substrates to be coated which were located remotely from the plasma; and (3) gas recirculation was used for the first time in CVD-diamond deposition systems.

At the high gas flow required for the proper operation of the LADS, a standard applicator using a quartz tube through a waveguide (similar to the NIRIM reactor) cannot be used. The plasma is extinguished as the gas flow through the tube increases. In order to stabilize the plasma, a bluff is inserted axially in the discharge tube at the center, just above the level of the waveguide. The bluff consists of a boron nitride cylinder around which the gas flows. The low-pressure region created downstream of the bluff stabilizes plasma formation even at very high power. The bluff also keeps the plasma from contacting the discharge tube wall. The substrates to be coated are located downstream from the applicator inside an oven. This allows good temperature control on the substrates. Two silicon wafers up to 20 cm in diameter can be placed in the oven in a position nearly parallel to the gas flow. As the reactant flow from the torch expands and impinges on the wafers, a rotating mechanism allows deposition uniformity of ±10% to be achieved. Small grain diamond is obtained with surface roughness of 30 nm without polishing. A closed circuit gas recirculation system is necessary for the large flows that are used in this device. As the gas goes through the plasma applicator and the deposition oven, a heat exchanger is used to cool the gas before a compression blower sends it back through the torch. A small amount of gas is constantly replenished, as in other standard reactors.

The diamond mass deposition rate that is achieved in the LADS device is of the order of 30 mg/hr. However, given the large area of the substrates used, the linear rate is limited to values below 0.2 μm/hr. This slow deposition rate and the fact that the only processes available lead to small grain diamond limited the application of the LADS device outside thin-film infrared optics.

2.3.5 The High Growth Rate Deposition Reactor

Angus et al. [2.28] pointed out a strong scaling of the mass deposition rate as a function of the power density for a variety of CVD-diamond deposition techniques. The scaling results at high power density were obtained from operation of acetylene and plasma torches (nonmicrowave) which operated at powers greater than 10 kW. In these units, very high deposition rates had been achieved [2.29,30]. Microwave reactors had been limited in power because of the lack of commercial availability of supplies at powers greater than 3 kW. As higher-power magnetrons became available, an effort to increase the power density in microwave discharges led to the discovery of a new plasma chemistry regime [2.8,9]. The new regime was characterized by a dramatic change in the light emission from the plasma and an increase in the diamond deposition rate of over one order of magnitude. Linear

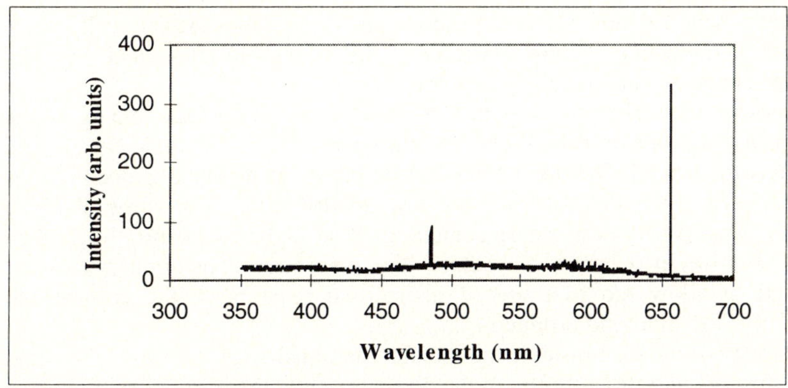

Fig. 2.4: The optical emission spectrum from a low power density microwave plasma at 1.5 kW, 30 Torr, and 1% methane in hydrogen.

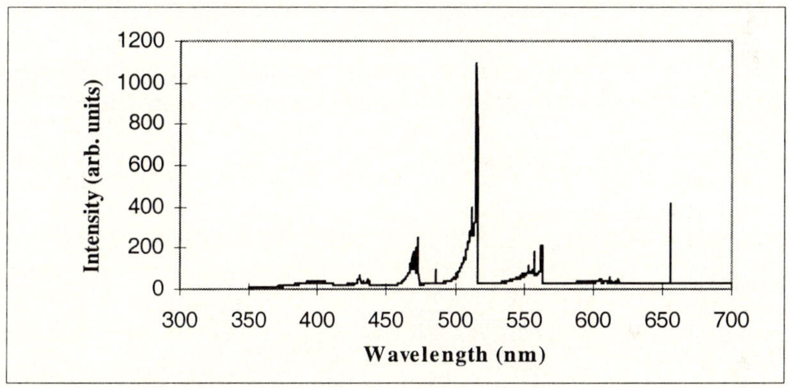

Fig. 2.5: The optical emission spectrum from a high power density microwave plasma at 8 kW, 110 Torr, and 5% methane in hydrogen.

deposition rates increased from approximately 1 µm/hr to over 10 µm/hr. The total mass deposition rate increased from about 6 mg/hr in standard reactors at 1.5 kW to over 60 mg/hr at 5 kW. The dramatic change in the emission signature of the plasma can be seen by comparing the spectra shown in Figs. 2.4 and 2.5. Figure 2.4 shows the emission spectrum from a discharge at 50 Torr, 1.5 kW, and 1% methane in hydrogen. Such discharge conditions are typical of diamond growth at low power densities. Only the H_α and H_β atomic hydrogen lines are clearly seen. The H_α line at 656.3 nm accounts for the characteristic pinkish color of the discharge. Figure 2.5 shows the spectrum from a discharge at 110 Torr, 8 kW, and 5% methane in hydrogen. Strong emission bands from C_2 molecules are observed at wavelengths near 470 nm, 510 nm, and 570 nm. These bands are the well known Swan bands of molecular carbon. Because of the strong emission near 510 nm, these plasmas have a characteristic green color.

At the new powers and pressures in these discharges, the power density on the substrate is of the order of 100 W/cm^2 and requires that the substrate be actively cooled instead of heated. Diamond films can be grown at methane concentrations that would previously yield soot or very low quality films. The films obtained show higher phase purity (very sharp Raman lines) at higher deposition rates than previously possible in microwave reactors. The new mass deposition rates were high enough to allow the first use of microwave reactors in the commercial production of CVD-diamond products.

The intensity of the C_2 emission and the diamond quality of the films deposited depend on the concentration of CH_4 in the discharge. It is therefore possible to use the emission signature from the plasma to control the quality of the diamond film that is grown [2.31]. This concept is shown in Fig. 2.6. A set of light-detecting diodes, each tuned to a different emission line through the use of interference filters, is connected to logarithmic amplifiers (used to increase the intensity bandwidth of the detectors). The output signal from the desired line is compared to a reference and the difference signal is used to drive the CH_4 flow from a mass flow controller. This inexpensive, *in-situ* control technique can be used very effectively in process transfer and reproducibility among production reactors.

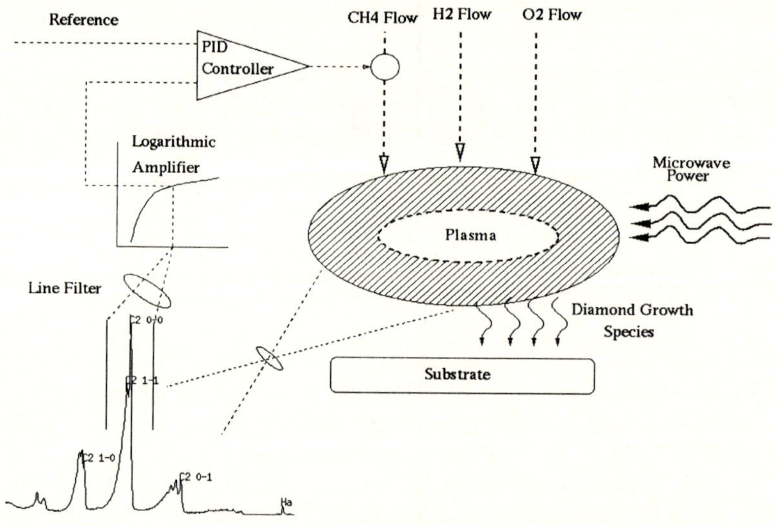

Fig. 2.6: A schematic diagram for active feedback control based on light emission from the microwave discharge. A line filter tuned to a desired line is fed to a comparator and the difference signal is used to drive the flow from a mass flow controller.

2.3.6 The Overmoded Cavity Plasma Reactor

With the development of high growth rate reactors, the power that could be coupled into fundamental cavity reactors operating at 2.45 GHz was near its limit. More aggressive cooling of the substrate would be necessary to further increase the deposition rate. It was not clear that this would be easily accomplished. In addition, as new applications of CVD diamond were introduced, a need for larger deposition areas and more economical diamond developed. A new reactor concept with a complete departure from the standard microwave coupling was developed [2.32]. The reactor cross-section is shown schematically in Fig. 2.7. The cavity radial dimensions are such that both the fundamental TM_{01} and the next radial mode TM_{02} are possible within the structure. Excitation of nonazimuthally symmetric modes is avoided by carefully maintaining symmetry in the launching structure.

Coupling into the cavity is done as follows: power traveling through a fundamental mode waveguide (not shown) is coupled to the bottom of the deposition chamber using a TEM transmission section which connects to the substrate holder. As the field enters the microwave cavity it travels radially outwards, and because of the cylindrical geometry of the configuration decreases in strength. A dielectric window is placed near the outer edge of the substrate holder where the electric fields are lower in order to avoid plasma breakdown at the window. As a result, the pressure-power stable operation of the discharge is greatly enhanced over that of the single-mode cavities discussed in Sects. 2.3.2 and 2.3.5. The window also serves as the vacuum seal since atmospheric pressure is present behind the window to avoid plasma formation in the TEM section in which the fields are high. As the waves travel into the cavity around the substrate holder (electrode), the chamber wall is shaped such that the electric fields increase towards the center of the cavity. The height of the domed section is tuned so that reflections at this section of the chamber produce a stable plasma directly on top of the substrate holder.

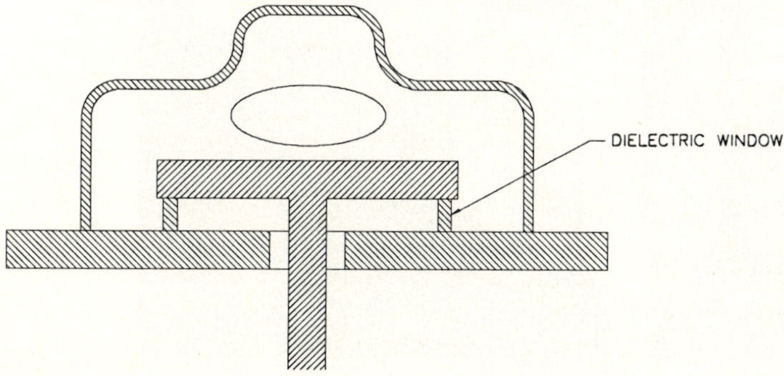

Fig. 2.7: The schematic diagram of an overmoded cavity reactor. The microwaves are coupled through a TEM section through the bottom. The window is not on the plasma line of sight.

In addition to the enhanced plasma stability of this design, the dielectric window is no longer in direct view of the plasma. This configuration avoids replacement of the window, since it does not become coated with byproducts of the plasma chemistry. Much greater power levels are possible in this design, where already up to 10 kW at 2.45 GHz have been coupled. As a result, larger plasmas than in single-mode cavities can be obtained. Higher total coupled power may yet be possible in the future, to further increase the deposition area and mass deposition rate while keeping the power density high. At power levels of 8 kW, $\pm 10\%$ uniformity diamond deposition has been demonstrated for substrates of 6.25 cm in diameter. The mass deposition rate is observed to be about 100 mg/hr in this case. Alternatively, high power density conditions can be relaxed and larger-diameter areas can also be coated. Such a configuration is used to deposit diamond on cutting tools, where a deposition area of 12.5 cm in diameter is possible.

2.3.7 The 915 MHz Microwave Frequency Reactor

The choice of microwave frequency in CVD-diamond applications has been dictated by the commercial availability of magnetron tubes. Tubes at 2.45 GHz are widely available, since this is the frequency used in microwave ovens. The next lower frequency with a large commercial application base is 915 MHz. At these

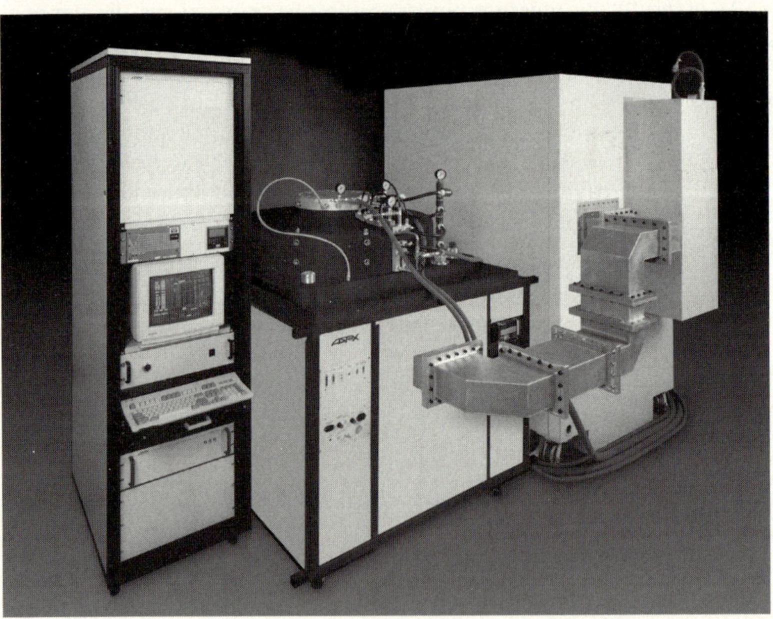

Fig. 2.8: A large-scale microwave-plasma deposition reactor prototype operating at 915 MHz. The device has already demonstrated CVD-diamond mass deposition rates of the order of 1 g/hr.

lower frequencies, magnetron tubes are available at power levels up to 100 kW continuous power. The overmoded reactors described in the previous section have been scaled up by the ratio of the wavelengths at the two frequencies (2450 MHz/915 MHz = 2.7). In order to keep the power per unit area approximately constant, the power required is of the order of 80 kW. Such scaling was used to construct a prototype unit at ASTeX. In this unit, diamond deposition on substrates of 15–20 cm in diameter and deposition uniformity of ±15% have already been accomplished. In addition, deposition rates near 1 g/hr have been obtained and microwave power in excess of 90 kW has been coupled. Diamond deposition over an area in excess of 30 cm diameter has also been achieved with these reactors, although with poorer radial uniformity. Rotating stages are now being developed to take advantage of the large plasma size possible in these reactors, so that deposition uniformity is not sacrificed. Using a rotating stage, diamond mass deposition rates in excess of 2 g/hr are expected.

A photograph of the reactor prototype is shown in Fig. 2.8, and an engineering drawing of the commercially available beta reactor is shown in Fig. 2.9. The reactor top rotates around a pivot for ease of access. For scale, the diameter of the deposition chamber is approximately 1 m. As a result of the improved plasma confinement in these reactors, because of their greater size, very high neutral pressures are required to limit the plasma size to the substrate size of interest.

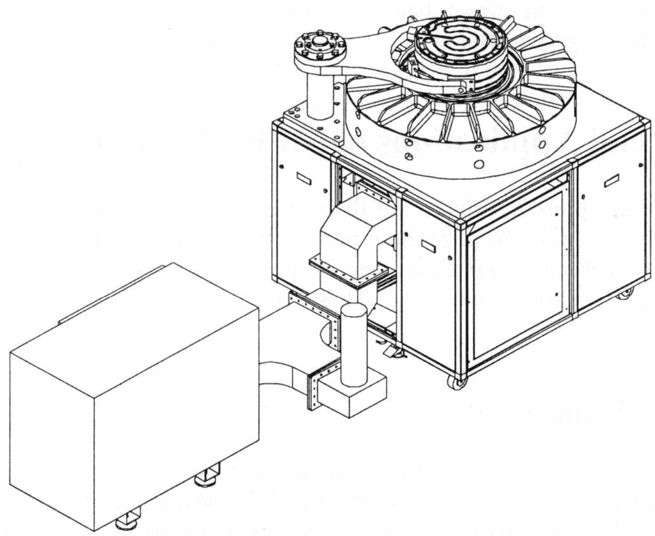

Fig. 2.9: An engineering drawing of a 100 kW microwave-plasma deposition reactor at 915 MHz. The chamber diameter is of the order of 1 m. Microwave power from the magnetron enclosure on the left is coupled to the bottom of the chamber as in Fig. 2.7.

packaging material exceeds about 10 K, it is likely that the junction temperature will be too high. With this assumption, diamond becomes the only suitable thermal substrate when the power density is about 500 W/cm^2.

2.4.2 Optics

Diamond is unique among all materials in its light propagation properties. Diamond is transparent from the UV just beyond the band gap edge at 220 nm all the way to microwave and rf frequencies. Intrinsic phonon absorption is only present in the 2–6 μm region. Such transmission properties make diamond ideally suited for long-wavelength transmission applications. Much of the early interest in diamond for optics was in the military where a highly impact resistant material capable of transmission in the 10 GHz region of the spectrum was sought. Diamond was used to coat optical elements in missile domes, to prevent rain and dust erosion of the elements. Recently, much progress has been made in the production of thick optical elements and domes for this application. Thick domes (over 2 mm thick and 6.4 cm diameter) can now be produced in microwave-plasma reactors [2.10, 11]. The intrinsic stress of the diamond and the high costs of polishing of the domes remain a challenge for the successful use of these films in this application.

With the development of the high power density processes in microwave-plasma reactors it has been possible to produce some of the highest-quality optical material available today. A comparison of the transmission spectra for high-quality CVD diamond produced at Raytheon and type IIa natural material is shown in Fig. 2.11 [2.10]. Other companies have been able to achieve similar results, also

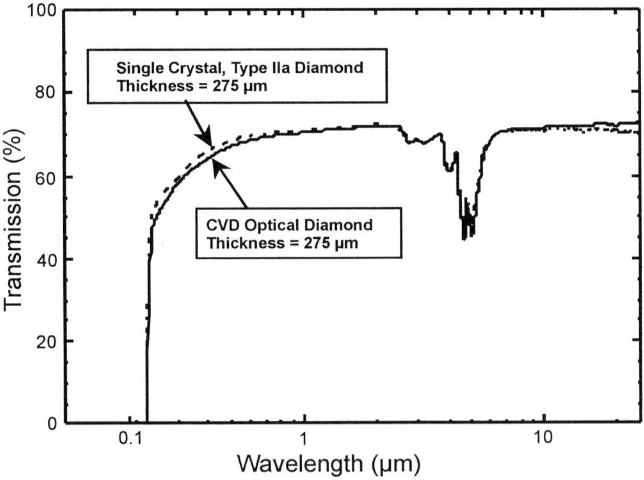

Fig. 2.11: Optical transmission spectra for CVD optical diamond and natural type IIa single-crystal material (from [2.10]).

using microwave technology [2.11, 12]. High power density reactors were critical to the production of such high quality material at practical deposition rates in the range of 2–5 µm/hr. For material which is transparent in the infrared and more lossy in the optical range, even higher deposition rates are possible.

Commercial applications of optical diamond in Fourier Transform Infrared instruments are gaining acceptance [2.35]. Beam splitters and other optical elements made with diamond are used to take advantage of the great transmission properties of diamond at the wavelengths of interest in this application. Because of its chemical inertness, optical diamond has also been used as window material in very aggressive chemical environments. A more complete review of the optical applications of CVD diamond is presented in Chap. 10.

2.4.3 Electronics

Much of the initial interest in CVD diamond centered around the development of electronic devices. However, it is unlikely that active devices based on this material will represent a significant commercial market in the near future. Boron doping of CVD diamond (p-type) to achieve semiconducting properties has been reported widely in the literature (see Chaps. 13 and 17). More recently, there are reports of nitrogen doping to achieve n-type material of interest in field emission displays [2.36].

Simple UV light detectors have now been developed using very high quality diamond substrates grown with the MPECVD technique [2.37]. Passive devices, such as pressure gauges, gas sensors, and so on, have also been reported in the literature [2.38]. However, among the passive devices that have the greatest commercial potential are surface acoustic wave (SAW) filters. The reader is referred to Chap. 14 for a review of this exciting application of CVD diamond. It is likely that, should this market be commercialized, the low deposition costs possible with MPECVD (see Sect. 2.5) will make this technique ideally suited for this application.

2.5 Economics of Diamond Deposition

As CVD diamond moves from the R&D laboratories into industrial applications, a good understanding of the costs associated with each step in the manufacture of diamond products is required. For the researcher, such an understanding is also crucial in order to be able to add an important piece of information to the choice of technology for a particular development. In this chapter, we focus only on the costs of the deposition step and we do not consider other costs, such as polishing, shaping, metallization, and packaging. Such costs will be similar with each technique as long as the material thickness uniformity and bow are similar. Today, post-processing costs are similar or slightly larger than deposition costs; however, both costs are rapidly coming down as the technologies evolve.

The elements of the deposition costs can be separated into fixed and variable. Fixed costs are those that do not change through time and include equipment depreciation, the cost of capital, building costs, and facilities. Variable costs are those that can change over time and include consumables, labor, and maintenance.

Once the fixed and variable costs associated with a factory or pilot production have been identified and quantified, the most important element that affects the cost of the material being produced is the production quantity. The amount of diamond produced will depend on the deposition rate that can be achieved, the yield of usable material per deposition run, and on the equipment uptime. Incorrect assumptions about these variables can make a big difference to the costs that are actually achieved in a real setting. A learning curve in which yield and uptime increase with time should be used when an accurate estimate of the costs is required.

We will try to keep the analysis general, so that it will become useful to many readers who usually are not familiar with calculations. A simple spreadsheet is the most convenient way to look at the results and the interaction among the variables in the calculation.

2.5.1 Fixed Costs Elements

Depreciation of equipment is often one of the most important costs to consider. Most companies use a straight-line depreciation schedule in which the annual depreciation costs are assumed to be the same during the depreciation period. General accounting practices allow some flexibility in the number of years that are assumed for depreciating a piece of equipment. The number of years depends on how fast the technology for the equipment is changing. In large-scale manufacturing machinery, ten years is common. For CVD-diamond equipment either 3 or 5 years should be assumed at this stage of the technology.

The cost of capital is often overlooked in these calculations. It takes into account the fact that money will typically have to be borrowed for the purchase of the equipment. Facilities costs which are specific to the installation of the equipment should be included in these costs, although the depreciation time for the facilities can be assumed to be longer.

Building and facilities costs include space rental and the installation and equipment costs for electrical, gas delivery, pressurized air, and cooling water services. In some areas, it is possible to rent facilities with all the required services. Otherwise, it is required to assess the expenses for each of the services. All facilities costs can be capitalized and depreciated over standard schedules.

2.5.2 Variable Costs Elements

Labor costs vary widely among geographic areas and therefore this cost area should be revised accordingly. In assessing these costs, it is necessary to estimate the number of personnel and management required for the deposition operation.

The number of personnel will depend on the degree of automation of the operation and the specifics of the product being manufactured. When equipment depreciation costs are high, as is the case here, multiple shifts may be necessary to avoid idle equipment time.

For our sample calculation, we include the salary of the individual and the costs of the benefits the worker may receive (paid vacation, health insurance, sick leave, etc.). We have not included any other overhead except that the salaries of supervisory personnel are also included. No general and administrative costs are considered in this calculation.

Consumables include gas, electricity, substrate and holder materials, and in the specific case of MPECVD, magnetrons (in other technologies the cost of material for filaments or consumable torch components needs to be included). Gas purity requirements depend on the diamond product being manufactured. In general, impurity levels greater than 50 ppm are unacceptable in most cases. In the case of hydrogen, high gas purity is easy to achieve at low costs. For other gases used in MPECVD, achieving high purity can be costly.

Maintenance costs can vary significantly depending on the deposition technology used. These costs depend on the frequency of replacement of key reactor components and the downtime associated with these replacements. In the case of MPECVD, these costs can be very low, as there are almost no replaceable components and infrequent maintenance is sufficient. We will assume that 2.5% of the reactor cost will be spent on miscellaneous maintenance-related hardware over the course of a year. Labor costs for maintenance will be included with the other labor costs.

2.5.3 Sample Deposition Cost Calculation

We consider an hypothetical CVD-diamond production factory that consists of a bank of ten diamond deposition reactors, each operating at 70 kW of microwave power at 915 MHz. The plant is assumed to be producing 400 μm thick, 15 cm diameter diamond substrates for thermal management applications, with a thermal conductivity in the range of 8–10 W/cmK. We will further assume that each substrate can be produced over a period of 35 hours (with 1 hour loading and unloading) so that four substrates can be grown in each reactor over a period of one week with two labor shifts. Regular maintenance will be assumed to take two weeks per reactor per year. A working year is assumed to have 50 weeks, so that only 48 weeks will be in actual production. We will further assume that we have reached 90% yield, so that the total plant output becomes 1728 diamond substrates, of 124 carats each.

Our hypothetical plant is capable of delivering a total of 700 kW of microwave power. Given a typical magnetron tube efficiency of 85% at the chosen frequency, a total cooling water capacity of the order of 1 MW is needed. We also need to estimate the cost of the gas delivery and exhaust systems capable of handling the gas load required (assumed to be of the order of 40 l/min of hydrogen) for this plant. We assume that the total cost for such facilities is $2 million. The purchase price for each reactor is taken to be \$600 000. The total capital investment required for the pilot plant is therefore \$8 million. Furthermore, the plant is assumed to be located in a 1000 m^2 facility with an annual rent of \$400 000.

Electricity and consumables costs are calculated based on a price of electricity of \$0.05/kWh, a high-purity hydrogen gas cost of \$10/m^3, and a high-purity methane gas cost of \$70/m^3. Each reactor is assumed to consume 4 l/min of hydrogen and 5% methane. In addition, 1.5 magnetrons per year are consumed per reactor, at a price of \$10 000 each.

The reactors are assumed to be completely computer-controlled, with minimal operator intervention required. Personnel is needed for loading and unloading functions, record keeping, and general reactor maintenance. It is assumed that two full time equivalent workers with an annual salary of \$40 000 each will be sufficient to operate this plant with ten reactors. In addition, a supervisor at a 20% level will be sufficient to oversee the diamond production operation. This supervisor will spend the rest of the time in other production-related areas (post-processing, cutting, polishing, etc.) which are not part of the deposition cost calculation. The supervisor will be assumed to have an annual salary of \$55 000. Employee benefits, vacation, and other costs are assumed to add an additional 30% to the gross labor. The annual operating cost summary is given in Table 2.2.

Table 2.2: Sample annual operating costs for a pilot plant with ten microwave reactors used to produce 214 300 carats/yr CVD-diamond substrates for thermal management substrates.

Fixed costs	\$
Capital equipment depreciation	1 200 000
Facilities depreciation	400 000
Cost of capital (6% interest)	96 000
Space rental	400 000
Total fixed costs	**2 096 000**
Variable costs	
Electricity	235 200
Gas	217 728
Magnetrons	150 000
Maintenance	150 000
Total consumables and maintenance	**752 928**
Labor	
Two full time employees (two shifts)	208 000
Supervisor (two shifts)	28 600
Total labor	**236 600**
Total annual operating cost	**3 085 528**

The cost per carat of diamond produced can now be calculated using the total operating cost from Table 2.2 and the throughput assumptions made earlier of 1,728 substrates, each 124 carats for a total of 214 300 carats. Under the assumptions made above, this plant produces diamond at a cost of approximately $14/carat. Although this is a hypothetical example, the deposition rates assumed are consistent with the values that have already been achieved in microwave-plasma deposition reactors. Other costs are estimated but are not far from being accurate. Therefore, we conclude that the deposition costs from the microwave technique are highly competitive with any other diamond deposition technique.

2.6 Conclusions

The microwave-plasma technique for CVD-diamond deposition is truly at the forefront of all techniques for the commercial and R&D applications of this material. Since its inception, this technique has demonstrated unique capabilities in the deposition of some of the highest-quality material yet reported.

The physics understanding of these reactors is such that the macroscopic plasma behavior and much of the microscopic phenomena are well characterized and understood. Plasma models that incorporate in a self-consistent way the electromagnetic fields inside the cavities of interest have been developed. These models closely predict the plasma shape, position, and behavior as a function of external variables such as power and pressure.

We have shown that the ionization fraction for most plasmas produced in these reactors (with the exception of magnetized plasmas) is extremely low. This weakly ionized plasma plays a limited role (if at all) in the chemistry of these discharges, as the reactions among neutral species dominate the chemistry of the diamond deposition. It is therefore important for the reader to understand that the main role of the electrons in these discharges is to efficiently couple the microwave power into gas heating through electron-neutral collisions. Increasing the power and pressure therefore leads to an increase in the gas temperature, but not necessarily to a significant change in plasma density. If the power density in the discharge is sufficiently high, a new gas chemistry regime can be reached, as evidenced by a significant change in the light emission spectrum from the discharge. The resulting large increase in diamond deposition rates is a result of the higher gas temperatures that drive the chemistry.

Commercially available MPECVD reactors have evolved rapidly over the past few years. The microwave power delivered, the substrate area that can be coated, and the total mass deposition rates have all been scaled up by several orders of magnitude. As a result, the diamond deposition costs that are achieved with this technique have dropped dramatically. The cost analysis of a realistic diamond-producing pilot plant indicates that the microwave technique is an economically attractive alternative for present and future commercial applications of CVD diamond.

References

2.1   M. Kamo, Y. Sato, S. Matsumoto, and N. Setaka, J. Cryst. Growth **62**, 642 (1983)

2.2   P.K. Bachmann and R.F. Messier, Chem Eng News **67**(20), 24 (1989)

2.3   P.K. Bachmann, W. Drawl, D. Knight, R. Weimer, and R.F. Messier, in Diamond and Diamond-like Materials, ed. A. Badzian, M. Geis, and G. Johnson, MRS Symposium Proceedings, Vol. EA-15, p. 99 (1988)

2.4   S. Yugo, T. Kimura, and T. Muto, Vacuum **41**, 1364 (1990)

2.5   X. Jiang and C.P. Klages, Diamond Rel. Mater. **2**(5–7), 1112 (1993)

2.6   B.R. Stoner and J.T. Glass, Appl. Phys. Lett. **60**, 698 (1992)

2.7   B.A. Fox, M.L. Hartsell, D.M. Malta, H.A. Wynands, C.-T. Kao, L.S. Plano, G.J. Tessmer, R.B. Henard, J.S. Holmes, A.J. Tessmer, and D.L. Dreifus, Diamond Rel. Mater. **4**, 622 (1995)

2.8   CVD Diamond. Thermal Management Substrates. Optical Windows. Protective Coatings Product Handbook, GEC–Marconi Materials Technology, Caswell, Towcester, Northamptonshire, England (1995)

2.9   C. Willingham, T. Hartnett, C. Robinson, and C. Klein, in Applications of Diamond Films and Related Materials. ed. Y. Tzeng, M. Yoshikawa, M. Murakawa, and A. Feldman, Materials Science Monographs, 73, Elsevier, Amsterdam (1991)

2.10 C.B. Willingham, Thomas M. Hartnett, Richard P. Miller, and Robert B. Hallock, Proc. SPIE, **Vol. 3060**, Window and Dome Technologies and Materials V (1997)

2.11 J.A. Savage, C.J.H. Wort, C.S.J. Pickles, R.S. Sussmann, C.G. Sweeney, M.R. McClymont, J.R. Brandon, C.N. Dodge, and A.C. Beale, Proc. SPIE, **Vol. 3060**, Window and Dome Technologies and Materials V (1997), p. 144

2.12 R.S. Sussmann, C.J. Wort, C.G. Sweeney, J.L. Collins, C.N. Dodge, and J.A. Savage, Proc. SPIE, **Vol. 2886**, Window and Dome Technologies and Materials IV, ed. P. Klocek (1994), p. 289

2.13 J.E. Graebner, T.M. Hartnett, and R.P. Miller, Appl. Phys. Lett. **64**, 2549 (1994)

2.14 E. Hyman, K. Tsang, I. Lottati, A. Drobot, B. Lane, R. Post, and H. Sawin, Surf. Coat. Technol. **49**, 387 (1991)

2.15 E. Hyman, K. Tsang, A. Drobot, B. Lane, J. Casey, and R. Post, J. Vac. Sci. Technol. **A 12**, 1474 (1994)

2.16 M. Capitelli, G. Colonna, K. Hassouni, and A. Gicquel, Plasma Chem. Plasma Process. **16**, 153 (1996)

2.17 K. Hassouni, S. Farhat, C.D. Scott, and A. Gicquel, J. Phys. III **6**, 1229 (1996)

2.18 N.A. Krall and A.W. Trivelpiece, Principles of Plasma Physics, McGraw-Hill, New York (1973)

2.19 S.C. Brown, Basic Data of Plasma Physics, M.I.T. Press, Cambridge, MA (1959)

2.20 Dieter M. Gruen, Chris D. Zuiker, Alan R. Krauss, and Xianzheng Pan, J. Vac. Sci. Technol. **A 13**, 1628 (1995)

2.21 P.K. Bachmann, D. Leers, and H. Lydtin, Diamond Rel. Mater. **1**, 1 (1991)

2.22 Chia-Fu Chen, Sheng-Hsiung Chen, Hsien-Wen Ko, and S.E. Hsu, Diamond Rel. Mater. **3**, 443 (1994)

2.23 P.A. Rizzi, Microwave Engineering, Prentice Hall, Englewood Cliffs, New Jersey (1988)

2.24 A.D. MacDonald, Microwave Breakdown in Gases, John Wiley, New York (1966)

2.25 A. Hatta, T. Yara, H. Makita, M. Yuasa, J. Suzuki, Y. Mori, T. Ito, T. Sasaki, and A. Hiraki, Diamond Films Technol. **5**, 29 (1995)

2.26 C.R. Eddy Jr., B.D. Sartwell, and D.L. Youchison, Surf. Coat. Technol. **48**, 69 (1991)

2.27 D.K. Smith, E. Sevillano, M. Besen, V. Berkman, and L. Bourget, Diamond Rel. Mater. **1**, 814 (1992)

2.28 J.C. Angus, F.A. Buck, M. Sunkara, T.F. Groth, C.C. Hayman, and R. Gat, MRS Bull., p. 38, October 1989

2.29 N. Ohtake, H. Tokura, Y. Kuriyama, Y. Mashimo, and M. Yoshikawa, Proceedings of the First International Symposium on Diamond and Diamond-Like Films, ed. J.P. Dismukes, A.J. Purdes, K.E. Spear, B.S. Meyerson, K.V. Ravi, T.D. Moustakas, and M. Yoder, Proceedings **Vol. 89-12**, The Electrochemical Society, NJ (1989) p. 93

2.30 Y. Hirose, Proceedings of the First International Conference on the New Diamond Science and Technology, Tokyo, Japan, October 24–26, 1988

2.31 E. Sevillano, L.P. Bourget, and R.S. Post, High Growth Rate Plasma Diamond Deposition and Method of Controlling Same, US Patent 5,518,759 (1996)

2.32 M.M. Besen, E. Sevillano, and D.K. Smith, Microwave Plasma Reactor, US Patent 5,556,475 (1996)

2.33 B. Nagy, Fourth International Conference on CVD Diamond & DLC Coatings and Thick Film Markets, Gorham Advanced Materials Institute, Atlanta, GA, March 11–13, 1996

2.34 K.J. Gray and P.M. Fabis, Diamond Rel. Mater. **6**, 191 (1997)

2.35 ASI Applied Systems, Millersville, MD, Product Data Sheets for DuraSamplIR™ and Comp™ Probes, (1996, 1997)

2.36 K. Okano, S. Koizumi, S.R.P. Silva, and G.A.J. Amaratunga, Nature **381**, 140 (1996)

2.37 M.D. Whitfield, R.D. McKeag, L.Y.S. Pang, S.S.M. Chan, and R.B. Jackman, Diamond Rel. Mater. **5**, 829 (1996)

2.38 J.L. Davidson, D. Wur, and W.P. Kang, Proceedings of the Third International Symposium on Diamond Materials. ed. J.P. Dismukes and K.V. Ravi, Proceedings, **Vol. 93-17**, The Electrochemical Society, NJ (1993) p. 1048

3. Combustion Flame Deposition of Diamond

Colin A. Wolden*, Zlatko Sitar, and Robert F. Davis

Department of Materials Science and Engineering,
North Carolina State University,
1001 Capability Drive, Raleigh, NC 27695–7919, USA

* Present address:
Department of Chemical Engineering, Colorado School of Mines,
Golden, CO 80401, USA
e-mail: cwolden@mines.edu

Springer Series in Materials Processing
Low-Pressure Synthetic Diamond Eds.: B. Dischler and C. Wild
© Springer-Verlag Berlin Heidelberg 1998

3.1   Introduction

Combustion synthesis is one of the competing chemical vapor deposition (CVD) technologies for diamond film growth. It was invented in 1988 by Hirose [3.1], who realized that acetylene flames produce copious amounts of atomic hydrogen and hydrocarbon radicals that are required for diamond growth. The discovery was subsequently confirmed at the Naval Research Laboratory [3.2] and has become the subject of intensive study. It has been argued that combustion synthesis is the most flexible of the CVD alternatives because of its scaleable nature, minimal utility requirements, and significantly reduced capital costs relative to plasma-aided processes [3.3]. In this chapter, the development of combustion CVD will be discussed from its inception, when conventional welding torches were used, to its present implementation with flat-flame burners. The unique features of the combustion CVD system are highlighted. The key experimental parameters, and their impact on deposition, are discussed in detail. The chemistry of combustion diamond CVD and its relation to the deposition mechanism is analyzed through a combination of *in-situ* diagnostics and detailed reactor modeling. Lastly, examples of the implementation of combustion synthesis for specific applications are provided.

3.2 Invention: The Welding Torch

3.2.1 Description of the System

Figure 3.1 shows the structure of an atmospheric oxy-acetylene torch in its unperturbed state and in the configuration used for diamond deposition. There are

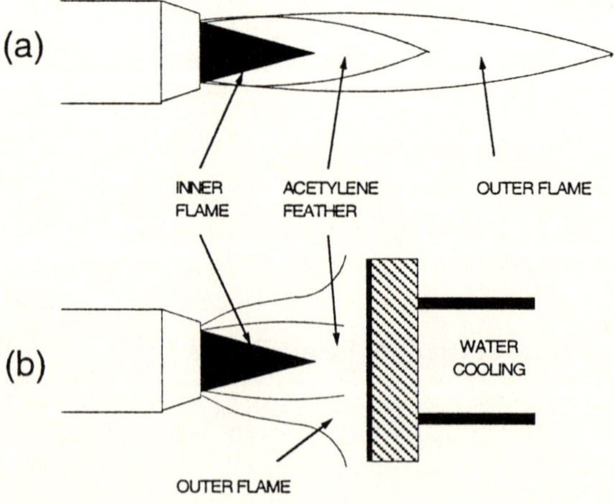

Fig. 3.1: Schematic diagrams of (**a**) the flame structure of an oxy-acetylene torch and (**b**) the flame in a diamond deposition configuration.

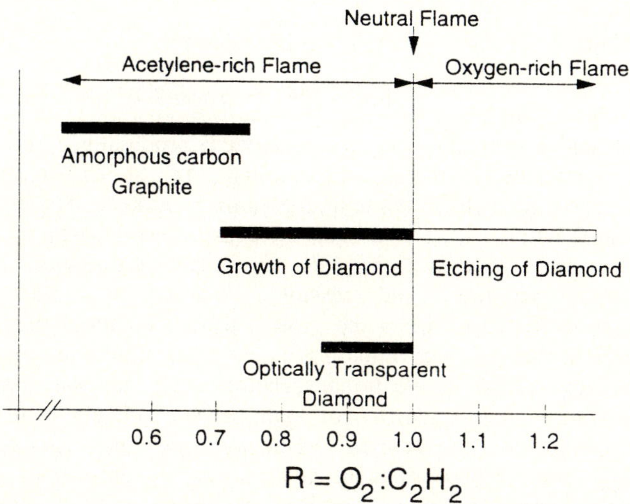

$$R = O_2 : C_2H_2$$

Fig. 3.2: Carbon deposition regions observed using atmospheric acetylene torches as a function of % acetylene supersaturation, S_{ac} (based on the findings of [3.4]).

three distinct regions in an acetylene flame: (i) the inner flame; (ii) the acetylene feather; and (iii) an outer diffusion flame. The substrate is placed in the feather region for diamond growth. The most important parameter in combustion synthesis is the volumetric ratio of oxygen to acetylene, $R = O_2 : C_2H_2$. At values of R near 1.0, a neutral flame is achieved. This is defined as the condition in which the feather region just disappears, since all the acetylene is consumed in the inner flame. The value of R at which a neutral flame occurs depends on both burner design and total flow rate. Because of these variations, Schermer and coworkers introduced the concept of acetylene supersaturation, S_{ac}, as a more universal measure of flame composition (see also Sect. 3.2.3). S_{ac} is defined as the percentage of additional acetylene compared to the acetylene flow required to achieve a neutral flame. The carbon deposition regimes as a function of composition are outlined in Fig. 3.2. The highest-quality diamond has been obtained in slightly rich acetylene flames, $S_{ac} = 0$–15 % [3.4].

The composition of acetylene flames differs much from that of the traditional H_2/CH_4 mixtures used in hot-filament and plasma CVD systems. However, the mechanism of diamond growth in combustion CVD is expected to be similar to that in other diamond growth systems. Bachmann et al. [3.5] found that when the gas-phase compositions used for diamond CVD from a variety of reactors were plotted on a ternary C–H–O diagram, all of the successful results fell into a narrow band centered on the CO tie line. At $R = 1.00$, the composition of an oxy-acetylene flame is positioned in the direct center of this growth regime. The chemistry of acetylene flames is discussed in more detail in Sect. 3.4, but its essence may be distilled to the following global reactions:

$$C_2H_2 + O_2 \rightarrow 2CO + H_2 \qquad\qquad \Delta H_R^\circ = -107 \text{ kcal/mol} \qquad (3.1)$$

$$H_2 \leftrightarrow H + H \qquad\qquad \textit{Reversible: } \text{function}(T,P) \qquad (3.2)$$

Nearly all of the oxygen and acetylene is rapidly consumed, producing carbon monoxide and hydrogen as described by the global reaction (3.1) [3.6]. The highly exothermic nature of this reaction results in flame temperatures > 3000 K. At these temperatures, a significant fraction of the hydrogen molecules decomposes into atomic hydrogen, as expressed by reaction (3.2) [3.7]. The degree of dissociation depends strongly on both temperature T and pressure P. Carbon monoxide is thermally stable and is not expected to affect the growth processes under these conditions. A small fraction of the acetylene remains unreacted or is converted into hydrocarbon radical species. Thus, conventional schemes that suggest that diamond growth proceeds through the addition of hydrocarbon radicals in the presence of a superequilibrium concentration of atomic hydrogen may also be applied to combustion systems. Where filament and plasma systems employ electrical energy to create atomic hydrogen, combustion systems rely on the chemical energy released by the oxidation of acetylene to generate heat and atomic hydrogen.

The welding torch arrangement developed by Hirose has been used in numerous subsequent studies [3.8–16]. Polycrystalline diamond has been grown at rates up to 200 µm/hr [3.8], and thick (150 µm) homoepitaxial layers have been deposited [3.9]. Atmospheric torches have successfully produced large individual crystals that approach 1 mm in diameter [3.10, 11]. The main drawbacks of the welding torch arrangement are the radial inhomogeneity of the deposited material [3.12], and the presence of significant amounts of nondiamond carbon [3.13]. Because of these limitations, welding torches are being replaced by flat-flame burners, which are more appropriate for large-area deposition. Nevertheless, the wealth of data from experiments with atmospheric torches has provided the foundation for the general understanding of combustion-grown diamond which is described below. A more comprehensive summary of experiments performed with atmospheric torches can be found in an earlier review [3.14].

3.2.2 Important Process Variables

Substrate Temperature

Substrate temperatures during the combustion CVD of diamond range from 950 K to 1650 K. Due to these high temperatures, substrates have been limited to materials such as molybdenum, alumina, silicon, and diamond. Evaluating the true substrate temperature is difficult in all diamond CVD environments, but especially so in combustion synthesis where extreme heat fluxes are present. The two techniques most commonly used are optical pyrometry and thermocouples placed on the back of the substrate. Pyrometry provides a good measure of the relative temperature of the substrate, but absolute values may be wrong by as much as

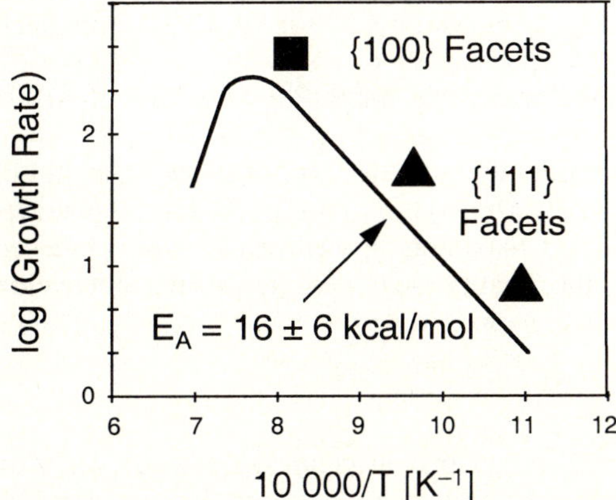

Fig. 3.3: The effect of substrate temperature on growth rate and morphology observed in combustion CVD of diamond. This diagram reflects data from [3.15–18].

100 K due to uncertainties in emissivities. Care must be taken to position the pyrometer in order to avoid emission from the flame itself. The accuracy of measurements by back-side thermocouples depends on both the quality of the contact and the magnitudes of the thermal gradients present. Both techniques work well for a single configuration, but caution is urged when comparing absolute temperatures obtained from different systems.

Substrate temperature has a dramatic effect on two important properties: growth rate and morphology. Several authors [3.15–17] have observed that the growth rate increases exponentially with temperature, as shown schematically in Fig. 3.3. This behavior is characteristic of combustion systems; however, the curve may be shifted to the right or left depending on operating conditions and the technique used to measure temperature. Weimer et al. [3.15] measured activation energies between 12 and 18 kcal/mol for diamond growth on individual crystal faces. Slightly higher values (16–23 kcal/mol) have been reported for polycrystalline films [3.16, 17]. As the substrate temperature increases, the growth rate reaches a maximum (Fig. 3.3). Increasing the substrate temperature beyond this point causes a rapid decline in both the quality and the growth rate [3.15, 17]. In atmospheric torches the maximum growth rate occurs at substrate temperatures between 1450 K and 1650 K [3.15, 18]. In low-pressure (40–50 Torr) flat-flame burners, the same growth rate behavior has been observed, but the maximum occurs at significantly reduced temperatures, in the range 1050–1250 K [3.17, 19]. Kim and Cappelli [3.19] suggested that the falloff at high temperature may be due to oxidation of diamond. An alternative possibility is that the growth precursor desorbs from the surface before having the opportunity to incorporate into the diamond lattice [3.20].

Substrate temperature also determines the diamond film morphology. The morphology of most films deposited by combustion CVD is typically dominated by {111} facets [3.18, 19, 21]. However, a number of researchers [3.17, 18, 21, 22] have observed that at higher temperatures the morphology consists of {100} facets parallel to the substrate.

As the substrate temperature increases, the growth habit changes from {111} faceting to cubic, as indicated in Fig. 3.3. The region of {100} faceting coincides with the substrate temperature at which maximum growth rates were observed [3.17, 18]. At higher temperatures the quality degrades, secondary nucleation increases, and no preferred texture is achieved.

Reactant Composition

In atmospheric torches, it has been observed that morphology also depends on reactant composition [3.23, 24]; it changed from octahedron growth at oxygen-rich conditions to cubic growth and finally to amorphous as the acetylene fraction was increased. The dependence on composition is in accord with observations of morphology variations with carbon concentration in plasma reactors [3.25, 26]. In flat-flame burners, similar behavior has been observed, although the range of R over which diamond is successfully formed is much smaller [3.17]. Growth rates generally increase linearly with acetylene concentration, although the exact behavior is somewhat system dependent.

3.2.3 Modifications of the Welding Torch

In the welding torch operation described in Fig. 3.1, diamond is often deposited in a small circle or as an annular ring. To increase the deposition area and uniformity, a number of modifications have been investigated, including turbulent-flame operation, multi-nozzle burners, and enclosed flames. The turbulent condition was achieved by increasing both the flow rate and the burner orifice [3.27, 28]. Despite some reports of improved quality [3.29], it was generally concluded that the changes were not dramatic, and not worth the costs associated with the increased flow rates [3.30]. Some success was achieved with a multi-nozzle configuration developed by Zhu and coworkers [3.31], although the morphology was not uniform over the deposition area.

During atmospheric operation, diffusion of oxygen or impurities from the ambient may affect film quality. Attempts to minimize the effect by surrounding the torch with a coannular flow of inert gas have been studied [3.32, 33]. Under these conditions, the outer flame (Fig. 3.1) was suppressed, but no dramatic changes were observed in the growth rate or quality. Another way to exclude the ambient is to simply enclose the torch in a vacuum chamber [3.34–36]. It has been observed that the outer diffusion flame disappears, as there is no oxygen present, and the deposition area increases by 20% [3.34, 35]. In addition, reducing the pressure to 300 Torr increases the deposition area by 200%, but the growth rate decreases significantly. Wang et al. [3.36] achieved some success using rastering techniques with an enclosed flame to coat larger areas. Despite the modifications discussed here, the welding torch design is limited to small-area ($< 1\,\mathrm{cm}^2$) applications.

3.3 Implementation: The Flat-Flame Burner

3.3.1 Reactor Design

To address the issues of increased deposition area and improved uniformity, a number of groups have turned to the use of flat-flame burners at both reduced [3.17, 19, 37–40] and atmospheric pressure [3.41–44]. Cooper and Yarborough [3.37] first demonstrated diamond deposition in a commercially available burner at pressures between 25 and 40 Torr. A schematic of a flat-flame burner CVD for reduced pressure operation is shown in Fig. 3.4. Premixed gas enters a water-cooled burner and exits through a matrix that creates a radially uniform velocity profile with only an axial component. In practice, this has been achieved by drilling 1 mm holes in a copper plate [3.19, 38] and by using a honeycomb made of thin-walled silica tubes [3.17]. A flat, circular flame is stabilized below the burner surface, as shown in Fig. 3.4. This is a very different structure than the conical flame that is associated with atmospheric torches (Fig. 3.1). In these one-dimensional flames, there is no acetylene feather or other distinguishing feature that allows the use of a concept such as acetylene supersaturation. Thus, reactant compositions are typically defined in terms of the volumetric ratio, R. The substrate is maintained at a certain temperature T_s, and positioned at a distance L from the burner surface. This design was developed in the 1950s to measure flame velocities near extinction/ignition limits [3.39, 40]. The burner's characteristics of

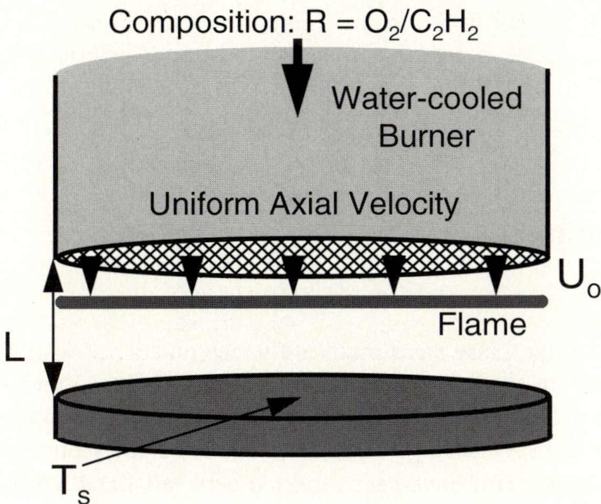

Fig. 3.4: A schematic of a flat-flame burner CVD system operated at reduced pressures.

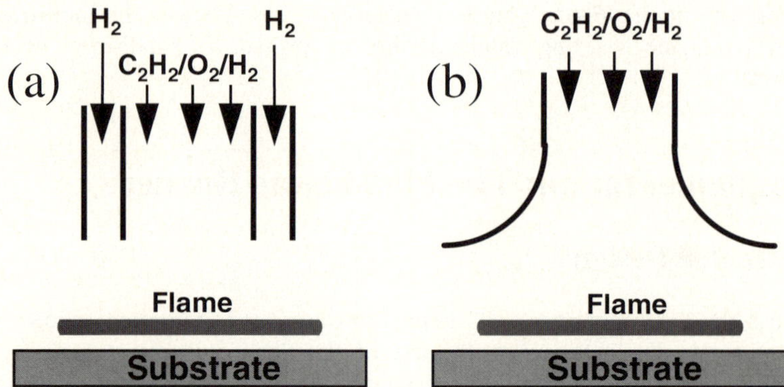

Fig. 3.5: Two designs of atmospheric flat-flame burners: (**a**) a coflow design, and (**b**) a trumpet bell design (abstracted from [3.41]).

stability and geometrical simplicity, which made it useful for that purpose, are equally valuable for chemical vapor deposition. Glumac and Goodwin [3.38] have used this design to deposit diamond uniformly over areas as large as 13 cm^2.

Atmospheric-pressure flat flames are less stable and require more complicated burner structures. Two designs that have been used include a coflow arrangement and a trumpet-bell design that are shown schematically in Fig. 3.5. Murayama and coworkers [3.42, 43] used the coflow configuration with hydrogen as the external gas to stabilize their flame. Workers at Sandia have scaled the trumpet-bell design to deposition areas greater than 20 cm^2 [3.44].

Comparison of Figs. 3.4 and 3.5 shows another important difference between the atmospheric and low-pressure operation. At low pressures the flame is stabilized at the burner surface, while atmospheric flames are substrate-stabilized and located 1–2 mm above the substrate. In low-pressure operation the substrate may be positioned at any distance from the burner, while in atmospheric systems the distance is constrained by the stability limits. Regardless of operating pressure, the unique feature of flat-flame burners is the radial uniformity achieved in both morphology and growth rate, as shown in Fig. 3.6.

3.3.2 Operating Conditions

Operating Pressure

As discussed above, operating pressure has a major influence on burner design and flame stability. The pressure also influences a number of critical issues, including growth rate, temperature control and deposition temperature. Growth rates in atmospheric systems range from 25 to 40 μm/hr [3.43, 44]. In comparison, growth rates in flat-flame burners at ~45 Torr have been reported between 4 and 5.5 μm/hr [3.17, 19]. Although the gas density is about 15 times greater in atmospheric

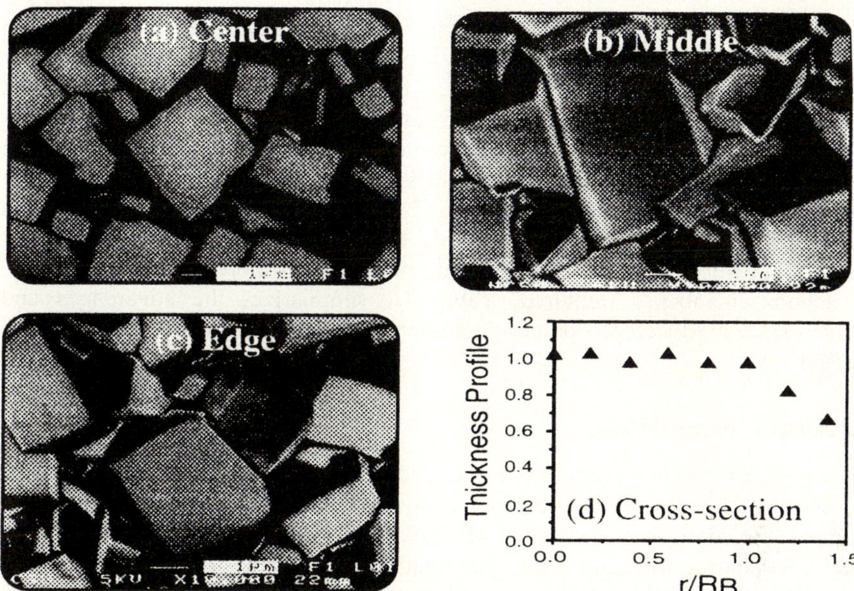

Fig. 3.6: A plot of diamond film thickness as determined by cross-sectional scanning electron micrographs (*inset*) as a function of radial position, showing the growth rate uniformity achieved in flat-flame deposition reactors (abstracted from [3.17]).

reactors, the growth rate is only a factor of ~7 times greater. More importantly, with regard to economics, the carbon conversion efficiency (the fraction of acetylene atoms converted to diamond) is similar in both cases.

As discussed above, growth temperatures in atmospheric systems are hundreds of degrees hotter than at low pressure. As such, substrate material selection is more limited for atmospheric operation. In addition, high growth temperatures lead to the formation of large compressive stresses in the films during cooling, due to

Table 3.1: A comparison of operating flat flames at atmospheric and reduced (40–50 Torr) pressures. This table reflects data and observations in [3.17, 19, 41–47].

Aspect	Reduced pressure	Atmospheric pressure
Growth rates	4–5.5 µm/hr	25–40 µm/hr
Carbon conversion efficiency	Similar ~0.005%	Similar ~0.005%
Vacuum chamber required	Yes	No
Deposition temperatures	400–1000°C	700–1200°C
Temperature control	Easy	Difficult
Acetylene flame stable?	Yes	No: requires H_2 addition
Safety	Advantage	
Burner design	Simple	Complex

differing coefficients of thermal expansion [3.41]. As discussed earlier, accurate substrate temperature control is required to achieve both the desired growth rate and appropriate texture [3.17]. The proximity of the flame and substrate in atmospheric reactors requires that energy must be removed at fluxes of the order of 500 W/cm^2. This requires the use of elaborate cooling strategies [3.18, 44, 45]. In contrast, low-pressure reactors are almost thermally neutral and films have been deposited without any cooling present [3.38]. Simple control strategies can maintain the substrate temperature within a few degrees of a desired setpoint for hours of deposition [3.46]. Atmospheric burner design is more complex; however, no vacuum chamber is required. Table 3.1 summarizes the advantages and disadvantages of different operating pressures for the combustion synthesis of diamond.

Reactant Composition

Low-pressure flat-flame burners are operated with a value of R near unity [3.17, 19]. At atmospheric pressure, pure oxy-acetylene flames are unstable. It has been found that the addition of hydrogen (18–25% by volume), with the C/O ratio kept at ~1, was required to stabilize the flame and achieve diamond growth [3.41–44]. Due to the homogenous nature of these flames, the composition range over which diamond is deposited is much smaller than observed with the welding torch design [3.4]. Most researchers [3.17, 38, 41–44] have reported a single-reactant composition that successfully produced high-quality diamond. However, a small range for diamond deposition has been reported for low-pressure systems [3.17, 19]. The reason for the narrow range is shown in Fig. 3.7, where the equilibrium

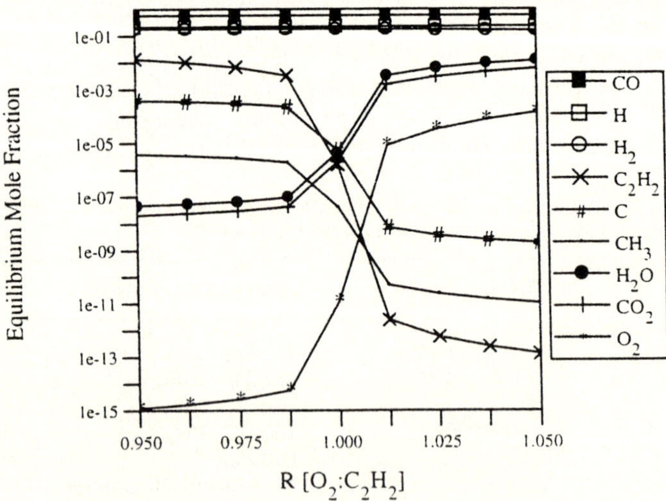

Fig. 3.7: Equilibrium mole fractions of selected species in an oxy-acetylene flame as a function of reactant composition $R = O_2 : C_2H_2$. Calculation performed at 40 Torr.

mole fractions of selected species are plotted as a function of composition. The three major species (CO, H_2, and H) are almost constant across this range of R. For the minor species, a very sharp transition is observed in the range $0.98 \leq R \leq 1.02$. The concentration of carbon-containing species (C_2H_2, CH_3, C) falls by four to seven orders of magnitude as the reactant composition varies by just a few percent. A concomitant increase in oxidation products (H_2O, CO_2, O_2) is also predicted. The position of the transition ($R = 1.00$) is in good correspondence with the composition employed for diamond growth, and the sharpness of the transition explains the narrow composition window for diamond growth.

3.4 Deposition Chemistry

3.4.1 Reactant Diagnostics

The chemistry of combustion diamond CVD has been primarily investigated using laser-induced fluorescence (LIF) [3.6, 48, 49] and mass spectroscopy [3.50, 51]. The two techniques are complementary. Mass spectroscopy has been used to provide quantitative measurements of the major stable species which are important for verifying kinetic models. In this technique, a quartz microprobe samples gases at the diamond growth surface. On the other hand, LIF is a nonintrusive optical technique which has been used to detect radicals such as C_2, C_2H, CH and OH. Researchers using plasma and hot-filament reactors have focused on C_1 species such as the methyl radical and atomic carbon as the dominant growth species. The evidence from combustion reactors suggest that C_2 species are also very important to diamond growth. Three different groups [3.6, 48, 49] have noted the strong relationship between C_2 concentration profiles and diamond deposition rates. The role of CH is less clear and OH appears to be of minor importance [3.49]. The work of Cappelli and Paul [3.48] suggested that C_2H also plays an important role in the deposition process. Using mass spectroscopy, Wolden et al. [3.50] noted that the strong compositional dependence observed in a flat-flame reactor was related to a sharp change in the acetylene concentration.

3.4.2 Reactor Modeling

An attractive feature of flat-flame reactors is that they may be accurately simulated using simple stagnation flow models that permit the inclusion of complex combustion kinetics. Under proper operating conditions [3.52], the axisymmetric stagnation flow may be treated in one dimension using a similarity solution [3.53]. A detailed kinetic mechanism containing 50 species and 218 reactions has been developed for the acetylene flame by Miller and Melius [3.54]. The model has been verified experimentally by Glumac and Goodwin [3.51] using mass spectroscopy and LIF at conditions that are slightly oxygen rich for diamond growth. Wolden and coworkers [3.7] performed a similar validation study at diamond growth conditions with a diamond-coated substrate present.

Several workers have used this approach to model flat flames at diamond growth conditions. Figure 3.8 shows the profiles of selected species between the burner and substrate at diamond deposition conditions [3.7]. Acetylene and oxygen are rapidly converted to hydrogen and carbon monoxide as the gas exits the burner. At the high temperatures involved, a large amount of atomic hydrogen is produced. The concentration of [H] decreases near the surface due to recombination reactions with the diamond surface [3.55, 56].

Based on their modeling results, Kim and Cappelli [3.57] postulated that optimum growth conditions were related to competition between oxidative pyrolysis and cyclization of postflame hydrocarbons. Simulations at Sandia National Laboratories indicated the importance of operating at high flame speeds to increase atomic hydrogen production [3.58]. Wolden et al. [3.50] showed that there is a strong coupling between the CH_x radical distribution and the rate of atomic hydrogen recombination on the diamond surface. The calculations of Goodwin et al. [3.59] confirmed that the chemical environment at the substrate is very similar to that in other CVD environments. For additional information on flame chemistry, the reader is directed to [3.7, 50, 51, 54–59].

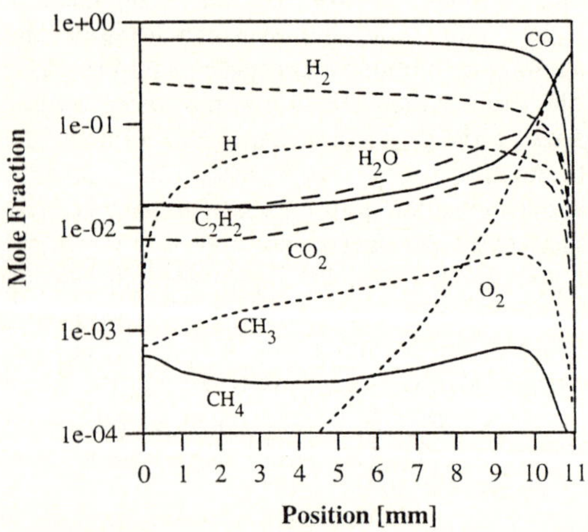

Fig. 3.8: Profiles of selected species between the burner (11 mm) and the substrate (0 mm) at diamond growth conditions: $R = 1.00$, $P = 47$ Torr, $T_s = 750°C$, $U_0 = 600$ cm/s (reproduced from [3.7]).

3.4.3 Alternative Fuels

Although acetylene is the most common fuel used for combustion synthesis, a number of alternatives have been examined, including hydrogen [3.60], ethylene [3.61], MAPP[1] gas [3.62] and methane [3.63]. The interest in other fuels is

---

[1] MAPP gas is a commercially available mixture of about half liquefied petroleum gas (mostly propylene), and half C3H4 (a mixture of the two isomers: methyl acetylene and propadiene).

Table 3.2: Adiabatic flame temperatures and flame speeds of selected fuels.

Fuel	Adiabatic temperature (K)[*]	Flame speed (cm/s)
Acetylene : C_2H_2	2910	141.0
Ethylene : C_2H_4	2565	68.3
Hydrogen : H_2	2525	264.8
Propylene : C_3H_6	2505	43.8
Methane : CH_4	2325	33.8

[*] Assumes base temperature = 25 °C; oxidizer = air; equivalence ratio = 1.0.

economic due to the relatively high cost of acetylene. Diamond has been deposited successfully in each case, but the growth rates were significantly less than with acetylene. The superior nature of acetylene is due to its high flame velocity and temperature. Table 3.2 compares these properties for a number of fuels. Examining Table 3.2, one finds that the flame temperature of acetylene is at least 350 K hotter than the alternatives. All of these fuels produce hydrogen, but the fraction of H_2 that dissociates into atomic hydrogen is critical for diamond deposition. The dissociation fraction has a strong exponential dependence on temperature, which explains why the other fuels are inferior to acetylene for diamond production.

3.5 Applications

Combustion CVD was invented several years after the discovery of hot-filament and plasma techniques, and consequently the majority of applied research has been performed in these conventional reactors. However, the conducted studies demonstrate that the combustion grown diamond is comparable to material produced in other systems. Some of these examples are described in this section, with the intention of demonstrating the flexible nature of combustion CVD.

3.5.1 Textured Film Growth

Thermal [3.64] and electronic [3.65] properties of diamond are adversely affected by the presence of grain boundaries. For many applications [3.65, 66], highly oriented {100} films have superior properties, since both the number and the angle of the grain boundaries are significantly reduced in these films relative to randomly oriented films [3.26, 46]. In addition, {100} growth provides a means of planarizing the growing surface [3.46], enabling the use of conventional silicon technology to manufacture device structures without expensive grinding and polishing steps [3.66]. Figure 3.9 shows a scanning electron micrograph of a highly oriented {100} film that was produced by a two-step process. First, oriented nucleation was achieved using bias-enhanced techniques on silicon in a conventional microwave-plasma reactor [3.67]. The substrate was transferred to a

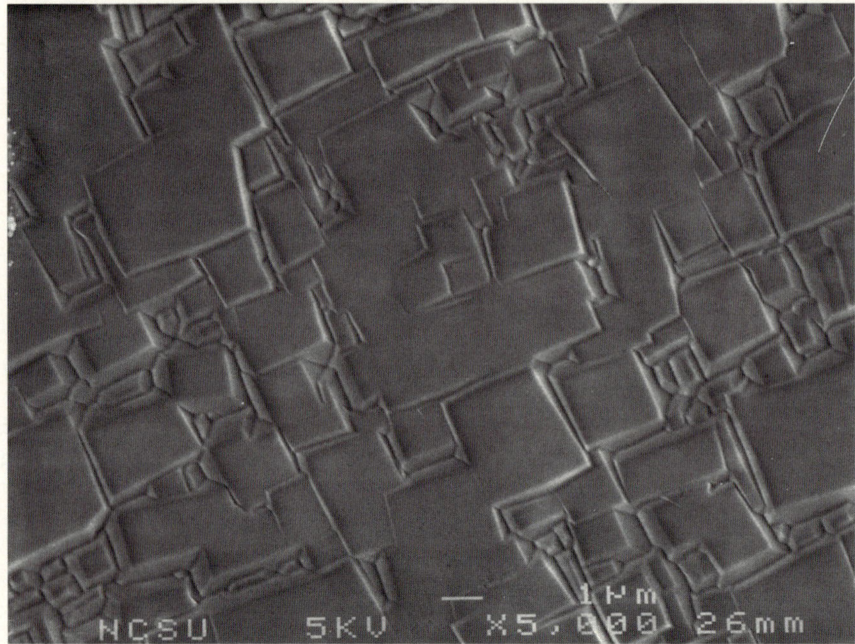

Fig. 3.9: A scanning electron micrograph of a highly oriented {100} film formed by a combination of bias-enhanced nucleation in a microwave reactor and textured growth in a low-pressure flat-flame CVD system (reproduced from [3.67]).

low-pressure flat-flame reactor for 3 hours of growth under conditions that favored a {100} texture [3.46]. Accurate substrate temperature control was requisite for achieving the film texture. It may be observed in Fig. 3.9 that individual grains have coalesced to form a very flat surface. These films have been used in the successful production of high-energy particle detectors [3.66].

3.5.2 Dopant Incorporation

Ravi and coworkers [3.68] demonstrated the high purity of combustion-grown diamond for electrical applications. Films having resistivities two orders of magnitude greater than natural diamond were produced. The doping of diamond imparts semiconductor properties that are required for a number of applications [3.65]. Most of the limited work on doping of combustion-grown diamond has been conducted at the Naval Research Laboratory [3.69–72]. Glesener and coworkers [3.69] fabricated Schottky diodes by depositing boron-doped diamond on molybdenum. Further studies [3.70, 71] indicated that high surface resistance was critical to the formation of good junctions. Boron was added as a dopant by bubbling part of the acetylene flow through a solution of diboric trioxide in acetone. Doverspike and coworkers [3.72] used an atomizer to entrain aerosol droplets that contained dopant species into the gas flow. Boric acid dissolved in methanol was used as a dopant solution. Recently, boron incorporation has been achieved without the use of a solvent. Trimethylborate is a high vapor pressure liquid that may be delivered using a bubbler and ordinary mass flow controllers, without the need for an additional solvent [3.73].

3.5.3 Low-Temperature Deposition

Low-temperature processing is required for compatibility with a number of opto-electronic materials. In addition, low-temperature deposition minimizes the stresses arising when diamond is grown on material with differing coefficients of thermal expansion. To date, most published low-temperature work has been performed in microwave plasmas [3.74, 75]. However, the deposition of diamond at rates of ~1 µm/hr on glass substrates at low temperature (~750 K) in a low-pressure combustion reactor was recently reported [3.47]. The low substrate temperature was achieved by increasing the burner–substrate distance to 20 mm. Continuous, adherent films were deposited on Vycor and Pyrex glasses. The best films were optically transparent, and no damage to the substrates was observed.

3.6 Conclusions

Although material deposited by combustion CVD has not been tested extensively, there is every indication that the quality is on par with diamond produced in hot-filament and plasma systems. Schottky diodes and high-energy particle detectors have been successfully fabricated using combustion-grown diamond. Issues of large-area deposition and uniformity have been addressed with flat-flame burners. Growth rates are exceeded only by atmospheric plasma torches. Important features such as textured growth, boron doping, and low-temperature deposition have all been achieved in combustion reactors. Oriented nucleation, which has been achieved by substrate biasing in microwave and hot-filament reactors, has not been demonstrated in a combustion CVD system.

Economically, combustion CVD is perceived as being a rather expensive technique, due to the cost and volume of acetylene required to produce diamond. A recent study [3.76] based on data from a low-pressure flat-flame reactor indicates that the cost of diamond produced by combustion CVD is somewhat more expensive than that from low-pressure microwave reactors. Combustion CVD requires less capital investment, but these savings are offset by high material costs associated with the large gas volumes that are processed. However, substantial savings would be realized by improvements in growth rates and carbon utilization efficiency.

Acknowledgments

The authors would like to thank Dr. John Prater of the US Army Research Office for stimulating and supporting our work in combustion CVD under contract DAAH04–93–D–0003. One of us (CAW) acknowledges his support as a National Research Council Postdoctoral Fellow.

References

3.1   Y. Hirose and N. Kondo, Program and Book of Abstracts: Japan Applied Physics 1988 Spring Meeting, Japanese Physical Society, Tokyo (1988), p. 434

3.2   L.M. Hanssen, W.A. Carrington, J.E. Butler, and K.A. Snail, Mater. Lett. **7**, 289 (1988)

3.3   K.V. Ravi, Diamond Rel. Mater. **4**, 243 (1995)

3.4   Y. Hirose, S. Amanuma, and K. Komaki, J. Appl. Phys. **68**, 6401 (1990)

3.5   P.K. Bachmann, D. Leers, and H. Lydtin, Diamond Rel. Mater. **1**, 1 (1991)

3.6   Y. Matsui, A. Yukki, M. Sahara and Y. Hirose, Jpn. J. Appl. Phys. **28**, 1718 (1989)

3.7   C.A. Wolden, Z. Sitar, and J.T. Prater, Combust. Flame, in press (1997)

3.8   K.A. Snail and L.M. Hanssen, J. Cryst. Growth **112**, 651 (1991)

3.9   G. Janssen, W.J.P. van Enckevort, J.J.D. Schaminee, W. Vollenberg, L.J. Giling, and M. Seal, J. Cryst. Growth **104**, 752 (1990)

3.10 T. Abe, M. Suemitsu, N. Miyamoto, and N. Sato, J. Appl. Phys. **73**, 971 (1993)

3.11 X.H. Wang, W. Zhu, J. von Windheim, and J.T. Glass, J. Cryst. Growth **129**, 45 (1993)

3.12 D.B. Oakes, J.E. Butler, K.A. Snail, W.A. Carrington, and L.M. Hanssen, J. Appl. Phys. **69**, 2602 (1991)

3.13 L.M. Hanssen, K.A. Snail, W.A. Carrington, J.E. Butler, S. Kellog, and D.B. Oakes, Thin Solid Films **196**, 271 (1991)

3.14 P.W. Morrsion Jr. and J.T. Glass, in Properties and Growth of Diamond, ed. G. Davies, INSPEC, London (1994), p. 380

3.15 R.A. Weimer, T.P. Thorpe, and K.A. Snail, J. Appl. Phys. **77**, 641 (1995)

3.16 K.A. Snail and C.M. Marks, Appl. Phys. Lett. **60**, 3135 (1992)

3.17 (a) C.A. Wolden, Z. Sitar, R.F. Davis, and J.T. Prater, Appl. Phys. Lett. **69**, 2258 (1996); (b) C.A. Wolden, Z. Sitar, and J.T. Prater, submitted to J. Appl. Phys. (1997)

3.18 J. Schermer, J.E.M. Hogenkamp, G.C.J. Otter, G. Janssen, W.J.P. van Enckevort, and L.J. Giling, Diamond Rel. Mater. **2**, 1149 (1993)

3.19 J.S. Kim and M.A. Cappelli, J. Mater. Res. **10**, 149 (1995)

3.20 C.A. Wolden and K.K. Gleason, Diamond Rel. Mater. **5**, 1503 (1996)

3.21 G.H. Ma, Y. Hirose, S. Amanuma, M. McClure, J.T. Prater, and J.T. Glass, in New Diamond Science and Technology, eds. R. Messier, J.T. Glass, J.E. Butler, and R. Roy, Materials Research Society, Pittsburgh, PA (1991), p. 587

3.22 K.V. Ravi and A. Joshi, Appl. Phys. Lett. **58**, 246 (1991)

3.23 K.V. Ravi, C.A. Koch, H.S. Hu, and A. Joshi, J. Mater. Res. **5**, 2356 (1990)

3.24 K. Hirabayashi and Y. Hirose, Diamond Rel. Mater. **5**, 48 (1996)

3.25 C. Wild, R. Kohl, N. Herres, W. Muller-Sebert, and P. Koidl, Diamond Rel. Mater. **3**, 373 (1994)

3.26 B.R. Stoner, S.R. Sahaida, J.P. Bade, P. Southworth, and P.J. Ellis, J. Mater. Res. **8**, 1334 (1993)

3.27 K.A. Snail, R.G. Vardiman, J.P. Estrera, J.W. Glesener, C. Merzbacher, C.J. Craigie, C.M. Marks, R. Glosser, and J.A. Freitas Jr., J. Appl. Phys. **74**, 7561 (1993)

3.28 J.J. Schermer, L.J. Giling, and P. Alers, J. Appl. Phys. **78**, 2376 (1995)

3.29 K.A. Snail, C.L. Vold, C.M. Marks, and J.A. Freitas Jr., Diamond Rel. Mater. **1**, 180 (1992)

3.30 J.J. Schermer, W.A.L.M. Elst, L.J. Giling, Diamond Rel. Mater. **4**, 1113 (1995)

3.31 W. Zhu, B.H. Tan, J. Ahn, and H.S. Tan, J. Mater. Res. **30**, 2130 (1995)

3.32 R.C. Aldredge and D.G. Goodwin, J. Mater. Res. **9**, 80 (1994)

3.33 T. Abe, M. Suemitsu, and N. Miyamoto, J. Cryst. Growth **143**, 206 (1994)

3.34 M. Murakawa and S. Takeuchi, Surf. Coat. Technol. **54/55**, 403 (1992)

3.35 K. Komaki, M. Yanagisawa, I. Yamamoto, and Y. Hirose, Jpn. J. Appl. Phys. **32**, 1814 (1993)

3.36 D.Y. Wang, Y.H. Song, J.J. Wang, and R.Y. Cheng, Diamond Rel. Mater. **2**, 304 (1993)

3.37 J.A. Cooper Jr. and W.A. Yarbrough, Diamond Optics, SPIE **1325**, 41 (1990)

3.38 N.G. Glumac and D.G. Goodwin, Mater. Lett. **18**, 119 (1993)

3.39 J. Powling, Fuel **28**, 25 (1949)

3.40 W.T. Bielder and H.E. Hoeschler, Jet Propulsion **27**, 1257 (1957)

3.41 K.F. McCarty, E. Meeks, R.J. Kee and A.E. Lutz, Appl. Phys. Lett. **63**, 1498 (1993)

3.42 M. Murayama, S. Kojima, and K. Uchida, J. Appl. Phys. **69**, 7924 (1991)

3.43 M. Murayama and K. Uchida, Combust. Flame **91**, 239 (1992)

3.44 D.W. Hahn, C.F. Edwards, K.F. McCarty, and R.J. Kee, Appl. Phys. Lett. **68**, 2158 (1996)

3.45 C.F. Edwards and D.W. Hahn, Proceedings of ILASS-95, Institute for Liquid Atomization and Spray Systems, Troy, MI (1995) p. 34

3.46 C.A. Wolden, S.K. Han, M.T. McClure, Z. Sitar, and J.T. Prater, Mater. Lett. **32**, 9 (1997)

3.47 C.A. Wolden, R.F. Davis, Z. Sitar, and J.T. Prater, Diamond Rel. Mater. **6**, 1862 (1997)

3.48 M.A. Cappelli and P.H. Paul, J. Appl. Phys. **67**, 2596 (1990)

3.49 R.J.H. Klein-Douwel, J.J.L. Spaanjaars, and J.J. ter Meulen, J. Appl. Phys. **78**, 2086 (1995)

3.50 C.A. Wolden, Z. Sitar, R.F. Davis, and J.T. Prater, J. Mater. Res. **12**, 2733 (1997)

3.51 N.G. Glumac and D.G. Goodwin, Combust. Flame **105**, 321 (1996)

3.52 G.H. Evans and R. Grief, Numer. Heat Transf. **14**, 373 (1988)

3.53 R.J. Kee, J.A. Miller, G.H. Evans, and G. Dixon-Lewis, in Twenty-second Symposium (International) on Combustion, The Combustion Institute, Pittsburgh, PA (1988), p. 1479

3.54 J.A. Miller and C.F. Melius, Combust. Flame **91**, 21 (1992)

3.55 C.A. Wolden, S. Mitra, and K.K. Gleason, J. Appl. Phys. **72**, 3750 (1992)

3.56 S.J. Harris and A.M. Weiner, J. Appl. Phys. **74**, 1022 (1993)

3.57 J.S. Kim and M.A. Cappelli, J. Appl. Phys. **72**, 5461 (1992)

3.58 E. Meeks, R.J. Kee, D.S. Dandy, and M.E. Coltrin, Combust. Flame **92**, 144 (1993)

3.59 D.G. Goodwin, N.G. Glumac, and H.S. Shin, in Twenty-sixth Symposium (International) on Combustion, The Combustion Institute, Pittsburgh, PA (1996), p. 1817

3.60 N.G. Glumac and D.G. Goodwin, Appl. Phys. Lett. **60**, 2695 (1992)

3.61 J.S. Kim and M.A. Cappelli, Appl. Phys. Lett. **65**, 2786 (1994)

3.62 S.J. Harris, H.S. Shin, and N.G. Glumac, Appl. Phys. Lett. **66**, 891 (1995)

3.63 J.S. Kim and M.A. Cappelli, Appl. Phys. Lett. **67**, 1081 (1995)

3.64 J.E. Graebner, S. Jin, G.W. Kammlot, B. Bacon, L. Seibles, and W.F. Banholzer, J. Appl. Phys. **71**, 5353 (1992)

3.65 B.A. Fox, B.R. Stoner, D. Malta, P. Ellis, R.C. Glass, and F.R. Sivazlian, Diamond Rel. Mater. **3**, 382 (1994)

3.66 Vlahovic et al., Proceedings of PANIC (Particle and Nuclear Interaction Conference), Williamsburg, VA (1996)

3.67 S.D. Wolter, B.R. Stoner, J.T. Glass, P.J. Ellis, D.S. Buhaenko, C.E. Jenkins, and P. Southworth, Appl. Phys. Lett. **62**, 1215 (1993)

3.68 K.V. Ravi, C.A. Koch, D.S. Olson, P. Choong, J.W. Vandersande, and L.D. Zoltan, Appl. Phys. Lett. **64**, 2229 (1994)

3.69 J.W. Glesener, A.A. Morrish, and K.A. Snail, J. Appl. Phys. **70**, 5144 (1991)

3.70 J.W. Glesener, A.A. Morrish, and K.A. Snail, Appl. Phys. Lett. **61**, 429 (1992)

3.71 J.W. Glesener, K.A. Snail, and A.A. Morrish, Appl. Phys. Lett. **62**, 181 (1993)

3.72 K. Doverspike, J.E. Butler, and J.A. Freitas Jr., Diamond Rel. Mater. **2**, 1078 (1993)

3.73 C.A. Wolden, Z. Sitar, and J.T. Prater, accepted for Diamond Rel. Mater. (1998)

3.74 R.K. Singh, D. Gilbert, R. Tellshow, P.H. Holloway, R. Ochoa, J.H. Simmons, and R. Koba, Appl. Phys. Lett. **61**, 2863 (1992)

3.75 Z. Ring, T.D. Mantei, S. Tlali, and H.E. Jackson, Appl. Phys. Lett. **66**, 3380 (1995)

3.76 M. Stonefield and C.A. Wolden, unpublished results (1997)

4. Plasma-Jet Deposition of Diamond

Mark A. Cappelli and Thomas G. Owano

High Temperature Gasdynamics Laboratory, Mechanical Engineering Department,
Stanford University, Stanford, CA 94305-3032, USA
e-mail: cap@leland.stanford.edu

Springer Series in Materials Processing
Low-Pressure Synthetic Diamond Eds.: B. Dischler and C. Wild
© Springer-Verlag Berlin Heidelberg 1998

4.1 Introduction

The deposition of diamond using moderate- and high-pressure plasma jets is a promising alternative to the more conventional low-pressure plasma discharges (microwave, RF and DC glow discharges) and hot-filament chemical vapor deposition methods. Introduced in the mid- to late-1980s, these so-called "thermal" plasma methods are now widely use in the synthesis of diamond films.

A plasma jet is a generic expression for a moderate to high pressure (~100 Torr – 1 atm) plasma discharge in which convection plays a significant role in transport processes. Plasma jets used in the deposition of polycrystalline diamond can be categorized by the methods in which the electrical discharge is sustained. The most commonly employed plasma jet is the direct-current (DC) arcjet discharge, which derives its name from the steady electric field that is used to drive high, relatively steady currents through the flowing, ionized, process gas. Other emerging technologies include the electrodeless arc discharges such as the radio-frequency (RF) inductively coupled plasma-jet and microwave (MW) plasma-jet sources. The RF plasma is supported by the time-varying electrical currents induced by the oscillating magnetic field. The MW plasma jet is supported by the time-varying electrical currents generated by high-frequency (typically 915 MHz or 2.45 GHz) electric fields. A schematic illustration of these three types of plasma jets used in diamond synthesis is presented in Fig. 4.1.

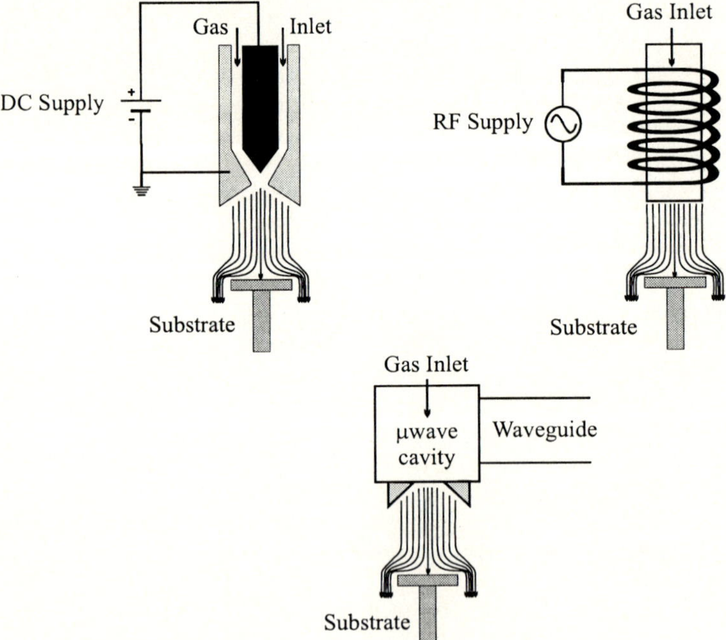

Fig. 4.1: A schematic of commonly employed plasma jets for diamond deposition.

In all cases, energy is transferred from the electric field to the free electrons, which subsequently transfer their energy to the process gas by ohmic dissipation and nonelastic collisions (dissociation, excitation, ionization). The time-averaged rate of power deposited (per unit volume), P_Ω, for a given time-averaged current density, J, is

$$P_\Omega = J^2 / \sigma \tag{4.1}$$

where, σ is the frequency-dependent plasma conductivity, which could depend on the temperature T, pressure p, and ionization fraction in the plasma. For a weakly ionized plasma in thermodynamic equilibrium, the plasma conductivity is given by

$$\sigma = \chi_e^{eq}\, \frac{kT}{p}\, \frac{e^2 v_m}{m_e(v_m^2 + \omega^2)}. \tag{4.2}$$

Here, χ_e^{eq} is the equilibrium electron mole fraction, v_m is the electron momentum-transfer collision frequency (which is inversely related to the electron mean free path), and ω is the field frequency. It can be seen that for $\omega \gg v_m$, as is often the case for microwave-sustained plasmas, and for all else equal, the ratio of the dissipated power for a high-frequency discharge to that of a low-frequency, or DC discharge, scales as $(\omega/v_m)^2$. Another interpretation is that less current density is required to deposit the same power in a high-frequency discharge than in a low frequency, or DC-discharge. This illustrates some of the advantages realized in using high-frequency discharge plasma sources in materials processing.

In a flowing gas, ohmic dissipation leads to an increase in the gas enthalpy and kinetic energy. In moderate- to high-pressure plasmas, relatively high gas temperatures can be achieved (3000–30 000 K). In molecular source gases, this leads to the production of a very reactive, partially ionized and dissociated gas stream. When hydrogen is introduced as a process gas, as in the case of diamond deposition, the temperatures achieved are sufficiently high to produce copious levels of atomic hydrogen, a necessary precursor in the synthesis of high-quality diamond.

In this chapter we shall discuss these plasma jets and their applications to diamond deposition, with a greater emphasis on DC arcjets, as they are the most commonly used plasma jet for diamond CVD. In the next section we shall present a brief overview of their operating principles and an historical perspective on their use in the synthesis of diamond films. In recent years, a growing number of researchers have studied the properties of these plasma flows in attempts to better understand the diamond synthesis process. Because of the growing number of papers published on this subject, it is not possible to cite and discuss each one in this review: however, we hope that this review will serve as a starting point for those interested in learning more about the use of plasma-jet processes in diamond CVD.

DC Arcjet Operating Characteristics

A schematic illustration of a typical DC arcjet in use as a source for diamond CVD is shown in Fig. 4.1. Electrical energy is converted to thermal and kinetic energy of a flowing gas mixture by an electric arc discharge. Like other CVD methods, a major constituent of the gas mixture is hydrogen, which is needed to provide a source of atomic hydrogen, and methane is most often introduced into the plasma jet to provide a source of carbon. The average plasma temperature is sufficiently high (~5000 K) to partially dissociate the gas. The plasma jet containing these reactive species impinges onto a cooled substrate surface (T_s ~ 1000–1500 K). Molybdenum is often used as a substrate, although other materials have also been employed [4.11]. In arcjet deposition of diamond, pretreatment of the substrate by polishing with diamond paste is not a necessary condition for film growth: however, it greatly enhances the nucleation density [4.18].

In most designs, the electric discharge is sustained between a concentric cathode rod and a surrounding cylindrical anode, creating an arc column (localized path of high current density) that is nearly fully ionized. Positive current (I) flows from the anode to the cathode, establishing the voltage drop (V) required to dissipate the total power ($P = IV$) in the arc. The current is generally concentrated in the hot region of the plasma due to the thermal pinch effect [4.1]. At sufficiently high current densities, self-induced magnetic fields further confine the motion of the charged particles within the arc column. The arc column diameter is strongly influenced by the conduction and convection of energy away from its core to the colder surrounding gas. The conductivity of the plasma in the core of a fully ionized arc varies as $T_{arc}^{3/2}$ [4.24]. As a result, an increase in arc current further increases the temperature, and this results in an increased conductivity which thereby reduces the arc voltage. Except for regions very near the electrodes,

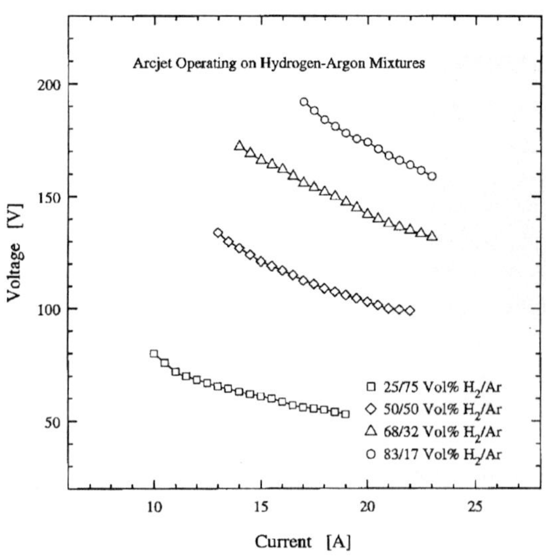

Fig. 4.3: Typical current–voltage characteristics of a supersonic DC arcjet used in diamond CVD [4.14].

there is a linear variation in the potential along the arc column from the cathode to the anode. In this respect, the arc column, or positive column, behaves as a resistor with a negative impedance. The actual length of the arc column is also determined by electromagnetic and gasdynamic forces acting on it. Increased mass flow rates tend to increase the arc length and reduce the diameter, thereby increasing the resistance to current flow and hence raising the operating voltage. The arc voltage established for any given operating current is therefore strongly influenced by the plasma gas mixture and by flow conditions such as the operating discharge chamber pressure and mass flow rates. An example of the sensitivity of the current–voltage characteristics of an arcjet used in diamond CVD to operating mixtures and mass flow rates is shown in Fig. 4.3 [4.14].

Electrode Processes

The thermionic emission of electrons occurs at the tip of the cathode. Like the arc column, the arc attachment at the cathode is highly constricted, implying that electron current is emitted from a small area (characterized by a spot diameter that is less than the cathode diameter). The spot at which the electron current is emitted is heated primarily by the concomitant ion bombardment of the cathode surface. This ion current, a consequence of the finite mobility of ions, is necessary to heat the cathode to temperatures that can support the high current densities ($\sim 10^6$–10^7 A/m^2). The relationship between the cathode current density and the cathode temperature, T_{cat}, is given by the Richardson–Duschman Law [4.25]:

$$J_{cat} = A T_{cat}^2 \exp-\left(\frac{\phi_w}{kT_{cat}}\right) \tag{4.4}$$

where $A = 1.2 \times 10^6$ A/m^2/K^2, and ϕ_w is the work function of the cathode material.

Selection of a cathode material with a low work function is desirable, so as to obtain the highest current densities possible for a given cathode temperature: 2% thoriated tungsten has a relatively low work function of only 3.35 eV, and is often employed as a cathode material. The arc constriction and ion bombardment generally results in a cathode tip that is in a molten state. It is interesting to note that a molten pool of tungsten (melting temperature of about 3400 K), 1 mm in diameter, can support an electron current of 11 A.

At extremely high current densities ($\sim 10^8$ A/m^2), field-enhanced thermionic emission may also be as important as thermionic emission [4.26]. The electrode heating process driven by ion bombardment also causes cathode erosion, contaminating the plasma with cathode material; a source of potential contamination of the diamond film. In some cases, the cathode may be water-cooled; however, in most DC arcjet sources, the primary source gas (i.e., mixtures of hydrogen and argon) is injected upstream of the cathode, providing in itself a means of cathode cooling beyond that achieved by conduction alone.

The electron current is collected at the anode, which, in most designs, is water-cooled to a temperature well below that of the arc. The strong property gradients and charged particle transport arising from these gradients that ensue near the anode due to this cooling give rise to relatively strong electric fields, and a potential drop between the anode and plasma that is often greater than that determined by the plasma arc column resistance alone. This potential fall is the so-called anode fall voltage. Electrons accelerated by this fall voltage gain energy and, when captured by the anode, transfer this energy (along with that released by the capture process – that associated with the anode work function) to the anode. This ubiquitous heating process may also lead to anode erosion and plasma contamination when the arc attachment to the anode is constricted.

Anode damage can be mitigated by methods such as magnetic stirring (applying a magnetic field to destabilize the constricted attachment mode and elongate the arc) [4.27]; the use of gasdynamic designs that force the anode to attach in a more diffuse mode [4.28]; or reducing the current density by elongating the arc in a segmented anode discharge [4.29]. All of these nontraditional variations have been incorporated into arcjet designs that have been used to deposit diamond films [4.14, 15, 29, 30].

In a DC arcjet, the primary source gas mixture is that gas mixture that flows through the discharge chamber with which the arcjet can be stably operated. The source gas mixture is the major factor that determines the operating voltage of the arc. A primary consideration in selecting the plasma gas-mixture is this operating voltage, and the availability of inexpensive and efficient (i.e., operating without a resistive ballast) power supplies that support this voltage. Since arc welders conventionally operate with argon, power supplies that are designed for voltages in the 20–80 V range are readily available. As a result, argon, a relatively inexpensive and inert gas, is often the gas of choice in most arcjet diamond CVD systems. Hydrogen is an additive to the argon as a source gas, and the addition of even small amounts of hydrogen to the argon (< 20% by volume) can significantly affect the current–voltage characteristics (see Fig. 4.3). Hydrogen may also be added downstream of the plasma jet, along with the carbon-containing reactants such as methane [4.31]. Setting the power supply constraints aside, a relatively straightforward analysis of the atomic hydrogen production efficiency in such an arcjet flow favors the use of pure hydrogen as the source gas, since there is a penalty paid for the seemingly unnecessary heating and ionization of the argon. However, it should be noted that no comprehensive studies exist that indicate that argon is entirely inert in such CVD systems. It is well known that argon atoms and ions have metastable electronically excited states that can participate via collisions in energy transfer to molecular species, thereby affecting the composition of the reactive mixture.

Recent Progress in the Study of DC Arcjet Diamond CVD

The relatively high growth rates afforded by DC arcjet diamond CVD provided some of the first convenient opportunities to study the morphological changes that occur during diamond crystal growth [4.32]. Scanning electron microscope images of the growth surface of diamond films deposited by a DC arcjet under varying methane to hydrogen ratios are shown in Fig. 4.4. The growth habit of DC arcjet

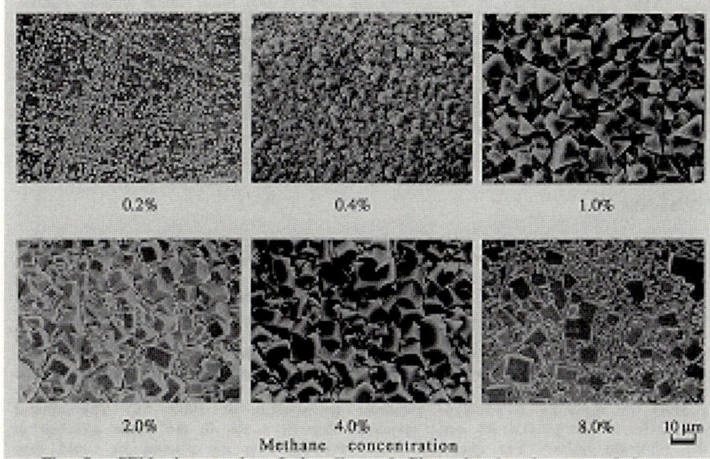

Fig. 4.4: SEM images of the growth surface of DC arcjet deposited diamond films for various percentages of methane in hydrogen [4.32].

deposited diamond at substrate temperatures of approximately 1000°C was found to change dramatically, from a predominantly {111} surface morphology at low volume fractions of methane in hydrogen (~1%) to a {110} and {100} surface morphology at relatively high volume fraction (~8%). Mixed morphologies were seen in between these values.

A similar trend was observed when the substrate-to-nozzle distance was varied (increased distances favored a shift from {111} to {110} or {100} morphologies). These results indicated that the growth of any particular crystallographic low-index plane was strongly correlated to the mole fraction ratios, [CH]/[H] and [C]/[H]. Using optical emission spectroscopy, strong variations in plasma composition were measured along the axial direction between the plasma-jet nozzle and substrate, in part because of the strong tendency for atomic hydrogen to diffuse towards the jet boundary more readily than heavier hydrocarbon fragments.

High-resolution scanning electron microscope analyses on DC arcjet deposited diamond films were performed by Ohtake et al. [4.33] and by Hirabayashi et al. [4.34]. These studies clearly revealed the presence of steps on both the {100} and {111} surfaces. In comparison to the {111} surface, the {100} was more resistant to subsequent plasma etching in air, leading to the suggestion that the {111} growth surface was more defective than the {100} surface, an argument that was supported by micro-Raman analysis of the deposits. A later study by the same authors revealed that the addition of oxygen to the arcjet the plasma gas dramatically altered the growth morphology [4.35]. With added oxygen, the films were smoother in texture and the {111} faces showed improved crystallinity. However, oxygen addition also reduced the overall growth rate. The authors concluded that the improved crystallinity arises from a reduction in the amount of active carbon and hydrocarbon radicals in the gas phase due to the formation of carbonmonoxide.

Consistent with the findings of many other studies of diamond CVD, the quality of the diamond deposited in DC arcjets degrades very rapidly with increasing methane to hydrogen volume flow ratios [4.7]. This suggests at first that the chemical mechanism controlling the synthesis of diamond over graphite in such high deposition rate arcjets is similar to that in other CVD systems, where the diamond growth rate can be orders of magnitude lower. Goodwin [4.36] has shown that such a condition is possible if the methyl radical (CH_3) is the growth species and if the surface growth mechanism follows that proposed by Harris [4.37]. However, in thermal plasmas such as DC arcjets, atomic carbon may be as abundant as CH_3. Recent theoretical calculations by Yu and Girshick [4.38] suggest that in many thermal plasma jets such as DC arcjet flows operating at modest pressures, atomic carbon can account for the observed growth rates.

In many of the first arcjet diamond CVD studies, the hydrocarbon source gas (usually methane) was mixed with the plasma carrier gas and introduced upstream of the cathode [4.2, 9, 32, 39]. Although upstream injection was attempted by Ohtake and Yoshikawa [4.7], it was found that the plasma was unstable and that the deposited films were of poor quality. As a result, their arcjet operating conditions specified downstream remote injection near the point at which the plasma exited the discharge nozzle. In the supersonic DC arcjet studies of Loh and Cappelli [4.14, 4, 40], diamond was deposited with either CH_4 or acetylene (C_2H_2). The point of injection was downstream of the nozzle and was selected to optimize the transport of either CH_3 radicals or C_2H_2 to the substrate, as it has been speculated that even acetylene can act as a precursor to diamond growth [4.41]. Recent experimental and modeling studies [4.42, 43] of diamond growth in a more conventional arcjet plasma (argon carrier gas and injected hydrogen) indicate that some increase in growth rate and quality can be obtained by optimizing the point of injection.

Extremely high diamond growth rates (~1 mm/h) have been achieved in DC arcjets that use liquid precursors as the carbon source [4.44]. In that study, the liquid precursors included a variety of ketones, alcohols, halogenated, and aromatic compounds, injected through the substrate opposite the direction of the impinging plasma jet. The highest growth rates were observed with ethanol and acetone.

Attempts have been made to enhance the diamond film growth rates in DC arcjets by applying a positive electrical bias to the substrates [4.45–47]. Using a 100 kW DC arcjet expanding to atmospheric pressure, Baldwin et al. [4.47] has shown that a six-fold increase in growth rate can be induced with only a modest secondary discharge (115 V, 3 A) while maintaining uniformity over a 1 cm diameter substrate (see Fig. 4.5). In such atmospheric-pressure discharges, there is considerable boundary-layer recombination of atomic hydrogen. The secondary discharge is believed to increase the electron temperature so as to maintain an elevated atomic hydrogen concentration in the near-surface region.

Measurements of the plasma temperature, velocity and plasma composition in arcjets under conditions of diamond growth are needed so as to better understand the energy balance in the system [4.48], and to validate detailed DC arcjet CVD

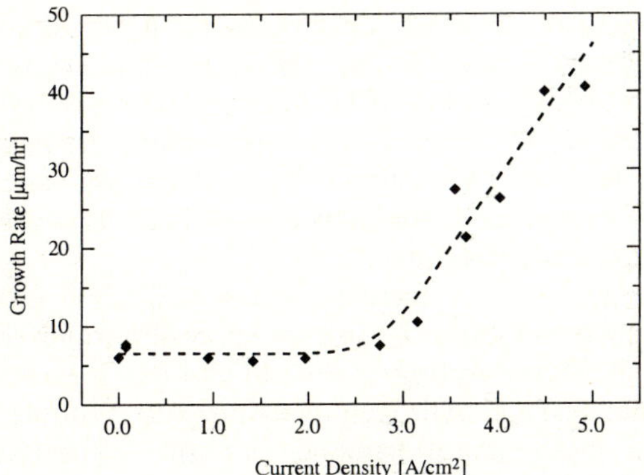

Fig. 4.5: Variation in growth rate with positive substrate bias during atmospheric pressure DC arcjet deposition of diamond.

computational flow models that have recently been developed [4.21, 49]. Diagnostics of DC arcjet flows employed in diamond CVD has been limited largely to optical emission spectroscopy [4.30, 50–52]. A limited number of mass sampling studies of the impinging plasma have been performed [4.23, 51, 53]. A few groups have employed laser-induced fluorescence to characterize flow properties [4.48, 54–56] in diamond-depositing DC arcjets.

Comparison is often made between DC arcjet emission and the emission from a CVD-diamond microwave plasma (see, for example [4.9]), where it is found that considerably greater emission is seen from molecular hydrogen. Spectral emission from the atomic hydrogen Balmer series indicates that there is a departure from local thermodynamic equilibrium in these plasmas. As a result, an estimate of the plasma temperature from the relative intensity of electronic transitions in atomic hydrogen is unreliable. Atomic hydrogen laser-induced fluorescence (LIF) measurements of plasma temperature and velocity have been made in supersonic arcjets used in diamond CVD [4.55]. A comparison of optical emission-based measurements with LIF measurements of temperature in a similar arcjet was reported by Jeffries et al. [4.56], and by Brinkman and Jeffries [4.48]. In those studies, it was found that emission-based measurements of arcjet exit temperature using the rotational and vibrational state distributions in electronically excited CH and C_2 radicals gave temperatures in the 3000–7000 K range, whereas LIF measurements of the rotational and vibrational state distributions in the electronic ground state gave temperatures in the 1200–2200 K range. The latter temperatures are consistent with those measured by Loh et al. [4.55]. It is apparent that the interpretation of plasma properties based on optical emission spectroscopy is difficult because of the nonequilibrium nature of the plume.

Arcjets have a distinct advantage over other diamond CVD methods in that the deposition of diamond can proceed simultaneously with the deposition of other ceramics and metals by introducing powders of various types into the plasma stream. Using this approach, Kurihara et al. [4.57, 58] deposited 40 µm thick diamond films on tungsten–molybdenum substrates by spraying an interlayer (70 µm thick) composed of tungsten carbide, followed by a composite diamond–tungstencarbide (40 µm thick) intermediate layer. Such a functional gradient material reduces the thermal stress between the diamond film and the substrate, and increased the adhesion strength by an order of magnitude over that which was obtained in the absence of the tungstencarbide interlayer.

The highest growth rate ever achieved in homoepitaxial diamond CVD was reported by Snail et al. [4.59]. That study employed a triple-arcjet design (referred to as a "triple torch plasma reactor"–TTPR) operating on argon/hydrogen mixtures as the plasma gas and with remote methane and hydrogen injection. The substrates consisted of single-crystal (0.25 mm thick) natural-diamond heat sinks (type IIa) that were cut into circular cross-sections and brazed on to threaded water-cooled molybdenum rods. Growth temperatures ranged from 1200°C to 1400°C, and growth rates were in the range of 100–200 µm/h, depending on the crystallographic plane. An SEM image of a {100} cylindrical shaped seed crystal following deposition for 30 min is shown in Fig. 4.6.

Bradley et al. [4.60] have presented technical and market strategies for DC arcjet CVD-diamond applications. Although in that paper they discuss a wide range of potential applications, it is apparent that DC arcjet processes will be most competitive in applications that make use of thick layers of diamond (so-called thick-film diamond) such as heat sinks for diode lasers, packaging for high-power electronics, and brazed, thick-film tool inserts. However, DC arcjet deposited diamond is beginning to show some advantages in thin-film applications as well. Olsen and Dawes recently compared the performance of tool inserts coated with thin-film diamond using DC arcjets and hot-filament methods [4.61]. Clearly, the range of potential operating conditions and DC arcjet designs have not been

Fig. 4.6: SEM images of a diamond crystal deposited on a cylindrical {100} diamond seed in a DC arcjet plasma: **(a)** 100 µm marker; **(b)** 1 µm marker [4.59].

fully explored, and future research will likely lead to more efficient DC arcjet reactors, higher growth rates, extended deposition areas, and high quality and deposition rates at reduced substrate temperatures.

4.2.2 Radio-Frequency Inductively Coupled Plasma Jet

The Radio-Frequency Inductively Coupled Plasma (RF-ICP) jet derives its name from the way in which RF energy is coupled to the plasma working gas by means of an intense oscillating magnetic field. This electrodeless discharge excites electrons in the gas which dissipate energy, via ohmic heating, into a flowing, high-pressure gas, thereby increasing the gas enthalpy and kinetic energy. In an RF-ICP jet, the plasma is confined to a dielectric walled chamber that contains a nozzle from which the plasma-jet discharges. The RF-ICP jet typically issues into a chamber that is maintained at a pressure equal to or lower than that of the RF-ICP chamber. Such high-enthalpy RF-ICP flows are often used in the spraying of metal and ceramic powders for aerospace coating applications, and are notable for their lack of electrode material contamination in the deposited coatings and their ability to operate on strongly oxidizing gases.

Historical Review of RF Plasma Diamond CVD

The first widely reported use of and RF-ICP jet system for diamond film deposition was by Matsumoto et al. in 1987 [4.62]. In this work, diamond microcrystals and polycrystalline films were synthesized in an atmospheric-pressure RF-ICP, with growth rates up to 60 μm/hr reported. This initial study was confirmed by others in 1990 [4.63, 64], and since then a relatively small number of research groups have pursued RF-ICP jet deposition of diamond thin films for both fundamental and applied studies.

Several groups have mapped the effects of various parameters on the quality and quantity of deposited diamond films [4.65–69], and others have delved into issues such as thermal management of the substrates [4.70, 71], the use of liquid precursors [4.72], enhanced film adhesion via multi-step metallization techniques [4.73], and techniques to increase the deposition area via lowered system pressures [4.74, 75]. More fundamental studies have used RF-ICP jet systems to study the diamond growth process via advanced laser-based diagnostics [4.76–78], gas sampling techniques [4.79, 80], fluid mechanics studies [4.81], and modeling of the growth kinetics [4.38, 65, 66].

RF Plasma Operating Characteristics

The RF-ICP jet systems are notable for several characteristics, including relatively large volume plasma production, an electrodeless discharge, and operation at atmospheric pressure. In an RF-ICP (shown schematically in Figs. 4.1 and 4.7), power is transferred to the ionized, and thus conducting, gas by means of a coil carrying an oscillating current. The current in this coil sets up a strong, pumping magnetic field in which eddy currents form in the cylindrical region of gas near the dielectric containment wall. The depth of penetration of this electromagnetic field into the gas is known as the skin depth, δ, given by:

$$\delta = 5.03/\sqrt{\mu\sigma f} \tag{4.5}$$

where δ is in cm, μ is the relative permeability, σ, as before, is the plasma electrical conductivity ($\Omega^{-1}cm^{-1}$), and f is the operating frequency (MHz). The frequency of the discharge and radius of the dielectric containment wall are chosen to optimize the power deposited into the plasma, and this leads to a typical plasma radius to skin depth ratio of approximately 1.5. Efficiencies of greater than 50% are possible for these systems.

Inside the RF-ICP discharge, the electromagnetic fields accelerate the electrons, and the electron energy is fed into the neutral gas particles by collisions. Ionizing collisions produce more electrons, and lead to a steady-state, hot, luminous, toroidal-shaped discharge. This off-axis maximum in temperature is typical of the discharge region of RF-ICP systems, but the center of the flowing plasma is rapidly heated by conduction and convection, and can lead to quite flat and uniform radial temperature profiles shortly downstream of the coil region. This trait of RF-ICP jets makes them an excellent candidate for relatively uniform, large-area deposition systems.

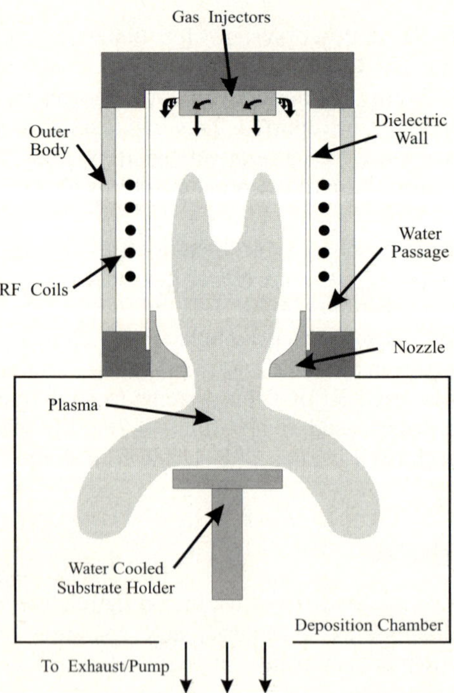

Fig. 4.7: A typical RF-ICP jet diamond deposition system.

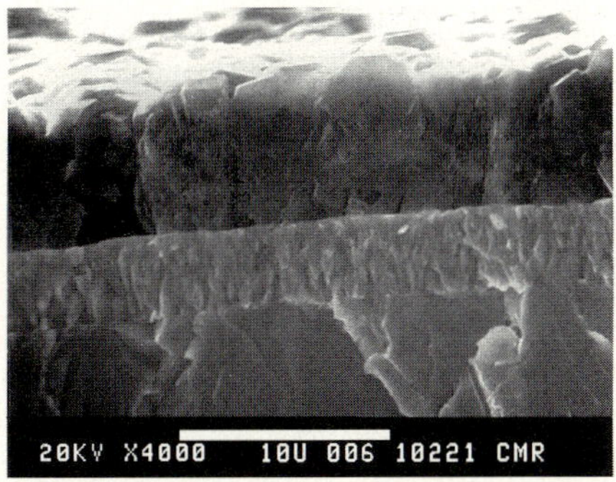

Fig. 4.8: A cross-sectional SEM of an RF-ICP jet deposited film.

The RF-ICP jet systems are considered "electrodeless" because the plasma discharge is separated from the load coil by means of a dielectric containment wall. Removing the physical contact between the coil and the plasma eliminates contamination of the deposition environment by electrode materials, and possible contamination of the environment by deterioration of the dielectric containment wall is typically minimized by water-cooling of the wall and the use of a cold, swirling sheath gas at the outer limits of the plasma discharge. In a moderate- to high-pressure (100–1000 Torr) RF-ICP jet diamond deposition system, frequencies of 3–13.56 MHz, dielectric wall radii of 1–4 cm, and power levels of 10–100 kW are typical.

Design Considerations and Optimal Process Design

RF-ICP jet systems are capable of producing high-quality diamond films at rates near 100 μm/hr over areas approaching 100 cm^2. A cross-sectional SEM of a diamond film produced by an RF-ICP jet system is shown in Fig. 4.8. In this figure, one can see a distinct interfacial layer between the approximately 10 μm thick diamond film and the molybdenum substrate. This interlayer has been identified by XRD analysis as Mo_2C [4.67], and is typical of the growth observed in these high-enthalpy systems. The layer has been observed to thicken as a function of deposition time, which indicates that carbon continues to diffuse into the molybdenum substrate from the nucleation side of the diamond film even after the film has completely overgrown the surface. The presence of this interfacial layer on carbide-forming substrates can strongly influence the adhesion properties of the deposited film.

The growth rate and quality of diamond films produced by RF-ICP jet systems are very sensitive to the operational environment. Several studies have been made to determine the effects of plasma power, substrate temperature, feed gas mixture, substrate material, and other deposition parameters on the deposited films. One such study by Baldwin et al. [4.66] examined the effect of the methane to hydrogen feed gas ratio on the growth rate of deposited films. The results of this study are shown in Fig. 4.9, and compared with model predictions at two different plasma freestream temperatures that bound the experimentally measured values.

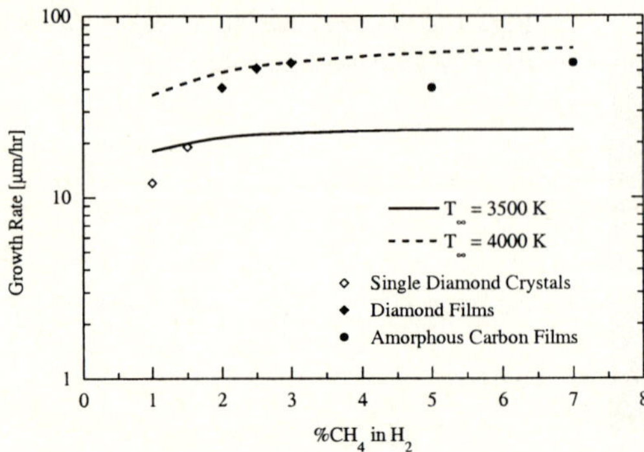

Fig. 4.9: Experimental and predicted growth rates as a function of % CH$_4$ in H$_2$ for a substrate temperature of 1100°C [4.66].

The experimental results compare favorably to the predicted values in the range over which continuous diamond films are produced. At low methane to hydrogen ratios, the computational model overpredicts the growth rate, due to its assumption of diamond on diamond growth and its neglect of nucleation effects.

In efforts to make more detailed comparisons between models of the growth environment and the RF-ICP jet deposition system, advanced optics-based diagnostic techniques have also been utilized to investigate the plasma properties and plasma chemistry [4.76, 82]. In moderate- to high-pressure RF-ICP jet-deposition systems, the benefits of high reactant densities are tempered by the presence of a collision-dominated, chemically reacting boundary layer above the deposition surface, in which important chemical species are both rapidly produced and destroyed. This aspect of high-pressure techniques is strikingly different from low-pressure techniques in which production and diffusion of chemical species control the deposition process. The application of a powerful nonlinear laser spectroscopy, degenerate four-wave mixing (DFWM), as a gas-phase optical diagnostic of these environments has allowed the chemistry of these boundary layers to be studied and compared to detailed models of the growth environment.

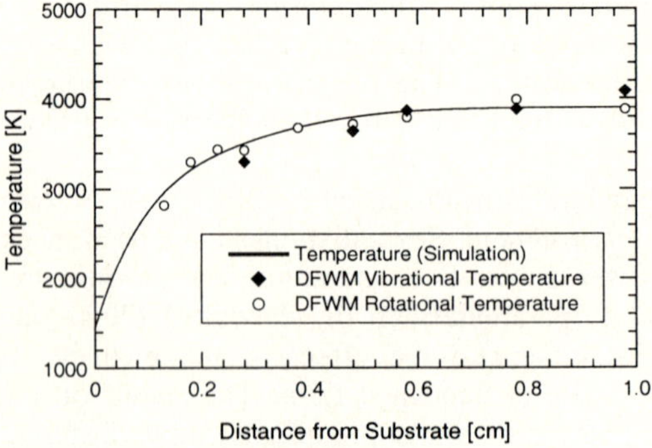

Fig. 4.10: Measured CH vibrational and rotational temperature profiles across the substrate boundary layer [4.82].

The CH radical was probed inside the boundary layer of an RF-ICP jet system with *in-situ* DFWM measurements of the CH $A^2\Delta{\leftarrow}X^2\Pi$ (0,0) system near 431 nm. A comparison of measured CH vibrational and rotational temperatures along the stagnation line of a deposition substrate with values from a computational model is shown in Fig. 4.10. Boundary conditions for the simulation are the measured freestream temperature of 3900 K, an estimated freestream velocity of 8 m/sec, and the measured substrate temperature of 1035°C. We can see in Fig. 4.10 the predicted thermal boundary layer (≈ 6 mm thick) with a steep falloff in temperature very close to the substrate. The measured CH vibrational temperatures are in close agreement with the predictions, and rotational temperature measurements are in good agreement with both the measured vibrational temperatures and the computational simulation.

Measurements of the relative CH mole fraction within the substrate boundary layer are compared with results of the computational simulation in Fig. 4.11. The CH mole fraction is approximately 30 ppm in the freestream (3900 K), and within the approximately 6 mm thick boundary layer it is predicted to first rise (due to production) as the plasma cools toward approximately 3700 K, reaching a peak of approximately 52 ppm at 2.5 mm from the substrate surface, then to be destroyed as the plasma cools further on its approach to the substrate surface.

A more direct insight into the RF-ICP boundary layer chemistry can be gained by comparing the CH mole fraction throughout the boundary layer to its equilibrium value at the measured local temperature. This is displayed in Fig. 4.12, where the CH is observed to remain near equilibrium until approximately 2 mm from the substrate surface, where it then rises to approximately 1000 times the equilibrium value close to the surface. The measured nonequilibrium is in close agreement with the predicted values. By studying the results of the numerical simulation, it has been shown that the nonequilibrium observed in the CH mole fraction is in fact driven by the atomic hydrogen chemistry. The timescale of

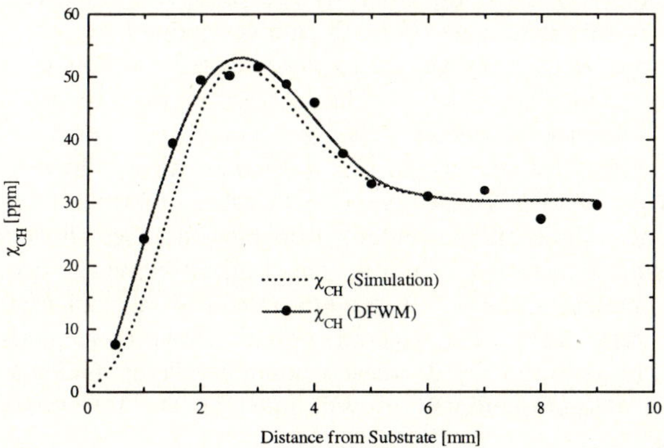

Fig. 4.11: The measured CH concentration profile across the substrate boundary layer [4.82].

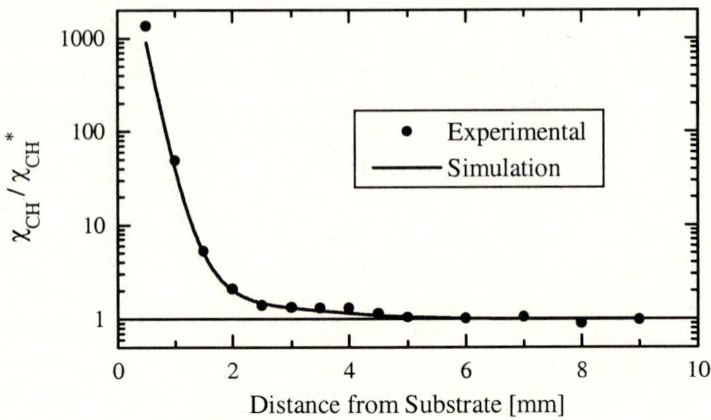

Fig. 4.12: A CH nonequilibrium profile in RF plasma-jet deposition of diamond [4.82].

transit through the boundary layer is insufficient for the atomic hydrogen recombination mechanisms to maintain atomic hydrogen concentration in equilibrium with the local gas temperature. The fast reactions between atomic hydrogen and the CH_x hydrocarbons maintain a partial equilibrium amongst these species, and thus an overpopulation of atomic hydrogen maintains an overpopulation of all CH_x hydrocarbons (notably CH_3, thought to be the major growth precursor). The interrogation of the RF-ICP jet environment using DFWM has demonstrated that the nonequilibrium behavior of atomic hydrogen chemistry is a critical component of the synthesis of diamond thin films at elevated pressures.

In an effort to enhance the nonequilibrium behavior of the substrate boundary layer in RF-ICP jet deposition systems, studies were performed [4.83] in which the boundary layer was purposely thinned by increasing the freestream plasma velocity. This increase in velocity was accomplished by accelerating the RF-ICP through smaller and smaller nozzle diameters. Growth rate (as judged by cross-sectional SEM) for the various freestream velocities investigated is plotted in Fig. 4.13, along with model predictions [4.83]. The corresponding Reynolds number range (based on substrate diameter) for the velocities shown is approximately $200 < Re_D < 5500$. We can see in Fig. 4.13 that the two lowest-velocity cases (particle growth) yield low effective growth rates (approximately 0.01–0.03 µm/h), while the next three higher-velocity film growth cases show a strong sensitivity to increasing freestream velocity. The highest-velocity case, however, indicates only a modest increase in growth rate and an apparent "saturation" of the growth rate curve. The overall growth rates and trends nevertheless show a strong link between the decreased boundary layer thickness and the increased experimental and predicted growth rates as the freestream Reynolds number is increased.

A detailed growth model of the RF-ICP jet diamond deposition system developed by Yu and Girshick [4.38] provides insight into this behavior. In this model, the growth by the methyl radical mechanism of Harris [4.37] is extended to

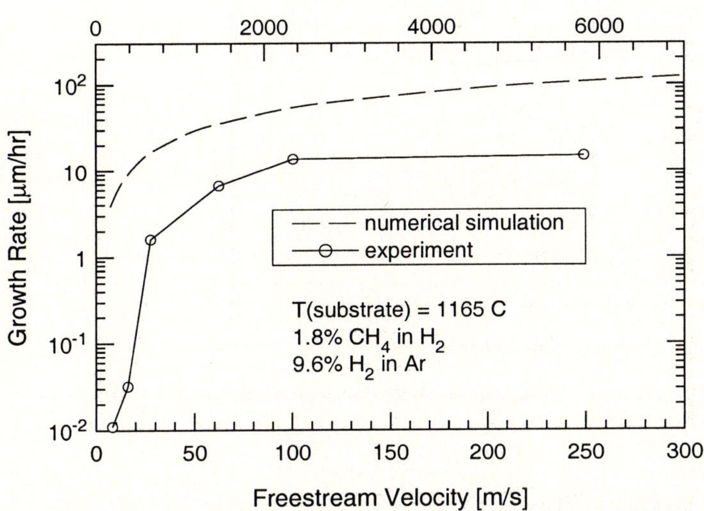

Fig. 4.13: Measured and predicted growth rates as a function of freestream velocity.

treat all CH_x ($x = 0$–3) species as possible growth precursors. Yu and Girshick utilized this model to predict radical concentrations at the surface as a function of boundary layer thickness, as well as the predicted growth rates and contributions of the various growth monomers. Figure 4.14 shows their model predictions for active gas-phase species mole fractions as a function of boundary layer thickness from 0.1 to 2.0 mm. In this figure, one can see that the radical concentrations at the growth surface are greatly increased as the boundary layer thickness decreases, and analysis of the model again indicates that the reduced timescale for recombination of atomic hydrogen through the boundary layer controls the chemistry of the growth environment.

A significant drawback of using small-nozzled RF-ICP jet systems for enhanced growth rates is the penalty incurred in the reduction of growth area due to the smaller jet size. Other investigators [4.13, 14] have sought to achieve a similar effect in RF-ICP jet systems without this penalty, by reducing recombination-inducing collisions via reduction of the system pressure to 100–250 Torr. This technique is successful in increasing growth rates and can also yield increased deposition areas from the expanded plasma plume, but requires the addition of a vacuum vessel and pumps to the RF-ICP jet system, and the inherent difficulties of loading and unloading pieces to be coated. The optimal techniques for RF-ICP jet deposition systems are still being researched, but it is clear that the needs and constraints of the particular coating application will be a determining factor in any system choice.

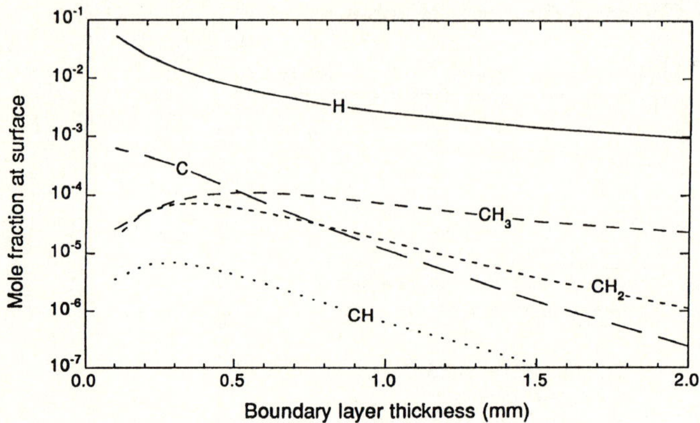

Fig. 4.14: Calculated active gas-phase species mole fractions at the substrate surface as a function of boundary layer thickness [4.38].

4.2.3 The Microwave-Plasma Jet

Despite its wide use in diffusion-dominated low-pressure diamond CVD, there have been only a few reports of microwave excitation of a gas to form a high-velocity plasma jet for diamond synthesis [4.84, 85].

Like the DC arcjet and RF plasma jet, the microwave- (MW-)plasma jet is a high-enthalpy convection-dominated plasma flow. It derives its name from the means by which microwave power is coupled into a flowing ionized and reactive gas mixture. MW-plasma jets (or MW-plasma sources in general) are often distinguished by the precise method by which the electromagnetic waves are coupled into the gas through either rectangular and/or coaxial waveguides, or other so-called microwave "applicators". A microwave applicator is simply an "antenna" that delivers or guides the radiation from the power source to the load. A good review of microwave coupling schemes and microwave-excited and sustained plasmas is given by Asmussen [4.86].

Background

In low-pressure microwave-plasma sources, microwaves are often delivered to a gas-filled resonant cavity via an applicator, where standing wave patterns are established, and a plasma is usually formed in the region of the most intense electric field. For a vacuum-filled cavity, these cavity modes can be calculated. However, the presence of the plasma greatly alters the standing wave patterns in the cavity, and so the spatial distribution of the electric field and the expected ohmic dissipation is often difficult to determine *apriori*.

Conventional wisdom suggests that, in comparison to RF plasma sources, smaller plasma volumes can be expected because of the increased operating frequencies (and hence shorter wavelengths) for the lowest cavity modes. This is a major drawback for many high-pressure plasma sources, in which diffusion is greatly reduced, since a large plasma volume is desired for large-area coverage. In addition to these cavity-mode frequency scaling limitations, the skin depth, describing the propagation depth of the microwave radiation into the plasma, is also greatly reduced, as it scales as the inverse frequency of the waves to the half power (see (4.6)).

Another factor that influences the coupling of radiation to the plasma, especially at lower pressures, is the requirement that the microwave frequency be at least the characteristic plasma frequency, ω_p, which is given as

$$\omega_p = \left(\frac{n_e e^2}{m_e \varepsilon_0}\right)^{1/2} .$$

(4.6)

At the higher frequencies associated with MW sources, this also indicates that the plasma will seek a configuration that establishes a relatively high plasma electron density (n_e) which, for a limited power, necessitates a smaller plasma volume.

Prior Research in MW-Plasma Jet Diamond Deposition

The problems of highly concentrated plasma generation limiting the possible deposition areas, such as DC arcs, can be mitigated somewhat by separating the plasma-generation region from the deposition region and expanding the plasma into low pressures through a nozzle to form a convection-dominated MW-plasma jet. In 1989, Mitsuda et al. [4.84] first reported on the use of a MW-plasma jet to deposit diamond films. In their design, a relatively high-power (2–5 kW) microwave plasma was transmitted through a rectangular waveguide to a coaxial applicator, which also formed the discharge chamber at atmospheric pressure. The center pin of the coaxial applicator serves as a source of intense electric fields and a means of stabilizing the formation of an intense arc plasma. This center "electrode" is surrounded by a coaxial grounded chamber into which a nozzle is formed. The hot high-pressure plasma issues out of this chamber as a high-velocity jet, which spreads out forming a larger area, and impinges on to a water-cooled substrate. The discharge was generally operated with argon as the main source gas, and the stability of the discharge was found to be greatly influenced by the operating power, partial pressure of molecular hydrogen, and flow rate. As expected, it was found to be most stable at higher specific powers (power per mass flow rate), and at lower molecular hydrogen partial pressures.

The MW-plasma jet design of Mitsuda is very similar to that of a DC arcjet, with the exception that the central electrode serves only as a source of an intense electric field for volume ionization of the gas, and not for the thermionic emission of electrons to sustain a high DC current. The downstream jet structure, although not reported, is expected to be very similar to that of a high-pressure arcjet, and, indeed, the diamond deposition characteristics (i.e., growth rates, growth rate dependence on the percentage of methane, etc.) reported are similar to those seen in other DC and RF plasma-jet methods.

Smith et al. [4.85] have reported the synthesis of large area diamond films using a convection-dominated low-pressure MW-plasma flow. Although the detailed design of the applicator and MW source is not reported, the MW plasma is described as being ejected as a directed jet of high velocity (50–500 m/sec) into a low-pressure (typically 50 Torr) furnace, in which the large-diameter substrates are located. Growth rates obtained with this MW-plasma jet are more like those of hot-filament and conventional low-pressure plasma CVD methods. However, coverage as large as 20 cm in diameter can be obtained at relatively modest powers (1–5 kW).

4.3 Summary

In this chapter, we have provided a review of the various plasma-jet methods used in the synthesis of diamond films. These include DC, RF, and MW-plasma-jet deposition methods, over varying ranges of operating conditions. The high plasma velocities provide enhanced reactive species transport to the growth surface, thereby offering greater deposition rates, and sometimes improved precursor utilization. All of these modalities show great promise for industrial scale-up, and are already the preferred commercial methods for the synthesis of large-area, thick-film diamond for a variety of applications. The future use of these technologies for the synthesis of thin-film diamond and diamond particles for a broader range of applications will depend on continued research in the improvement of growth rates, deposition areas, and diamond deposition efficiencies.

Acknowledgments

Research at Stanford on the plasma-jet synthesis of diamond during the past ten years has been generously supported by the Office of Naval Research, Olin Aerospace, the Daimler–Benz Corporation, the National Science Foundation, the Electric Power Research Institute, and the Department of Energy.

References

4.1   G.M. Giannini, Sci. Am. **197**, 80–88 (1957)

4.2   K. Kurihara, K. Sasaki, M. Kawarada, and N. Koshino, Appl. Phys. Lett. **52**, 437 (1988)

4.3   Y.A. Carts, Laser Focus World **126**, 41 (1990)

4.4   F. Akatsuka, Y. Hirose, and K. Komaki, Jpn. J. Appl. Phys. **27**, L1600 (1988)

4.5   A. Hirata and M. Yoshikawa, Diamond Rel. Mater. **4**, 1363 (1995)

4.6   D.G. Goodwin, J. Appl. Phys. **74**, 6895 (1993)

4.7   N. Ohtake and M. Yoshikawa, J. Electrochem. Soc. **137**, 717 (1990)

4.8   N. Ohtake, Y. Kuriyama, M. Yoshikawa, H. Obana, M. Kito, and H. Saito, Int. J. Japan Soc. Prec. Eng. **25**, 5 (1991)

4.9   K. Kurihara, K. Sasaki, and M. Kawarada, Fujitsu Sci. Tech. J. **25**, 48 (1989)

4.10 N. Ohtake, Y. Mashimo, and M. Yoshikawa, Proceedings of the 1991 International Conference on New Diamond Science and Technology (Materials Research Society), Washington, DC (1991), p. 173

4.11 R. Furakawa, H. Uyama, and O. Matsumoto, IEEE Trans. Plasma Sci. **18**, 930 (1990)

4.12 P. Klocek, J. Hoggins, P. Taborek, and T. McKenna, SPIE Proceedings Vol. 1325, Diamond Optics III (The International Society for Optical Engineering) (1990), p. 63

4.13 R. Li, H. Shi, Z. Yan, S. Tang, and H. Zhu, in Applications of Diamond Films and Related Materials, ed. Y. Tzeng, M. Yoshikawa, M. Murakawa, and A. Feldman, Elsevier, Amsterdam (1991), p. 207

4.14 M.H. Loh and M.A. Cappelli, AIAA Paper 92-3534, 28th Joint Propulsion Conference, Nashville, TN (1992)

4.15 M.H. Loh and M.A. Cappelli, Surf. Coat. Technol. **54/55**, 408 (1992)

4.16 Z.P. Lu, J. Heberlein, and E. Pfender, Plasma Chem. Plasma Process. **12**, 35 (1991)

4.17 Z.P. Lu, L. Stachowicz, P. Kong, J. Heberlein, and E. Pfender, Plasma Chem. Plasma Process. **11**, 387 (1991)

4.18 Z.P. Lu, J. Heberlein, and E. Pfender, Plasma Chem. Plasma Process. **12**, 55 (1992)

4.19 S. Matsumoto, I. Hosoya, and T. Chounan, Jpn. J. Appl. Phys. **29**, 2082 (1990)

4.20 N. Ohtake, Y. Mashimo, and M. Yoshikawa, Proceedings of the 1991 International Conference on New Diamond Science and Technology (Materials Research Society), Washington, DC (1991), p. 173

4.21 N. Ohtake, M. Ikegami, and M. Yoshikawa, Diamond Films Technol. **2**, 1 (1992)

4.22 S.K. Baldwin, Jr., Ph.D. Thesis, Mechanical Engineering Department, Stanford University (1996)

4.23 K.R. Stalder and R.L. Sharpless, J. Appl. Phys. **68**, 6187 (1990)

4.24 L. Spitzer and R. Harm, Phys. Rev. **89**, 977 (1953)

4.25 S. Dushman, Rev. Mod. Phys. **2**, 381 (1930)

4.26 W. Schottky, Ann. Phys. **44**, 1011 (1914)

4.27 G.L. Cann, US Patent Nos. 4,471,003 and 4,487,162 (1984)

4.28 M.A. Cappelli and M.H. Loh, US Patent No. 5,358,596 (1994)

4.29 J.J. Beluens, Doctoral Thesis, University of Eindhoven (1992)

4.30 R.L. Woodin, L.K. Bigelow, and G.L. Cann, in Applications of Diamond Films and Related Materials, ed. Y. Tzeng, M. Yoshikawa, M. Murakawa, and A. Feldman, Elsevier, Amsterdam (1991), p. 439

4.31 T.G. Owano, M. Zhao, C.H. Kruger, and S.K. Baldwin, Paper AIAA 97-0689, 35th Aerospace Sciences Meeting, Reno, NV (1997)

4.32 K. Kurihara, K. Sasaki, M. Kawarada, and N. Koshino, Mater. Res. Soc. Symp. Proc. **162**, 115 (1990)

4.33 N. Ohtake, M. Yoshikawa, K. Suzuki, and S. Takeuchi, Applications of Diamond Films and Related Materials, ed. Y. Tzeng, M. Yoshikawa, M. Murakawa, and A. Feldman, Elsevier, Amsterdam (1991), p. 431

4.34 K. Hirabayashi, N.I. Kurihara, N. Ohtake, and M. Yoshikawa, Jpn. J. Appl. Phys. **31**, 355 (1992)

4.35 N. Ohtake and M. Yoshikawa , Jpn. J. Appl. Phys. **32**, 2067 (1993)

4.36 D.G. Goodwin, Appl. Phys. Lett. **59**, 277 (1991)

4.37 S.J. Harris, Appl. Phys. Lett. **56**, 2298 (1990)

4.38 B.W. Yu and S.L. Girshick, J. Appl. Phys. **75**, 3114 (1994)

4.39 K. Kurihara, K. Sasaki, and M. Kawarada, in Proceedings of the 1st International Symposium, FGM, Sendai (1990), p. 65

4.40 M.H. Loh and M.A. Cappelli, in Proceedings of the 3rd International Symposium on Diamond Materials Vol. 93-17, The Electrochemical Society, Honolulu, HI (1993), p. 17

4.41 S. Skokov, B. Weiner, and M. Frenklach, J. Phys. Chem. **90**, 8 (1994)

4.42 S.W. Reeve, W.A. Weimer, and D.S. Dandy, Appl. Phys. Lett. **63**, 2487 (1993)

4.43 D.S. Dandy and M.E. Coltrin, Appl. Phys. Lett. **66**, 391 (1994)

4.44 Q.Y. Han, T.W. Or, Z.P. Lu, J. Heberlein, and E. Pfender, in Proceedings of the 2nd International Symposium on Diamond Materials (The Electrochemical Society), Washington DC **91**(8), 115 (1991)

4.45 S. Matsumoto, I. Hosoya, and T. Chounan, Jpn. J. Appl. Phys. **29**, 2082 (1990)

4.46 N. Ito, M. Yamamoto, S. Nakamura, and T. Hattori, J. Appl. Phys. **77**, 6636 (1995)

4.47 S.K. Baldwin, T.G. Owano, and C.H. Kruger, in Proceedings of the 12th International Symposium on Plasma Chemistry, Minneapolis, MN4 (1995), p. 4

4.48 E.A. Brinkman and J.B. Jeffries, Paper AIAA 95-1955, 26th Plasmadynamics and Lasers Conference, San Diego, CA (1995)

4.49 C.D. Moen and H.A. Dwyer, paper presented at the Spring Meeting of the Western States Section/The Combustion Institute (1994)

4.50 M. Kawarada, K. Kurihara, K. Sasaki, A. Teshima, and K. Koshino, SPIE Proceedings Vol. 1146, Diamond Optics II (The International Society for Optical Engineering) (1989), p. 28

4.51 M.A. Cappelli and M.H. Loh, Diamond Rel. Mater. **3**, 417 (1993)

4.52 S.K. Baldwin, Jr., T.G. Owano, and C.H. Kruger, Appl. Phys. Lett. **67**, 194 (1995)

4.53 K.R. Stlader and W. Homsi, Appl. Phys. Lett. **68**, 3710 (1996)

4.54 G.A. Raiche, G.P. Smith, and J.B. Jeffries, in Proceedings of the 1991 International Conference on New Diamond Science and Technology (Materials Research Society), Washington, DC (1991), p. 251

4.55 M.H. Loh, J.G. Liebeskind, and M.A. Cappelli, Paper AIAA 93-2227, 29th Joint Propulsion Conference, Monterey, CA (1993)

4.56 J.B. Jeffries, G.A. Raiche, and M.S. Brown, paper presented at the Spring Meeting of the Western States Section/The Combustion Institute (1994)

4.57 K. Kurihara, K. Sasaki, and M. Kawarada, in Proceedings of the 1st International Symposium, FGM, Sendai (1990), p. 65

4.58 K. Kurihara, K. Sasaki, M. Kawarada, and Y. Goto, in Applications of Diamond Films and Related Materials, ed. Y. Tzeng, M. Yoshikawa, M. Murakawa, and A. Feldman, Elsevier, Amsterdam (1991), p. 461

4.59   K.A. Snail, C.M. Marks, Z.P. Lu, J. Heberlein, and E. Pfender, Mater. Lett. **12**, 301 (1991)

4.60   D.S. Bradley, S.M. Jaffe, and L.K. Bigelow, Workshop on the Industrial Applications of Plasma Chemistry, August 25–26, Proceedings Volume B, Thermal Plasma Applications (IUPAC) (1995), p. 77

4.61   J.M. Olsen and M.J. Dawes, J. Mater. Res. **11**, 1765 (1996)

4.62   S. Matsumoto, H. Hino, and T. Kobiyashi, Appl. Phys. Lett. **51**, 737 (1987)

4.63   M.A. Cappelli, T.G. Owano, and C.H. Kruger, J. Mater. Res. **5**, 2326 (1990)

4.64   T.G. Owano, C.H. Kruger, and M.A. Cappelli, Carbon **28**, 748 (1990)

4.65   T.G. Owano and C.H. Kruger, Plasma Chem. Plasma Process. **13**, 433 (1993)

4.66   S.K. Baldwin, Jr., T.G. Owano, and C.H. Kruger, Plasma Chem. Plasma Process. **14**, 383 (1994)

4.67   R. Hernberg, T. Lepisto, T. Stenberg, and J. Vattulainen, Diamond Rel. Mater. **1**, 255 (1992)

4.68   G. Verven, T. Priem, S. Paidassi, F. Blein, and L. Bianchi, Diamond Rel. Mater. **2**, 468 (1993)

4.69   S. Matsumoto, I. Hosoya, Y. Manabe, and Y. Hibino, Pure Appl. Chem. **64**, 751 (1992)

4.70   M.T. Bieberich and S.L. Girshick, Plasma Chem. Plasma Process. **16**, 157 (1996)

4.71   Q.D. Zhuang, H. Guo, J. Heberlein, and E. Pfender, Diamond Rel. Mater. **3**, 319 (1994)

4.72   I. Hosoya and S. Matsumoto, J. Chem. Vapor Depos. **1**, 210 (1992)

4.73   C. Tsai, J. Nelson, W.W. Gerberich, J. Heberlein, and E. Pfender, J. Mater. Res. **7**, 1967 (1992)

4.74   M. Kohzaki, K. Uchida, K. Higuchi, and S. Noda, Jpn. J. Appl. Phys. **32**, L438 (1993)

4.75   K. Eguchi, S. Yata, and T. Yoshida, Appl. Phys. Lett. **64**, 58 (1994)

4.76   D.S. Green, T.G. Owano, S. Williams, D.G. Goodwin, R.N. Zare, and C.H. Kruger, Science **259**, 1726 (1993)

4.77   T.G. Owano, C.H. Kruger, D.S. Green, S. Williams, and R.N. Zare, Diamond Rel. Mater. **2**, 661 (1993)

4.78   J. Larjo, J. Vattulainen, and R. Hernberg, Appl. Phys. B (Lasers and Optics) **62**, (1996)

4.79   J.W. Lindsay, J.M. Larson, and S.L. Girshick, Diamond Rel. Mater. **6**, 481 (1997)

4.80   S.L. Girshick and J.M. Larson, Pure Appl. Chem. (in press); also in Proceedings of the 9th International Conference on High Temperature Materials Chemistry, State College, Pennsylvania, May 19–23, 1997, ed. K.E. Spear (Electrochemical Society Proceedings Volume 97-39), pp. 537–546.

4.81   S.L. Girshick, C. Li, B.W. Yu, and H. Han, Plasma Chem. Plasma Process. **13**, 169 (1993)

4.82   T.G. Owano, E.H. Wahl, C.H. Kruger, and R.N. Zare, Paper AIAA 95-1954, 26th AIAA Plasmadynamics & Lasers Conference, San Diego, CA (1995)

4.83   T.G. Owano, C.H. Kruger, and D.G. Goodwin, in preparation (1998)

4.84 Y. Mitsuda, T. Yoshida, and K. Akashi, Rev. Sci. Instrum. **60**, 249 (1989)

4.85 D.K. Smith, E. Sevillano, M. Besen, V. Berkman, and L. Bourget, Diamond Rel. Mater. **1**, 814 (1992)

4.86 J. Assmussen, J. Vac. Sci. Technol. A, **8**, 883 (1989)

5. Hot-Filament Deposition of Diamond

Claus-Peter Klages and Lothar Schäfer

Fraunhofer-Institut für Schicht- und Oberflächentechnik,
Bienroder Weg 54E, D-38108 Braunschweig, Germany
e-mail: schaefer@ist.fhg.de

Springer Series in Materials Processing
Low-Pressure Synthetic Diamond Eds.: B. Dischler and C. Wild
© Springer-Verlag Berlin Heidelberg 1998

5.1 Introduction

The historical development of diamond growth under conditions of its thermodynamical metastability can be traced back to the early 1950s, when Eversole at the laboratories of Union Carbide achieved the first overgrowth of diamond from the gas phase on the surface of diamond powder particles, probably before the first successful high-pressure/high-temperature synthesis of diamond. However, the method used at that time was far from being economically viable, owing to a very low yield and the necessity for repeated removal of simultaneously deposited graphite by a treatment in a pure hydrogen atmosphere. A very important breakthrough in the efforts to grow diamond by a chemical vapour deposition process was the finding that this carbon phase can be deposited continuously and largely free from graphite in the presence of a superequilibrium concentration of atomic hydrogen. A comprehensive account of this development, with particular consideration of the contributions by Russian scientists and many references to the original literature, was published recently [5.1].

Although this achievement can most probably be dated back to the early 1970s, it took another decade, until – due to a series of papers from the Japanese NIRIM (National Institute for Research in Inorganic Materials) – its importance was recognized by the scientific–technical community. Among the multitude of diamond CVD processes which have been developed since about 15 years, the hot-filament CVD process (HFCVD) – together with the microwave-plasma CVD method (MWPCVD), one of the two first methods described in detail by the Japanese scientists – has continued to be one of the most used diamond deposition processes, due to its simplicity and up-scaling potential.

5.2 Process Variants of Diamond HFCVD

5.2.1 The First Descriptions of the HFCVD Method

The essential key in the HFCVD process, turning the pyrolysis of a hydrocarbon from a graphite-forming process to a diamond CVD method, is the presence of a heated refractory metal wire in the immediate vicinity of the substrate. The elementary principle was depicted by Matsumoto et al. in 1982 [5.2, 3], and is shown in Fig. 5.1.

The NIRIM scientists used a coiled tungsten filament, heated to about 2000°C by passing an electrical current, mounted near the substrate. The temperature of the deposition chamber was between 700°C and 1000°C, as measured by a thermocouple in contact with the silica substrate holder. Using a mixture of hydrogen and methane (about 1 vol%) as a feed gas (10–100 cm^3/min), diamond was formed at chamber pressures between roughly 1 and 10 kPa. The deposit consisted of cubo-octahedral or multiply twinned isolated particles or closed films of diamond, depending on the substrate conditions. Although a substrate scratching

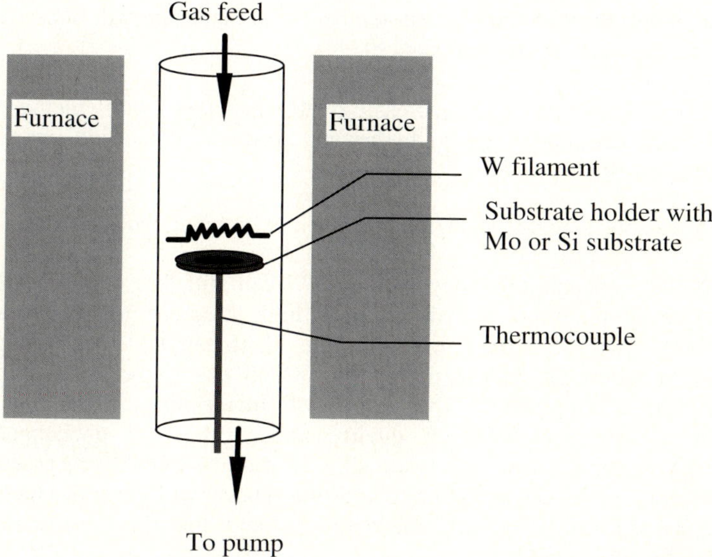

Fig. 5.1: A schematic view of the deposition chamber as used by Matsumoto et al. [5.2, 3].

with diamond powder is not explicitly mentioned in the papers, it is highly probable that this kind of pretreatment was already in use in order to achieve a dense nucleation and early coalescence of the crystallites to closed films.

The early Japanese papers left no doubt that diamond was actually formed by the methods described: scanning electron micrographs showed the cubo-octahedral morphology typical of CVD-grown diamond, electron diffraction data were consistent with ASTM values and the Raman peak at 1334.5 cm^{-1} was very close to the position reported for diamond single crystals.

Linear growth rates of roughly 1 µm/h were achieved under the conditions quoted above; this growth velocity can still be considered as typical for HFCVD processes employing hydrocarbon/hydrogen mixtures forming good-quality diamond. In the literature, numerous investigations of the growth parameter influences on diamond growth rates and film qualities are reported. In [5.4] five important factors have been summarized:

1. A proper substrate pretreatment – for example by ultrasonic irradiation in a diamond powder suspension – increases the nucleation density considerably, leading to faster formation of a closed film and better utilization of growth species in the early growth phase.

2. As a function of substrate temperature, a maximum growth rate is achieved at about 850°C.

3. The growth rate increases with the filament temperature; higher growth rates are achieved at lower filament–substrate distances.

4. Diamond can be deposited with increasing growth rates with methane concentrations from 0.1 to 2 vol% in hydrogen: however, the non-diamond carbon components will also increase.

5. As a function of gas pressure, the growth rate has a maximum at about 10 kPa.

However, methane concentrations beyond 10 vol% can still lead to high-rate (19 μm/h), high-quality diamond growth, if very high filament temperatures (2500–2600°C) are used [5.5, 6]. This example demonstrates the strong interrelation between optimum growth parameters which has to be taken into account in process developments. The effect of increasing carbon feed concentration on growth rate and film quality under otherwise unchanged conditions is a very common effect in presumably all diamond CVD processes. Therefore, a compromise between growth rate and film quality will generally have to be made, depending on economic considerations and the film properties required for a specific application.

5.2.2 Source Gas Compositions

Mixtures of 0.1 to a few vol% methane and hydrogen are still the most used gas atmospheres for hot-filament deposition of diamond. Additions of oxygen-containing source gases in HFCVD of diamond are possible only to a limited extent – in contrast to plasma-activated deposition methods – due to the tendency for oxidation and volatilization of the filament material under oxidizing conditions. Nevertheless, additions of oxygen, water vapour or oxygen-bearing organic compounds to the HFCVD gas feed can have a number of advantages, if the quantitative gas phase composition is carefully chosen. This was shown first by Hirose and Terasawa, using a range of oxygen- and nitrogen-containing organic compounds [5.7]. Films with good crystallinity and high quality (according to Raman spectroscopy and electron diffraction) were obtained at growth rates roughly an order of magnitude higher than typical for oxygen-free gas feeds. Compared with methane/hydrogen mixtures, much higher gas pressures, up to nearly one atmosphere, and gas phase concentrations of the carbon-bearing species were possible in these experiments. These results were later essentially confirmed by other authors, achieving growth rates up to 10 μm/h by using ethanol and acetone (1–8 vol%) in hydrogen at pressures up to 13 kPa [5.8].

A comparison of CH_4–H_2 and CH_4–H_2–O_2 mixtures, carried out by Kawato and Kondo [5.9], showed that an increased growth rate and extended pressure and concentration ranges (40 kPa, 4 vol%) are also obtainable with additions of molecular oxygen. Numerous investigations have been performed in an attempt to understand the reasons for the effects of oxygen-bearing gases observed in diamond HFCVD and MWPCVD (see, for example, [5.9–14]). The results of these studies can only be summarized here, without further comment. Oxygen effects have been attributed to:

- a reduction of the acetylene concentration, leading to fewer non-diamond carbon phases
- promotion of hydrogen dissociation and methane decomposition
- formation of active growth sites by improved hydrogen abstraction
- efficient removal of non-diamond carbon phases by hydroxy radicals.

Aside from restrictions due to the effect of oxidizing atmospheres on filament lifetimes, there are more general limits to the C,H,O-concentrations that can be used in diamond HFCVD, similar to probably all other diamond CVD processes. A compilation of diamond-forming gas phase compositions from a large number of papers shows that – within a ternary C,H,O-concentration diagram – those concentrations leading to diamond growth are restricted to an essentially wedge-shaped area extending from the H-corner along the CO line of the triangle [5.15]. Smaller carbon concentrations generally do not lead to the formation of deposits, due to the dominance of etching processes over carbon deposition. At concentrations higher than those situated in the diamond wedge, sp^2-coordinated carbon phases tend to dominate.

Additions of molecular nitrogen or nitrogen-containing compounds have in recent years become the subject of increased interest as a key to the growth of fiber-textured diamond films (see Sect. 5.2.6). Homoepitaxial growth experiments of Cao et al. [5.16] have demonstrated that nitrogen greatly enhances the growth rate of {100} facets. This is attributed to nitrogen compounds breaking the dimers of the (2×1) reconstructed {100} surface, thus favouring the <100> growth direction.

5.2.3 Filament Materials

While tungsten was exclusively used as a filament material in the early HFCVD work, several other materials have been tried subsequently. Apart from tungsten, only tantalum and rhenium have so far gained greater importance. A selection of the relevant literature is given in references [5.17–22]. Criteria to be considered in choosing a filament material are, among others:

- economic aspects; that is, price and availability of the material
- chemical and dimensional stability under carburizing and possibly oxidizing conditions
- hydrogen dissociation capability of the filament surface and growth rates achievable
- stability against surface poisoning by carbon deposition
- the possibility of destruction of growth species on the filament surface
- the amount of filament impurities incorporated in the diamond film.

The use of tantalum filaments for diamond CVD was the subject of a Swiss patent that was already filed in 1984 [5.17]. A good reason to use tantalum instead of tungsten is its much smaller carburization rate. While both materials, W and Ta, are thermodynamically unstable with respect to their carbides under typical diamond-forming conditions, the carbide-forming reaction, which severely deforms and embrittles the filaments, proceeds much more slowly through a tantalum wire. Therefore, under comparable conditions, Ta keeps its mechanical strength – due to a core of uncarburized material – longer than a W filament of the same thickness. A 0.8 mm thick W filament at a temperature of 2000°C, to take an example from the work of the Vienna group [5.18], was completely carburized after 23 hours (5.3 kPa, 0.5 vol% CH_4), while under the same conditions a Ta filament of the same thickness had only a carbide scale of 70 µm thickness. The advantage of using Ta instead of W as a filament material for the diamond coating of cutting tools because of the to much higher filament temperatures possible was clearly demonstrated [5.19]. In contrast to Ta and W, Re does not form a carbide. Therefore rhenium filaments, if used under proper conditions, can retain their ductility and do not show significant dimensional changes [5.20]. However, the much higher price of Re has to be taken into account if deposition of diamond on greater than a laboratory scale is considered.

As far as growth rates achievable with different filament materials and their hydrogen dissociation capabilities are concerned, results reported in the literature are conflicting. Comparing tantalum and tungsten filaments, Sing et al. [5.21] found no differences in diamond crystal growth rates, concluding that dissociation of the gas mixture is independent of the filament material. In contrast, distinct differences between rhenium, tantalum and tungsten filaments in HFCVD of diamond were found by Okoli et al. [5.22]. For well-faceted good-quality diamond films, different methane concentrations were needed for the three materials at a filament temperature of 2200°C. Additionally, the growth rates under these conditions differed by a factor of two, indicating a material-dependent catalytic effect on the dissociation of the gas phase. However, at filament temperatures of 2400°C tantalum and rhenium (the tungsten filaments used were not stable at this temperature) showed equal growth rates. According to the results of Sommer and Smith [5.20], who found a reduction of filament activity by non-reactive carbon layers, material- and temperature-dependent formation of carbon contamination layers has to be taken into account in interpreting these findings. For methane containing gas phases, the comparison of atomic hydrogen concentrations measured by two-photon laser-induced fluorescence with calculated equilibrium values at equal filament temperatures and diameters yielded values below the equilibrium concentrations calculated for the filament temperature. Among the materials investigated (W, Ta, Ir), tungsten yielded the highest concentrations [5.23]. The deviation of molecular hydrogen dissociation from thermal equilibrium, indicating kinetic limitations, is also confirmed by higher degrees of dissociation in pure hydrogen gas phases, where no carburization or carbon impurities affect the filament surface. For oxygen-containing systems, McNamara and Gleason [5.24] also detected significant differences in heterogeneous reactions for tantalum and rhenium filament surfaces.

The incorporation of filament material, due to temperature-dependent evaporation from the filament surface, has been detected by X-ray microfluorescence [5.25], energy-dispersive X-ray spectroscopy [5.26, 27] and *in-situ* ellipsometry [5.28]. For more detailed studies, secondary ion-mass spectroscopy (SIMS) [5.29–32] and Rutherford backscattering (RBS) [5.31–34] have been used.

Quantitative determinations of filament-related impurities have been carried out for rhenium, tantalum and tungsten filaments. For filament temperatures between 1900°C and 2350°C, the amount of Re incorporated in diamond films increases by three orders of magnitude, to a maximum of about 0.1 at% [5.29]. At filament temperatures of about 2250°C, Gheerhaert et al. measured a nearly constant tungsten concentration around 0.1 at% with RBS [5.33]. For tantalum filaments at 2300°C, maximum Ta concentrations of 0.03 at% were detected using SIMS [5.31].

For carbide-forming filament materials, the degree of filament material impurities depends strongly on the carburization state of the filaments. This is caused by the fact that, at the relevant temperatures, the carbides have lower vapour pressures than the corresponding uncarburized materials [5.31]. Therefore, high amounts of filament impurities are often found at the diamond substrate interface when the deposition process is started with uncarburized filaments [5.26, 32, 34].

Filament impurities in HFCVD-diamond films are obviously inevitable. However, they can be reduced by:

- low filament temperatures, which, on the other hand, reduce gas phase activation and growth rate
- the use of precarburized filaments
- increased methane concentrations [5.32] in the gas phase, which, on the other hand, affect the phase purity of deposited diamond.

5.2.4 Bias-Assisted HFCVD

While HFCVD with negative-biased substrates has been the subject of increasing interest in recent years with respect to enhanced uniform and selected-area diamond nucleation [5.35], electron-assisted HFCVD applying positive bias to the substrate has been used to investigate its influence on diamond film quality via micro-Raman spectroscopy [5.36], to promote selected-area deposition of diamond films [5.37] and to produce smooth diamond films containing considerable amounts of non-diamond carbon with enhanced growth rates [5.38]. Lee et al. [5.39, 40] investigated bias-controlled HFCVD under positive as well as negative substrate bias relative to the filament. Under high-current conditions, the created DC plasma leads to a deterioration of film quality in both cases. The best results with respect to film quality have been found for negative bias and low-current conditions, this being attributed to minimum electron and ion bombardment of the diamond film and to additional hydrogen dissociation by the electron flux from the surface. For positive-biased substrates, higher defect densities and multiple interfacial layers with incorporated filament material were found [5.40, 41]. Better-faceted diamond crystals at slightly lower deposition rates and unchanged nucleation densities were found under conditions of positive substrate bias values. Okoli et al. [5.42] attributed this effect to additional atomic hydrogen formation in the generated discharge. They also found an increase in the growth rate for deposition after formation of a closed diamond film. Bannholzer and Kehl [5.43] attributed the increase of the growth rate in electron-assisted HFCVD to the increase in substrate temperature that they detected upon applying different positive substrate bias values. These different and somewhat confusing results might be understandable when one considers that the deposition conditions under bias-assisted HFCVD are very sensitive to the details of the experimental arrangement of substrates and filaments.

According to our investigations [5.44], ion bombardment of negative-biased substrates yields high nucleation densities of 10^{10} cm^{-2}, comparable to bias-nucleation in microwave-plasma CVD processes. Negative substrate bias during diamond deposition, however, causes the deposition of non-diamond carbon, due to the destruction of the diamond structure by ion impact. Under high-current conditions at positive substrate bias, the electron flux to the substrate causes a discharge in the vicinity of the substrate. Atomic hydrogen generated in this plasma increases the diamond growth rate through the formation of additional growth species close to the substrate surface. (Because of their short lifetimes, HFCVD-diamond growth species generated at a greater distance from the surface do not contribute to diamond growth.) The additional atomic hydrogen also contributes to improved etching of non-diamond deposits, thus improving diamond phase purity.

5.2.5 HFCVD with Forced Convection

Experiments with forced convection of the source gases were first performed in the authors' laboratory in order to investigate and possibly overcome the detrimental effect of filaments on the near-substrate concentration of growth species as encountered in experiments with multi-filament arrays [5.45] (see Sect. 5.3.1). As acetylene is inactive for diamond growth in HFCVD, the formation of C_2H_2 will hinder diamond growth by consuming methyl growth species. In order to avoid the supposed conversion of methane and methyl to acetylene at the surface of the hot filaments, the source gas was fed directly on to the substrate surface using a nozzle of a few millimeters in diameter, ending a few centimeters from the filament array. With this modification the Peclet number, which characterizes the relative efficiencies of convective and diffusive transport, resp., was increased to values larger than 1, corresponding to the dominance of convective transport. Maximum growth rates around the stagnation point have been found at Peclet numbers of about 5. For Peclet numbers greater than 10, the growth rates decrease to zero. This is not due to a lack of atomic hydrogen, as no saturation of atomic hydrogen generation at the filaments was found for Peclet numbers up to 70. The effect may, rather, be attributed to the fact that the thermal equilibrium ratio for the generation of methyl radicals at the substrate boundary layer is not reached at high gas velocities [5.46]. Similar experiments were performed by Kondo et al. in what was termed an "advanced HFCVD method" [5.47]. Experiments with local supplies of methane and acetylene were also carried out at the Pennsylvania State University [5.48].

A semi-quantitative explanation of the effects involved in multi-filament arrays and HFCVD with forced convection was given in [5.49]. This explanation is consistent with the results of a comparative modelling of diffusion- and convection-controlled CVD methods [5.50]. Although considerably increased growth rates around the stagnation point can be achieved by enhanced transport of atomic hydrogen from the filaments to the substrate boundary layer, the process is presently not expected to gain economic importance, due to the non-uniform deposition over large-areas. However, important conclusions concerning the mechanism of diamond growth have been deduced from experiments of this kind. According to the experimental results and computer simulations of gas kinetics and flow dynamics [5.45–53], methyl has been identified as the major diamond growth species in HFCVD. In order to avoid methyl consumption, the formation of acetylene should be minimized. For high diamond growth rates, kinetic and flow dynamic conditions should be chosen such that, at atomic hydrogen concentrations that are as high as possible, the methyl generation reaction reaches thermal equilibrium within the methyl diffusion length or boundary layer, resp., in front of the substrate.

5.2.6 Textured and Heteroepitaxial Growth of Diamond by HFCVD

From an applications point of view, textured and heteroepitaxial diamond films differ from randomly oriented films mainly with respect to a reduced influence of grain boundaries and the possibility of producing smooth diamond surfaces. Because of their great importance for electronic and optical applications, plasma-activated methods have generally been used for their deposition [5.54].

In randomly oriented diamond films, the surface roughness usually increases with the film thickness. Wild et al. were able to grow fiber-textured diamond films by choosing growth conditions in which the fastest growth direction is almost perpendicular to {100} planes [5.55]. Under these conditions, thick smooth diamond films with <100> texture have been prepared. The influence of gas phase composition, substrate temperature and filament temperature has been examined in HFCVD in order to control texture formation [5.16, 56, 57]. Improvement of texture formation has also been demonstrated by applying additional hydrogen treatment, alternating with diamond growth [5.58], and by careful *ex-situ* etching pretreatment [5.59]. As shown for microwave-plasma CVD [5.60], textured diamond growth in HFCVD is also attributed to the competition between growth rates of different facets [5.61]. Additionally, Clausing et al. [5.61] found a correlation between different defect densities and growth facets, indicating that the control of texture might lead to a reduction of defects in the diamond films.

For the deposition of heteroepitaxial diamond films, textured growth has to be combined with oriented diamond nucleation. Besides oriented seeding [5.62, 63], resulting in low nucleation densities, negative substrate bias relative to the filaments has been used for *in-situ* oriented nucleation with densities above $10^9\,\mathrm{cm}^{-2}$ [5.64–66]. Minimum misorientations of about 7° between the diamond and silicon (001) planes were measured by high-resolution transmission electron microscopy [5.65]. This value is considerably higher than those reported for plasma CVD methods. Additionally, the heteroepitaxial area is smaller. It has to be taken into account that a multi-filament array introduces non-uniform biasing conditions over large areas, whereas it is well known that oriented nucleation is very sensitive to biasing conditions [5.67]. In the authors' laboratory, as a first attempt to obtain heteroepitaxial HFCVD-diamond films, the established microwave-plasma oriented bias nucleation [5.68] was combined with <100>-textured HFCVD growth. Textured deposition was achieved uniformly with growth rates of about 2 µm/h over an area of several hundred cm². Epitaxial growth up to 140 µm thickness has been demonstrated by this method (Fig. 5.2). According to rocking curve measurements, the misorientation in the films is improved to about 4°.

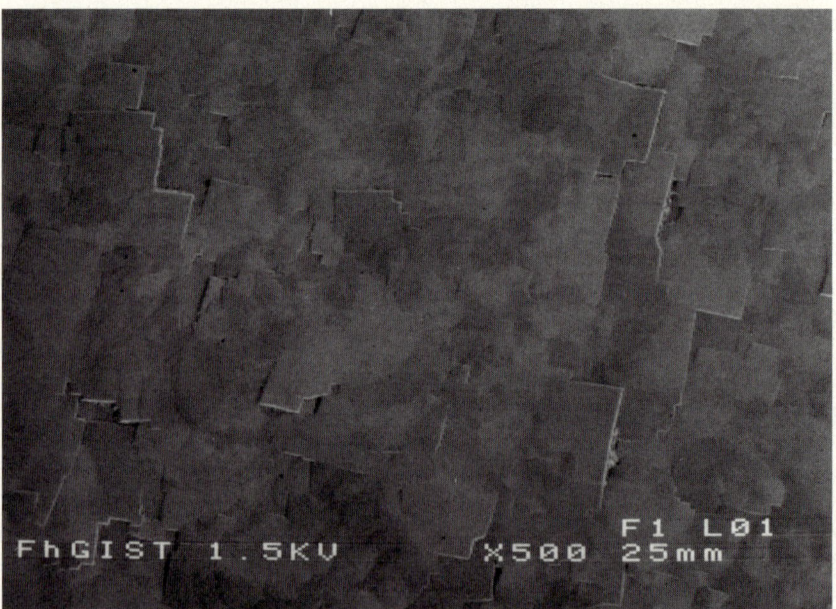

Fig. 5.2: Heteroepitaxial HFCVD-diamond film grown after MWPCVD bias nucleation. The area presented is 175 × 175 µm².

5.3 Application Aspects of Diamond HFCVD

5.3.1 Process Scale-Up

Compared with the total number of papers that have meanwhile been published so far on diamond CVD processes and diamond film properties, there are few publications on the upscaling of HFCVD, due – among other reasons – certainly to economic interests involved in upscaled diamond deposition processes and possibly also a supposed lack of scientific challenges. From a scientific–technical viewpoint, the development of a large-area diamond multi-filament CVD process is by far not a completely trivial tasks, owing to a more complicated role of the hot filament than just delivering hydrogen atoms by dissociation of molecular H_2 on its surface and possibly heating the substrate by radiation and thermal conduction. Experiments performed with hanging parallel arrays of thick straight tantalum filaments have revealed a surprising deviation of film thickness profiles from what would be expected from a superposition of profiles for individual filaments [5.45]. There are now good reasons to conclude that this effect is due to a decomposition of methyl radicals – which are essential growth species for diamond formation

Fig. 5.3: A three-dimensional filament arrangement for diamond deposition on many micromachining shaft tools.

(see Sect. 5.2.5) – at the relatively large surface of the hot wires. In order to suppress this effect as much as possible, a careful optimization of inter-filament and filament–substrate distances and filament diameters is necessary.

To reduce the decomposition of methyl radicals by the filament surfaces, thin multi-filament arrangements may be used. Uniform multi-filament deposition with filament diameters of 0.5 mm and 0.2 mm, respectively, has been reported [5.30, 51]. Possibly due to the poor mechanical stability of the thin filaments, maximum deposition areas up to about 50 cm^2 have been achieved by adequate filament holders compensating for filament deformation during the deposition process. For large-area deposition. the authors have investigated multi-filament HFCVD with mechanically more stable filaments, with thicknesses up to 2 mm [5.52]. Diamond deposition with effective filament arrays of 400 cm^2 has been demonstrated. Based on these results, a HFCVD process with an effective filament array of 3200 cm^2 is now being tested.

An additional advantage of thicker, stable filaments is the possibility of adjusting gas phase activation to three-dimensional geometries (Fig. 5.3). This possibility makes three-dimensional multi-filament arrangements superior to flat-filament arrangements and plasma processes with respect to diamond deposition on tools and parts with cylindrical geometries.

5.3.2 Applications of HFCVD-Grown Diamond Films

Because of its peculiarities, the HFCVD process is mostly used for diamond deposition on cutting tools and wear parts [5.69–71]. For these applications, filament impurities in the diamond coating (see Sect. 5.2.3) have no known negative effect on performance. Excellent results have been reported, especially for dry cutting applications [5.71, 72] (see also Chap. 12). Three-dimensional HFCVD of <100>-textured diamond was recently demonstrated in the authors' laboratory (Fig. 5.4). The strongly reduced surface roughness is presumably one reason for the high cutting performance [5.71] and might also be utilized for low-friction applications on non-flat wear parts.

Furthermore, the flexibility of HFCVD with respect to the activation area may be used not only for coating large numbers of cutting inserts, but also for uniform deposition on shaft tools and grinding tools [5.71] with extended dimensions (Fig. 5.5).

In MWPCVD, the deposition of well-adhering films on complex-shaped tools is rendered difficult by field enhancement at corners and edges, leading to non-uniform growth conditions. Owing to the absence of electric fields, HFCVD is therefore better suited to deposition on three-dimensional objects. Moderate activation temperatures in HFCVD are advantageous for temperature-sensitive substrate materials. By applying adequate interlayers and simple substrate cooling [5.71, 73], high-speed steel cutting inserts have been coated with adherent diamond films at temperatures below 550°C, without loss of hardness of the steel insert.

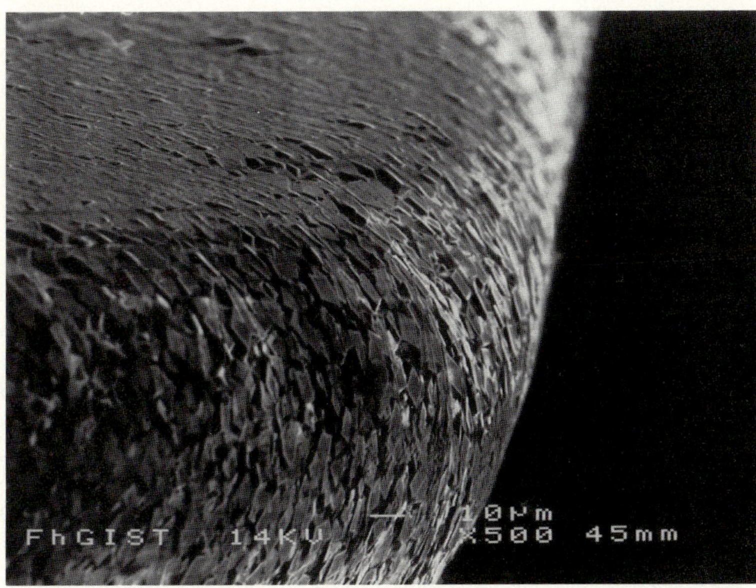

Fig. 5.4: A <100>-textured diamond film over the cutting edge of a cemented carbide insert.

Fig. 5.5: Diamond deposition on a cylindrical grinding wheel 90 mm in diameter and 20 mm in height.

Although the purity of HFCVD-diamond films is generally lower than that of high-quality plasma CVD films, thermal conductivities of more than 10 W/cmK have been reported [5.74, 75]. This value is sufficient for many thermal management applications [5.76]. Compensating lower deposition rates by large-area multi-filament deposition is therefore pursued as an alternative to high-power plasma CVD for cost-effective production of thick films for thermal management applications (see also Chap. 9).

Insulating and chemically resistant diamond films on relatively large-dimensioned substrates are required for level sensor systems in chemically aggressive environments. Matthée et al. [5.77] therefore modified the HFCVD process to deposit insulating films with electrical resistivities up to 5×10^{12} Ωcm on 2 mm diameter tungsten electrodes that were 15 cm in length. Even larger dimensions of chemically resistant, highly conducting diamond films are required for electrochemical applications [5.78]. The application of HFCVD-diamond films with conductivities above 1 $(\Omega \text{cm})^{-1}$ in electrochemical processing environments has shown, that the material properties can meet the requirements with respect to long-term chemical resistance. The further development of large-area HFCVD deposition processes is also promising to reduce the fabrication costs for this application to an acceptable level.

For active electronic devices such as diamond transistors, the requirements on material perfection are much higher than for the previously discussed applications, due to the sensitivity of electronic transport to point defects and other imperfections, which might be introduced into the film by the hot filaments. The requirements on device dimensions and film thickness, on the other hand, are not so extreme. Therefore high-quality plasma-activated CVD should be the preferred deposition methods for diamond in electronic applications.

5.4 Conclusions

Hot-filament chemical vapour deposition (HFCVD) is an established deposition process for diamond films. Considerable research and development in numerous research groups and in industry has contributed to a better understanding of the mechanism, and to the evaluation of the capabilities and limitations of HFCVD-diamond growth. The advantages of HFCVD over plasma CVD methods are the adaptability to product geometries and the moderate activation temperatures for deposition on temperature-sensitive substrates. The lower degree of gas phase activation limits the linear growth rates of HFCVD, which might, however, be compensated for by multi-filament large-area deposition processes. The importance of gas-phase kinetics and transport as well as the relevant growth species have been elucidated and identified, respectively. The engineering of material properties, such as morphology, thermal and electrical properties, has been demonstrated. Thus tailoring of diamond films for specific applications via a proper process control is possible. The testing of prototypes of HFCVD-diamond products has yielded very good results for various tool applications, and for applications in which diamond is used as an insulating protective coating in chemically adverse environments. Today, HFCVD processes are available for job lot production of shaft tools and inserts, with productivities of 20 shaft tools or more than 80 CVD-diamond inserts per hour, respectively. However, further technological and product development has to be carried out to transfer the HFCVD-diamond technology to industrial production of high-quality products on a large scale.

References

5.1 R.C. DeVries, A. Badzian, and R. Roy, RS Bull. Feb. 96, 65 (1996)
5.2 S. Matsumoto, Y. Sato, M. Kamo, and N. Setaka, Jpn. J. Appl. Phys. **21**, L183 (1982)
5.3 S. Matsumoto, Y. Sato, M. Tsutsumi, and N. Setaka, J. Mater. Sci. **17**, 3106 (1982)
5.4 S. Zhou, Z. Zhihao, X. Ning, and Z. Xiaofeng, Mat. Sci. Eng. **B 25**, 47 (1994)
5.5 J. Brückner and T. Mäntylä, Diamond Rel. Mater. **2**, 373 (1993)
5.6 D.M. Li, T. Mäntylä, R. Hernberg, and J. Levoska, Diamond Rel. Mater. **5**, 350 (1995)
5.7 Y. Hirose and Y. Terasawa, Jpn. J. Appl. Phys. **25**, L519 (1986)
5.8 M. Murakawa, S. Takeuchi, H. Miyazawa, and Y. Hirose, Surf. Coat. Technol. **36**, 303 (1988)
5.9 T. Kawato and K. Kondo, Jpn. J. Appl. Phys. **26**, 1429 (1987)
5.10 Y. Saito, K. Sato, H. Tanaka, K. Fujita, and S. Matsuda, J. Mater. Sci. **23**, 842 (1988)
5.11 C. Chen, Y.C. Huang, S. Hosomi, and I. Yoshida, Mat. Res. Bull. **24**, 87 (1989)
5.12 J.A. Mucha, D.L. Flamm, and D.E. Ibbotson, J. Appl. Phys. **65**, 3448 (1989)
5.13 S.J. Harris and A.M. Weiner, Appl. Phys. Lett. **55**, 2179 (1989)
5.14 D.N. Belton and S.J. Schmieg, J. Appl. Phys. **69**, 3032 (1991)
5.15 P.K. Bachmann, D. Leers and H. Lydtin, Diamond Rel. Mat. **1**, 1 (1991)
5.16 G.Z. Cao, J.J. Schermer, W.J.P. van Enckevort, W.A.L.M. Elst, and L.J. Giling, J. Appl. Phys. **79**, 1357 (1996)
5.17 SE 8 403 428-9, 27.06.84
5.18 S. Okoli, R. Haubner, and B. Lux, J. Phys. IV **1**, 923 (1991)
5.19 H. Matsubara and T. Sakuma, J. Mater. Sci. **25**, 4472 (1990)
5.20 M. Sommer and F.W. Smith, J. Mater. Res. **5**, 2433 (1990)
5.21 B. Singh, Y. Arie, A.W. Levine, and O.R. Mesker, Appl. Phys. Lett. **52**, 451 (1988)
5.22 S. Okoli, R. Haubner, and B. Lux, J. Phys. IV, Colloque C2 (1991), C2-923
5.23 L. Schäfer, U. Bringmann, C.-P. Klages, U. Meier and K. Kohse-Höinghaus, in Diamond and Diamond-Like Films and Coatings, ed. R.E. Clausing et al., Plenum Press, New York (1991), p.†643
5.24 K.M. McNamara and K.K. Gleason, J. Electrochem. Soc. **140**, L22 (1993)
5.25 A.M. Bonnot, Phys. Rev. B **41**, 6040 (1990)
5.26 C.P. Beetz, Jr. and T.A. Perry, private communication
5.27 A. Venter and J.H. Neethling, Diamond Rel. Mater. **3**, 168 (1993)
5.28 R.W. Collins, Y. Cong, H.V. Nguyen, I. An, K. Vedam, and T. Badzian, J. Appl. Phys. **71**, 5287 (1992)
5.29 F. Jansen, M.A. Machonkin, and D.E. Kuhman, J. Vac. Sci. Technol. **A8**, 3785 (1990)
5.30 C.-F. Chen, T.-M. Hong, and T.C. Wang, Scripta Metall. Mater. **31**, 413 (1994)
5.31 M. Griesser, G. Stingeder, M. Grasserbauer, H. Baumann, F. Link, P. Wurzinger, H. Lux, R. Haubner, and B. Lux, Diamond Rel. Mater. **3**, 638 (1994)

5.32 D. Li, M. Griesser, R. Hernberg, T. Mäntylä, and M. Grasserbauer, presented at Diamond Films '96, Tours, France, September 8–13, 1996

5.33 E. Gheeraert, A. Deneuville, M. Brunel, and J.C. Oberlin, Diamond Rel. Mater. **1**, 504 (1992)

5.34 H.-J. Hinneberg, M. Eck, and K. Schmidt, Diamond Rel. Mater. **1**, 810 (1992)

5.35 W. Zhu, F.R. Sivazlian, B.R. Stoner, and J.T. Glass, J. Mater. Res. **10**, 425 (1995)

5.36 A.M. Bonnot, in Proceedings of the 1st International Symposium on Diamond and Diamond-Like Films, ed. J.P. Dismukes et al., The Electrochemical Society, Proceedings Volume 89-12, 579 (1989)

5.37 T. Inoue, H. Tachibani, K. Kumagai, K. Miyata, K. Nishimura, K. Kobashi, and A. Nakaue, J. Appl. Phys. **67**, 7329 (1990)

5.38 G. Popovici, C.H. Chao, M.A. Prelas, E.J. Charlson, and J.M. Meese, J. Mater. Res. **10**, 2011 (1995)

5.39 Y.H. Lee, P.D. Richard, K.J. Bachmann, and J.T. Glass, Appl. Phys. Lett. **56**, 620 (1990)

5.40 Y.H. Lee, G.-H. Ma, K.J. Bachmann, and J.T. Glass, Mat. Res. Soc. Symp. Proc. **162**, 119 (1990)

5.41 G.-H.M. Ma, Y.H. Lee, and J.T. Glass, J. Mater. Res. **5**, 2367 (1990)

5.42 S. Okoli, R. Haubner, and B. Lux, Diamond Rel. Mater. **1**, 955 (1992)

5.43 W. Banholzer and R. Kehl, Surf. Coat. Technol. **47**, 51 (1991)

5.44 M. Fryda, X. Jiang, L. Schäfer, and C.-P. Klages, Diamond Films '95, September 10–15, 1995, Tours, France

5.45 L. Schäfer, M. Sattler, and C.-P. Klages, in Proceedings of the 2nd Internatinal Conference on the Applications of Diamond Films and Related Materials, MYU, Tokyo, ed. M. Yoshikawa et al. (1993), p. 133

5.46 R. Kröger, L. Schäfer, C.-P. Klages, and R. Six, Phys. Status Solidi A **154**, 33 (1996)

5.47 E. Kondo, T. Ohta, T. Mitomo, and K. Ohtsuka, J. Appl. Phys. **72**, 705 (1992)

5.48 W.A. Yarbrough, K. Tankala, and T. DebRoy, J. Mater. Res. **7**, 379 (1992)

5.49 C.-P. Klages, M. Sattler and L. Schäfer, in Proceedings of the 3rd International Symposium on Diamond Materials, During the 183rd Meeting of the Electrochemical Society, Honolulu, Hawai (1993)

5.50 D.G. Goodwin, J. Appl. Phys. **74**, 6895 (1993)

5.51 G. Zhao, E. Charlson, B.-Y. Liaw, A. Khan, R. Roychoudhury, E.J. Charlson, T. Stacy, and J. Meese, Diamond Films Technol. **5**, 67 (1995)

5.52 L. Schäfer, M. Sattler, and C.-P. Klages, in Applications of Diamond Films and Related Materials, ed. Y. Tzeng et al., Elsevier, Amsterdam (1991), p. 453

5.53 E. Kondoh, K. Tanaka, and T. Ohta, J. Appl. Phys. **74**, 4513 (1993)

5.54 T. Suzuki and A. Argoitia, Phys. Status Solidi A **154**, 239 (1996)

5.55 C. Wild, W. Müller-Sebert, T. Eckermann, and P. Koidl, in Applications of Diamond Films and Related Materials, ed. Y. Tzeng et al., Elsevier, Amsterdam (1991), p. 197

5.56 Y.-J. Baik and K.Y. Eun, Thin Solid Films **214**, 123 (1992)

5.57 W.B. Alexander, P.H. Holloway, L. Heatherly, and R.E. Clausing, Surf. Coat. Technol. **54/55**, 387 (1992)

5.58 D. Ganesan and S.C. Sharma, J. Mater. Res. **10**, 1764 (1995)

5.59 X. Zhang, T. Shi, J. Wang, and X. Zhang, J. Cryst. Growth **155**, 66 (1995)

5.60 C. Wild, P. Koidl, W. Müller-Sebert, H. Walcher, R. Kohl, R. Locher, and R. Samlenski, Diamond Rel. Mater. **2**, 158 (1993)

5.61 R.E. Clausing, L. Heatherly, L.L. Horton, E.D. Specht, G.M. Begun and Z.L. Wang, Diamond Rel. Mater. **1**, 411 (1992)

5.62 P. Yang, W. Zhu, J.T. Glass, US Patent Number 5,298,286 (1994)

5.63 M.W. Geis, H.I. Smith, A. Argoitia, J. Angus, J.E. Butler, C.J. Robinson, and R. Pryor, Appl. Phys. Lett. **58**, 2485 (1991)

5.64 F. Stubhan, M. Ferguson, H.-J. Füßer, and R.J. Behm, Appl. Phys. Lett. **66**, 1900 (1995)

5.65 J. Yang, Z. Lin, L.-X. Wang, S. Jin, and Z. Zhang, Appl. Phys. Lett. **65**, 3203 (1994)
5.66 Q. Chen, L.-X. Wang, Z. Zhang, J. Yang, and Z. Lin, Appl. Phys. Lett. **68**, 176 (1996)
5.67 X. Jiang, K. Schiffmann, and C.-P. Klages, Phys. Rev. B **50**, 8402 (1994)
5.68 M. Fryda, unpublished results
5.69 R. Haubner and B. Lux, Diamond Rel. Mater. **2**, 1277 (1993)
5.70 A. Feldmann et al. (eds.), Applications of Diamond Films and Related Materials: Third International Conference, NIST Special Publication **885** (1995)
5.71 C.-P. Klages, M. Fryda, T. Matthée, L. Schäfer, and H. Dimigen, in 14th International Plansee Seminar '97, Plansee Proceedings 3, 1 (1997)
5.72 T. Leyendecker, O. Lemmer, S. Esser, and M. Frank, 4. Werkstoffwissenschaftliches Kolloquium Innovative Werkstofftechnologie, 1996, E. Lugscheider (Editor), Verlag Mainz, Wissenschaftsverlag Aachen, pp. 23–29
5.73 L. Schaefer, A. Bluhm, M. Sattler, R. Six, and C.-P. Klages, in Applications of Diamond Films and Related Materials: Third International Conference, ed. A. Feldmann et al., NIST Special Publication **885**, 399 (1995)
5.74 J.E. Graebner, Diamond Films Technol. **3**, 77 (1993)
5.75 D.T. Morelli, C. Uher, and C.J. Robinson, Appl. Phys. Lett. **62**, 1085 (1993)
5.76 P.J. Boudreaux, in Applications of Diamond Films and Related Materials: Third International Conference, ed. A. Feldmann, NIST Special Publication **885**, 603 (1995)
5.77 Th. Matthée, L. Schäfer, A. Schmidt, and C.-P. Klages, Diamond Rel. Mater. **6**, 293 (1997)
5.78 G.M. Swain, Adv. Mater. **6**, 388 (1994)

6. Low-Temperature Diamond Deposition

Akimitsu Hatta[1],[*] **and Akio Hiraki**[2]

[1] Department of Electrical Engineering, Osaka University, Suita 565, Japan

[2] Department of Electronic and Photonic Systems Engineering,
Kochi University of Technology, Tosayamada-cho, Kochi 782, Japan

* Present address:
  Department of Electronic and Photonic Systems Engineering,
  Kochi University of Technology, Tosayamada, Kochi 782-8502, Japan
  e-mail: ahatta@ele.kochi-tech.ac.jp

Springer Series in Materials Processing
Low-Pressure Synthetic Diamond Eds.: B. Dischler and C. Wild
© Springer-Verlag Berlin Heidelberg 1998

6.1 Introduction

The success of diamond growth from the vapor phase has been an important innovation in expanding the area of application of this fascinating material. The chemical vapor deposition (CVD) method has enabled us to obtain crystalline diamond at easily controllable temperatures such as 800–900°C. For actual use of CVD-diamond films in various application areas, however, the growth temperature is still too high. Many efforts have been devoted to lowering the growth temperature to extend the limitations of material for the substrate at such a high temperature. Such high temperatures of the substrate during growth induce stress due to the difference in the expansion rates. Low-temperature growth is an essential technique required for practical application areas, especially for electronics applications. For electronics applications, the substrate temperature is limited not only by the melting point of the material but also by the diffusion of impurities. While a silicon substrate itself is still stable at the conventional CVD growth temperature, silicon devices will easily lose their structures of impurity profiles, heterojunctions, and metal interfaces. One of the objectives of low-temperature CVD of diamond is the temperature at which electronics devices can be active; for instance, from 200°C to 400°C. Another objective is diamond coating on low melting point materials, especially on organic polymers.

The trend of research on low-temperature deposition has been reviewed thoroughly by Muranaka et al. [6.1]. Several papers have reported successful diamond growth at temperatures below 200°C. Nakao et al. succeeded in the fabrication of diamond films on an Al substrate at 140°C using a DC-plasma CVD method [6.2]. Ihara et al. confirmed that diamond particles could be formed at a substrate temperature of 135°C by a hot filament assisted CVD method with water cooling [6.3]. Muranaka et al. reported the fabrication of diamond films on silicon substrates at 130°C by a microwave-plasma CVD method using a $CO/O_2/H_2$ gas mixture and cooling the substrate by ventilation [6.4]. Yara et al. reported the fabrication of microcrystalline diamond films at below 100°C [6.5].

In this chapter, the technical trends of low-temperature deposition of diamond films by the plasma CVD method are introduced. The following are indispensable problems for fabrication of diamond films at low temperature: heating of the substrate by the plasma; forced cooling and measurement of temperature; and growth and nucleation at low temperature. These problems for low-temperature growth are discussed, and some actual techniques to achieve low-temperature deposition are suggested.

6.2 Substrate Heating by Plasma

In plasma, the substrate is automatically heated up by many kinds of energies. There are several kinds of energy flow to the substrate; that is radiation in the infrared, visible, or ultraviolet regions, electromagnetic waves, charged or neutral species with kinetic energy, or the release of chemical or electronic excited states. Some of these energies will enhance the surface reaction necessary for diamond growth, while some of them will only heat up the substrate. In conventional microwave-plasma CVD for diamond growth, the substrate temperature automatically exceeds 500°C. In order to keep the substrate at a proper temperature, it is necessary to cool down the substrate rather than to intentionally heat up. For the low-temperature deposition of diamond, it is important not only to cool down the substrate but also to decrease the extra energy which only heats up the substrate without any contribution to diamond growth. If the extra heating is successfully suppressed, it will result in a lower or more easily controlled temperature.

Here, the dominant energy for the automatic heating of the substrate should be clarified. The thermal flux due to neutral gas molecules, Γ_g, is estimated from the energy loss due to their collisions with the substrate surface. The neutral molecules are heated in the plasma. By means of the general kinetic theory of gas molecules, Γ_g is expressed as

$$\Gamma_g = \frac{1}{4} n_g \bar{v} \cdot \eta \cdot \frac{1}{2} m \overline{v^2}$$
$$= \frac{3}{8} \eta p \sqrt{\frac{3RT_g}{M}}. \tag{6.1}$$

Here, n_g is the gas density at pressure p, m and $\bar{v}$ denote the mass and the mean velocity of the molecules, respectively, M is the atomic mass number, and η is a coefficient of energy transfer from the molecule to the substrate in a collision. The coefficient 1/4 appears in the calculation of the collision frequency of a molecule against a wall of unit area. In the case of conventional microwave plasma, Γ_g becomes 11 W/cm^2 under the assumption of $p = 50$ kPa, $T_g = 1000$ K, $M = 2$ (H$_2$ molecule), and $\eta = 0.5$. The molecules are heated by collisions with electrons which are accelerated in the electric field up to several eV; that is, to several tens of thousands of degrees Kelvin. The actual gas temperature, T_g, has been measured as several thousand degrees Kelvin [6.6]. The actual value of η depends on the difference between the gas temperature and the substrate temperature. η will decrease when the substrate temperature approaches the gas temperature. Therefore, the substrate temperature will saturate at a proper value when a balance is reached in the temperature gradient. In any event, the potential heating power of the hot gas is of the order of 10 W/cm^2.

The kinetic energy of such charged particles is much larger than that of neutral molecules because of acceleration in the electric field in the plasma. In the conventional microwave plasma, however, the ionization ratio of the gas molecules is quite small; for instance, the ion (or electron) density is of the order of 10^{10}/cm^3, while the molecule density is 10^{16}/cm^3. The energy flux to the substrate due to the charged particles, Γ_c, is expressed as

$$\Gamma_c = V_{sh} I_{is} + T_e \frac{I_e}{e}. \tag{6.2}$$

The first and second terms on the right-hand side are the energy fluxes due to ions and electrons, respectively. V_{sh} and T_e denote the sheath potential between the plasma and the substrate and the electron temperature, respectively. The electron current, I_e, is equal to the ion saturation current, I_{is}, if an electrically floated substrate is assumed, where I_{is} is limited by the electron density. By using typical values for the conventional microwave plasma, $V_{sh} = 20$ V, $T_e = 5$ eV, and $I_{is} = 10$ mA/cm^2, Γ_c becomes 250 mW/cm^2.

Heating by photons of any wavelength will be negligible in comparison with Γ_g. If a substrate is taken out of the plasma column, it can be kept at almost room temperature during deposition. Dielectric heating of the substrate by the microwave radiation can be also ignored, because the microwave power is almost completely used up for discharge. Therefore, the dominant thermal flux to heat up the substrate in the conventional microwave-plasma is Γ_g; that is, heating by hot gas molecules. The most effective way to cool down the substrate is to decrease the gas pressure.

6.3 Low-Pressure Plasma CVD

Liou et al. first attempted to lower the substrate temperature by decreasing the gas pressure from 50 to 6 Torr and also decreasing the microwave power from 1 kW to 260 W, resulting in a temperature as low as 365°C. A low-power RF glow discharge at the low pressure of 1–3 Torr was also examined for low-temperature deposition of diamond [6.7]. By reducing the pressure for easy control of the substrate temperature, it becomes difficult to ignite or maintain the plasma due to an increase in the loss of electrons and ions to the wall. A magnetic field is effective for low-pressure discharge because of electron trapping along magnetic lines of force. It is also advantageous to use electron cyclotron resonance (ECR) heating by the microwave and to use the R-wave propagation mode of the microwave from the higher magnetic field side [6.8], which is called ECR discharge.

In the plasma, the reaction process is triggered by electron impact to the source molecules. The collision frequency between the electrons and the gas molecules governs the production rate of the reactive species that will be necessary for diamond growth. For example, the dissociative collision frequency for a kind of source gas, ν_d, is

$$\nu_d = n_e n_g < \sigma_d \overline{v_e} > \tag{6.3}$$

where n_e and n_g denote the electron density and the gas molecule density, respectively, σ_d is the dissociation cross-section of the gas molecule, and $\overline{v_e}$ is the electron velocity. In the conventional microwave-plasma without a magnetic field, the electron density saturates at the cut-off density for the microwave frequency even if the incident power is increased, where the cut-off density is 7×10^{10}/cm^3 for the frequency of 2.45 GHz. When the gas pressure – that is the molecule density – is decreased, the collision frequency – that is, the production rate of the reactive species – will decrease. By using a magnetic field, however, the electron density can exceed the cut-off density by the additional power input through the R-wave mode propagation.

An increase in the electron density will compensate for the decrease in molecule density. In addition, electrons are heated by ECR heating, resulting in the higher reaction probability $< \sigma_d v_e >$, due to the dependence of σ_d on the electron energy, which has a threshold value of about 10 eV for dissociation [6.9].

Figure 6.1 shows the magnetoactive microwave-plasma CVD system. It is almost the same apparatus as for an ECR plasma reactor, which is widely used for the fabrication of microelectronics devices [6.10]. For an operational condition, a higher gas pressure and a higher magnetic field than for the conventional ECR plasma reactor are employed, such as 10 Pa and 0.25 T, respectively, while the ECR plasma is produced at pressures below 0.1 Pa in magnetic fields of about 0.1 T at the peak. This requires a magnetic field several times higher than that of the ECR condition to suppress reflection of the microwave power at pressures higher than those for conventional ECR plasma. The microwave power is absorbed in the plasma mainly due to collisional damping of the wave in the higher magnetic field region.

The operating pressure, 10 Pa, is still lower by two orders of magnitude than for conventional microwave-plasma CVD. This decrease in pressure also enables uniform plasma production over a large diameter, resulting in a large deposition area more than 6 inches wide. The typical growth conditions for the magnetoactive microwave-plasma are shown in Table 6.1.

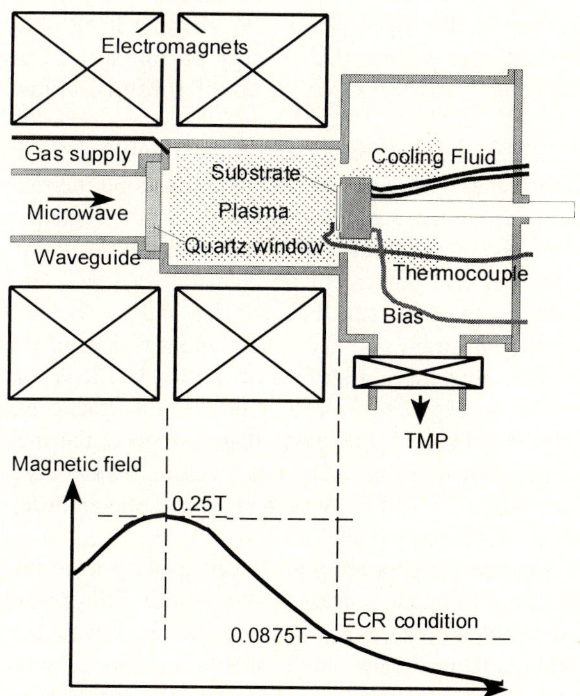

Fig. 6.1: The magnetoactive microwave-plasma CVD system. The quartz window for microwave is positioned at the peak of the magnetic field. The substrate holder is positively biased (usually +30–40 V) from the grounded chamber.

Table 6.1: Typical conditions for diamond growth by magneto-active microwave plasma CVD at low temperature.

Source gas	CH_3OH, or CH_4+CO_2 (15–25 sccm)
Dilution gas	H_2 (75–85 sccm)
Total pressure	10 Pa
Microwave power	1.3 kW
Substrate temperature	200–600°C (surface)
Substrate bias	+30–40 V

In the magnetoactive microwave-plasma at 10 Pa, the heating of the substrate by the hot gas molecules, Γ_g, is less than in the conventional microwave-plasma by almost two orders of magnitude, due to the lower gas pressure when the same values of η, T_g, and M are assumed. The substrate temperature can be controlled easily by additional heating or cooling through the contacts with the substrate holder.

6.4 Forced Cooling and Measurement of Temperature

Even if the extra energy heating the substrate was successfully reduced by lowering the pressure, some other kinds of heating would be inevitable. At the very least, the surface reaction of diamond growth in the vapor phase is exothermic itself; that is, chemically or electronically excited species lose their energy on the surface to become diamond structures. Forced cooling is necessary to keep the substrate at low temperature.

Figure 6.2 shows a substrate holder with forced cooling by a circulating fluid, such as water or ethylene glycol. The material of the holder should have good thermal conductivity to transfer the heat flux from the substrate to the fluid, although the thickness of the wall between the fluid and the substrate should ensure the atmospheric pressure of the circulating fluid. For reliable control of the substrate temperature, most attention should be paid to the contact between the substrate and the holder. The thermal conductivity between the substrate and the holder greatly depend on the roughness of both the back of the substrate and surface of the holder. It is not adequate without any filling, as shown in Fig 6.2a even if both of the surfaces are mirror-polished. They should be fixed with a filling paste that has good thermal conductivity, such as silicon grease, as shown in Fig. 6.2b. The resulting temperature of the substrate is quite different for Figs. 6.2a and b where both of the temperatures were measured in the same way as shown in Fig. 6.2b.

It is not easy to measure the accurate temperature with forced cooling from the back, because the temperature decreases from the surface to the holder. An optical pyrometer in the visible emission range is not useful at low temperature. A radiation thermometer in the infrared region and a thermocouple are available for a wide range of temperatures. For semiconductor substrates, such as Si, however, the substrate is transparent in the measurement range of the wavelength. The thermometer will show

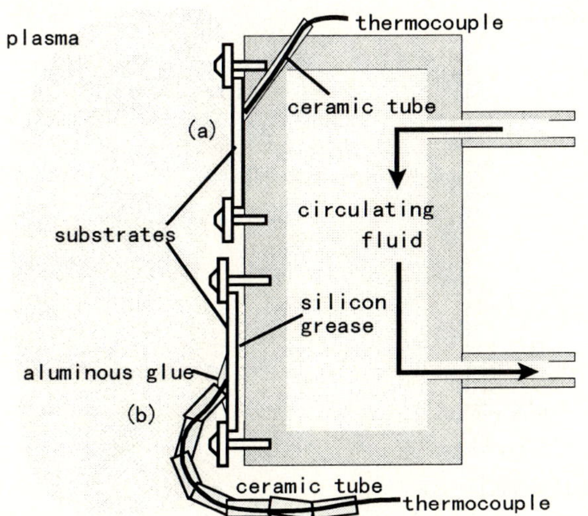

Fig. 6.2: A substrate holder with forced cooling by a circulating fluid, such as water with ethylene glycol: (**a**) is a conventional way and (**b**) is a recommended way to keep the substrate at a lower temperature while monitoring the temperature from the surface.

the temperature not of the Si substrate itself but of the holder under the substrate. To use a thermocouple, special attention should be paid to the thermal contact with the substrate. If the thermocouple is fixed on the back of the substrate, as shown in Fig. 6.2a, it thermally contacts the holder rather than the substrate. The temperatures of diamond deposition reported in some literature would be not correct, but much lower than the actual temperature because they are usually measured from the back contact. The thermocouple should touch only the surface of the substrate and should be covered with insulator to avoid irradiation to the plasma, as shown in Fig. 6.2b. The substrate temperature measured as shown in Fig. 6.2a was almost equal to the temperature of the holder.

6.5 Nucleation at Low Temperature

Nucleation is one of the key processes in the fabrication of diamond films, except for homoepitaxial growth. The proper conditions for nucleation are not the same as those for crystal growth. Some pretreatment methods have been employed to enhance nucleation before starting the growth procedure. Polishing with diamond paste or ultrasonically scratching with diamond powder is the most popular way to obtain an adequate nucleation density on Si or some other kinds of material. The bias-enhanced nucleation process is the other promising method to realize heteroepitaxial growth of diamond on Si or SiC. At low temperatures, however, the nucleation density and/or the quality of the nuclei produced by these methods are not sufficient for the fabrication of high-quality diamond films.

It has been reported that the nucleation density decreases with a decrease in the substrate temperature on ultrasonically scratched Si substrates in a microwave-

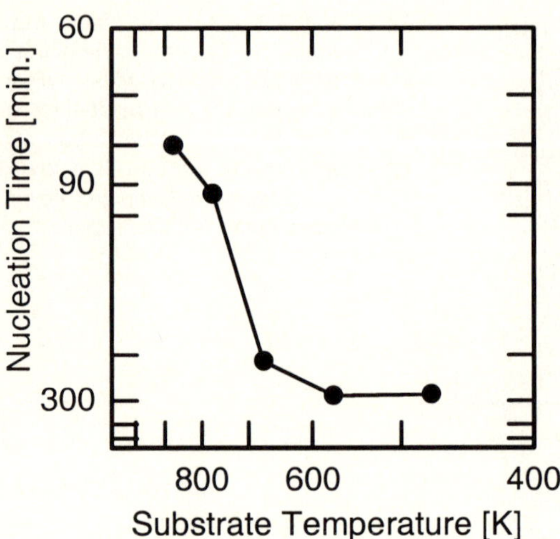

Fig. 6.3: The dependence of the nucleation time on the substrate temperature on ultrasonically scratched Si substrates in magnetoactive microwave-plasma CVD. The nucleation time is plotted on an inverted axis to show the nucleation rate.

plasma CVD [6.4]. The decrease in nucleation density will result in a decrease in the effective growth rate due to the lateral growth in the initial stage, and, furthermore, will result in difficulty in making continuous films. Figure 6.3 shows the dependence of the nucleation time on the substrate temperature on ultrasonically scratched Si substrates at the proper conditions for diamond growth in magnetoactive microwave plasma CVD [6.11]. The nucleation time is the delay of the growth in the initial stage, which is an intercept on the growth time in the plot of film thickness versus deposition time. When the substrate is cooled below 400°C, the nucleation time on ultrasonically scratched Si substrates becomes very long, so the deposition time will be mostly spent not on the growth but on the nucleation. It seems that diamond nucleation is more difficult than growth at low temperature in the vapor phase. The bias-enhanced nucleation process is also influenced by the substrate temperature and it is not applicable for low-temperature deposition.

Instead of such a self-nucleation process from the vapor phase, seeding with diamond powder on the substrate becomes the key technique as a promising pre-treatment for low-temperature deposition. The requirements of the seeding process are high density, hopefully to cover the whole surface of the substrate, and good adhesion. The authors have developed seeding methods using nanocrystal diamond powder, because the finer powder will perform better for both the seeding density and adhesion. The nanocrystal diamond powder, which is synthesized by an implosion process, is commercially available, and its size is 3–5 nm. By using a colloidal solution of the nanocrystal diamond powder purified by acid, very high density seeding of 2×10^{11} cm^{-2} has been obtained on silicon substrates [6.12]. This very high density seeding with diamond powder enabled the fabrication of an extremely thin and continuous diamond film within a short growth time, below 10 min, by a conventional microwave-plasma CVD method. It is expected that the loss of effective growth time due to lateral growth in the initial stage will be eliminated.

Fig. 6.4: SEM images of films fabricated at 200–250°C for 24 hours by magnetoactive microwave-plasma CVD from a $CH_3OH(15\%)/H_2$ gas mixture in a same batch: **(a)** on an Si substrate pretreated by ultrasonic scratching with diamond powder in a conventional way; and **(b)** on an Si substrate seeded with nanocrystal diamond powder.

The nanocrystal seeding method has been employed for low-temperature deposition of diamond films [6.11]. It has improved the effective growth rate by eliminating the loss time for nucleation and lateral growth, which was more than 5 hours at 200°C. Figures 6.4a,b are SEM images of fabricated diamond films at 200°C on ultrasonically scratched silicon substrate and a nanocrystal seeded one, respectively, by magnetoactive microwave-plasma CVD, in a same batch. On the nanocrystal-seeded substrate, a well-faceted polycrystalline diamond film was even grown at 200°C, while no crystalline particles could be found on the scratched one. It is clear that diamond can be grown on the seeds with fairly good crystallinity even at low temperature, as shown in Fig. 6.4b. On the ultrasonically scratched Si substrate, however, the nucleation of high-quality diamond could not be obtained. Nanocrystal seeding is the essential technique for low-temperature deposition of diamond films.

6.6 Film Properties

Diamond films deposited at low temperature have been qualified by Raman scattering spectroscopy, scanning electron microscopy, X-ray diffraction, and so on. The Raman spectra showed significant features for diamond films deposited at low-temperature.

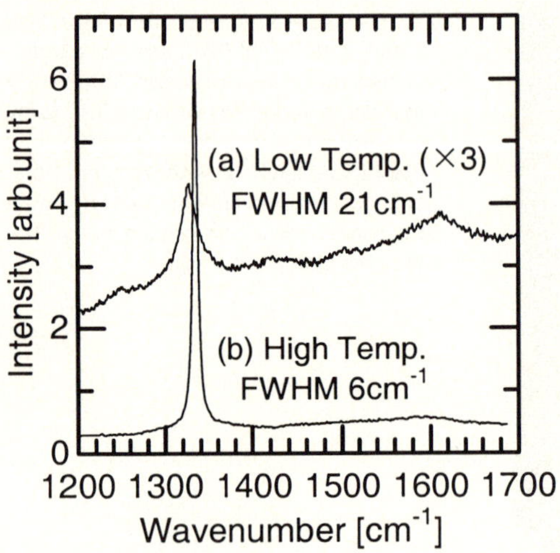

Fig. 6.5: Raman spectra of fabricated films: **(a)** at 200°C by magnetoactive microwave-plasma CVD at a proper condition from a CH_3OH (15%)/H_2 gas mixture on the nanocrystal-seeded Si substrate; and **(b)** at 930°C by conventional microwave-plasma CVD from a gas mixture of CO(10%)/H_2 on an Si substrate pretreated by ultrasonic scratching.

In first-stage trials of low-temperature deposition, the films fabricated at temperatures ranging from 400°C to 500°C showed a sharp diamond peak with a small shift of the peak position [6.13], while the background intensity was sometimes very large due to laser-induced luminescence [6.14].

Degradation in quality is significant for films fabricated at temperatures below 200°C. Films fabricated below 200°C by DC-plasma CVD showed submicron grains about 0.1 μm in size, with a round shape, composed of nanocrystalline diamond about 20 nm in size and with an amorphous carbon phase, which were characterized by diffraction patterns in TEM observation because no diamond peak was found in the Raman spectra [6.2]. Films fabricated at 130°C by microwave-plasma CVD [6.4] and at 135°C by hot-filament CVD [6.3] clearly showed a diamond peak in the Raman spectra. In the case of microwave-plasma CVD, although the inclusion of a nondiamond phase was successfully suppressed by additional oxygen incorporation in the source gas, the films fabricated at 130°C showed a ball-like morphology and the diamond peak in the Raman spectra was broad [6.4].

The causes of the degradation can be separated into two categories, with reference to the nucleation and the subsequent crystal growth. The nucleation was improved by the nanocrystal seeding process, as shown in Fig. 6.4b. Figure 6.5a shows the Raman spectrum of a diamond film fabricated at 200°C by magnetoactive microwave-plasma CVD. Although the film shows well-faceted morphology, as shown in Fig. 6.4b, the diamond peak is broadened to 21 cm^{-1} in FWHM and shifted to the lower-wavenumber side. For comparison, Fig. 6.5b shows a film fabricated at 930°C by conventional microwave-plasma CVD. The spectrum for the film deposited at low temperature includes a swell centered at 1600 cm^{-1} due to the inclusion of nondiamond-phase carbon. Even if the nuclei are improved, the inclusion of a nondiamond-phase and/or defects during the following crystal growth results in the degradation of the crystal.

Fig. 6.6: SEM images of films fabricated by magnetoactive microwave-plasma CVD for 8 hours from a $CH_3OH(15\%)/H_2$ gas mixture at (a) 200°C and (b) 180°C respectively, on Si substrates seeded with nanocrystal diamond powder.

Figures 6.6a,b show SEM images of films fabricated at 200°C and 180°C, respectively, by magnetoactive microwave-plasma CVD on an Si substrate seeded with nanocrystal diamond. At 180°C, the film is composed of microcrystalline particles, possibly of diamond, about 10 nm in size. This may be due to secondary nucleation during the growth phase, caused by many inclusions of nondiamond-phase and/or defects.

6.7 Growth Rates at Low Temperature

Although the mechanism of diamond growth in the vapor phase has not yet been clarified, it is thought that selective etching of nondiamond-phase carbon, with partial retention of the diamond phase, is essential [6.15]; there are two competing processes: deposition and etching. Without the etching, the diamond phase will be buried in the nondiamond-phase carbon film, growing faster than the diamond crystal. The competition between the deposition of activated hydrocarbon species and the etching of the hydrogenated amorphous phase and the graphite phase must be well balanced. We should consider not only deposition but also etching.

The success of diamond growth in the vapor phase by hot-filament CVD [6.16] and microwave-plasma CVD [6.17] was due to the production of sufficient hydrogen radicals for the etching. If the etching rate is not adequate, amorphous carbon will be deposited quickly, as in the deposition of amorphous silicon from silane gas. At the proper temperature, the etching of nondiamond-phase carbon by the atomic hydrogen can overcome the total growth rate of the amorphous carbon. Due to the difference between the etching rates of the crystal diamond phase and the amorphous phase, the atomic hydrogen can completely remove the amorphous phase, leaving a propostion of the carbon atoms as a diamond phase on the growing surface.

When the substrate is cooled to below 400°C, however, the atomic hydrogen cannot perform adequate etching of nondiamond-phase carbon. It is necessary to match the etching rate to the growth rate of the nondiamond-phase carbon. To enhance the etching at low temperatures, oxygen has been incorporated into the source gases [6.13, 18]. A small amount of additional oxygen has a strong influence on diamond growth [6.19, 20]. Oxygen-containing carbon sources, such as CO [6.21], CO_2 [6.22], CH_3OH [6.23], and C_2H_5OH [6.7] are popular for low-temperature growth because they can also supply oxygen atoms. Without oxygen, the concentration of the carbon source should be reduced to slow down the deposition of carbon. In the case of hot-filament CVD from a CH_4/H_2 gas mixture, the CH_4 concentration is below 1% at a temperature of 135°C [6.3].

The deposition process consists of the production of chemically activated species of carbon and hydrocarbon, transportation from the plasma to the substrate, and sticking and migration on the growing surface. The production of the reactive species is governed by the plasma condition or filament temperature. It is almost independent of the surface condition. On the contrary, the sticking and migration on the growing surface is drastically influenced by the substrate temperature. In general, the sticking coefficient will increase with a decrease in the substrate temperature. The increase in the sticking probability will result in a decrease in its migration length on the surface. Without any etching by hydrogen or oxygen, the hydrocarbon deposition rate will increase with a decrease in the substrate temperature. Actually, at substrate temperatures below 200°C, polymer-like films are formed very quickly at the proper conditions for diamond growth at high temperature. In order to overcome the fast growth of organic polymer films, the etching effect of the nondiamond-phase should be enhanced by increasing the oxygen concentration.

If the nondiamond phase can be successfully removed by hydrogen and/or oxygen, the decrease in migration length will result in a small probability of the formation of diamond structure. The reactive species will more likely stick before finding any proper site for diamond structure, by sp^3 bonding, to the growing surface. The actual growth rate will decrease due to the decrease in the migration length on the surface. To improve the actual growth rate of high-quality diamond crystal at low temperatures, it is necessary to increase the oxygen concentration for enhanced etching and the carbon source concentration for adequate deposition, at the same time, as reported in experimental results [6.13].

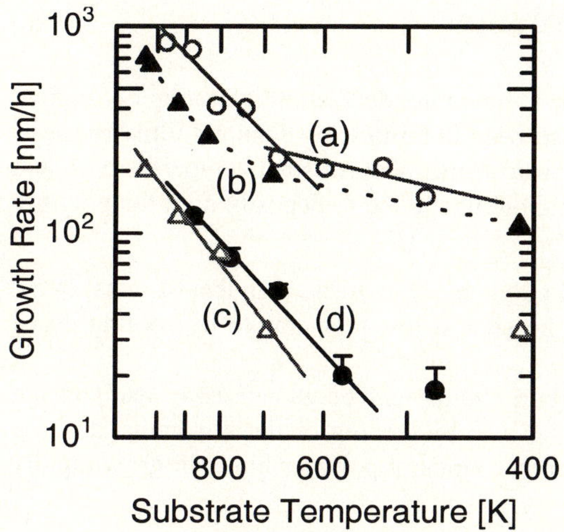

Fig. 6.7: The dependence of the growth rate on the substrate temperature: **(a)** by hot-filament CVD from CH_4/H_2; **(b)** by microwave-plasma CVD from CO/H_2; **(c)** by the microwave-plasma CVD from $CO/O_2/H_2$; and **(d)** by magnetoactive microwave-plasma CVD from CH_3OH/H_2.

Figure 6.7 shows the dependence of the diamond growth rate on the substrate temperature, as found in the literature. Some results would seem to include ambiguities in the measurement of the temperature and the estimation of growth rate. The substrate temperatures were measured from the back of the substrate, except for the data of Fig. 6.7d. In the case of Fig. 6.7d, the substrate temperature was measured on the top surface of the substrate, and the nucleation time could be eliminated by using high-density seeded substrate, with nanocrystal diamond. Figure 6.7a is for a case of hot-filament CVD from $CH_4(0.5\%)/H_2$ [6.24], and Figs. 6.7b,c are cases of microwave-plasma CVD from $CO(7\%)/H_2$ and $CO(8\%)/O_2(2.2\%)/H_2$, respectively [6.4]. The difference in the growth rate of Figs. 6.7b,c was thought to be due to the inclusion of a nondiamond phase in the crystal from the gas mixture without additional oxygen. Figure 6.7d is for a case of magnetoactive microwave-plasma CVD from $CH_3OH(15\%)/H_2$. Figure 6.7a shows an activation energy of 5 kcal/mol in the temperature region above 400°C. It becomes about 1 kcal/mol at temperature below 400°C. Figures 6.7c,d show activation energies of 7 kcal/mol and 5 kcal/mol, respectively, except for the lówest-temperature data. These values are smaller than those of conventional temperature deposition; for instance, 22–24 kcal/mol in the range of 740–930°C by hot-filament CVD [6.25]. The low-temperature growth is governed by some kinds of reaction stages that differ from those present in high-temperature growth.

6.8 Diamond Coating on Polymer

One of the important applications of low-temperature deposition is coating on organic polymer substrates. Even if researchers succeed in fabricating diamond films on metal or Si at a low temperature, which is below the heat-resistant temperature of the polymer material, the method cannot be directly applied to deposition on the polymer itself, for the following reasons:

(1) Because the thermal conductivity of polymer is much less than that of metal or Si, the surface of the substrate cannot be kept at low-temperature by cooling down from the back surface.

(2) The polymer substrate of hydrocarbon chains will be etched away itself in the environment for diamond growth – that is by etching of the nondiamond phase by hydrogen and/or oxygen – some other kinds of polymer film will grow rapidly at the low temperature.

To keep the substrate at a low temperature, it is effective to lower the gas pressure, as mentioned before, and to use a thinner film for the substrate. The difficulty of diamond coating on polymer concerns not only the substrate temperature but also the etching of the substrate.

At temperatures below 200°C, the deposition of a nondiamond phase is very fast, due to higher sticking coefficient and the polymerization of the hydrocarbon species. For the selective growth of diamond, the etching of the nondiamond-phase should be enhanced to match the fast deposition rate. This means that in an environment for the low-temperature growth of diamond, polymer substrate cannot exist by itself. Some protective layer is necessary in the initial stage of diamond growth, and this will be an interlayer between the polymer substrate and the coated diamond film.

With a protective layer of amorphous Si, a diamond film can be fabricated on polyimid film [6.26]. Figure 6.8 shows a 125 µm thick polyimid substrate, after an a-Si coating of 0.3 µm, and after an diamond coating of 0.6 µm. With an a-Si coating below 0.3 µm in thickness, the substrate was damaged by the plasma. The successfully fabricated diamond film was continuous, though it consisted of microcrystalline particles, due to poor quality of a nucleation by a conventional ultrasonic scratching method. The diamond-coated polyimid film shows a mirror-like surface due to the fine crystallinity, and it can be flexibly bent to a radius of about 10 mm.

6.9 Future Prospects

The remaining challenge for low-temperature deposition of diamond is improvement of the growth rate and film quality. The diamond growth rate at low-temperature is still too slow to use for actual application. One of the possibilities for improving the growth process is pulse modulation of the plasma [6.27].

The diamond growth rate due to magnetoactive microwave-plasma CVD increases with an increase in the microwave power. The high microwave power perhaps

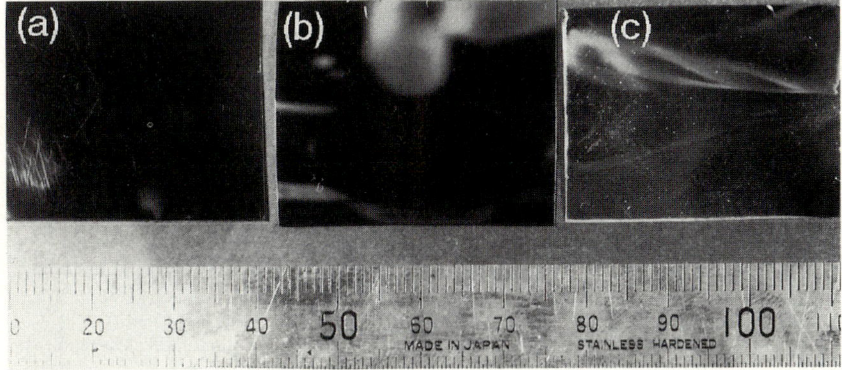

Fig. 6.8: (a) A 125 μm thick polyimid substrate, (b) after amorphous Si coating by sputtering, and (c) after 0.6 μm diamond deposition.

yields a high growth rate because of the high-density production of reactive species necessary for diamond growth. At a high power level such as 5 kW, however, it is not easy to keep the substrate at low temperature. It is necessary to reduce the microwave power [6.13] without decreasing the growth rate. Intermittent discharge on a time scale of several tens of seconds has been attempted as a means of keeping the substrate at a low temperature [6.18].

To maintain an adequate growth rate, however, a high production rate of the reactive species is necessary. The pulse modulation of plasma is quite interesting because of the time averaged lower power and the higher controllability of the reaction process in the vapor phase. The time-averaged power can be reduced with a decrease in the duty ratio (the ratio of the working duration to the period), resulting in a lower and more easily controlled substrate temperature. Even if the time-averaged power is reduced, the density of the reactive species can be kept high because of the difference in the time constant for the production and loss of each reactive species. A pulse-modulated controlled process will also improve the film quality by controlling the density ratio of each reactive species [6.28].

References

6.1　Y. Muranaka, H. Yamashita, and H. Miyadera, Diamond Rel. Mater. **3**, 313 (1994)
6.2　S. Nakao, M. Noda, H. Kusakabe, H. Shimizu, and S. Maruno, Jpn. J. Appl. Phys. **29**, 1511 (1990)
6.3　M. Ihara, H. Maeno, K. Miyamoto, and H. Komiyama, Appl. Phys. Lett. **59**, 1473 (1991)
6.4　Y. Muranaka, H. Yamashita, and H. Miyashita, J. Appl. Phys. **69**, 8145 (1991)
6.5　T. Yara, M. Yuasa, M. Shimizu, H. Makita, A. Hatta, J. Suzuki, T. Ito, and A. Hiraki, Jpn. J. Appl. Phys. **33**, 4404 (1994)
6.6　T. Lang, J. Stiegler, Y. von Kaenel, and E. Blank, Diamond Rel. Mater. **5**, 1171 (1996)
6.7　I. Watanabe, T. Matsushita, and K. Sarahara, Jpn. J. Appl. Phys. **31**, 1428 (1992)
6.8　F.F. Chen: Introduction to Plasma Physics, Plenum Press, New York (1974) chap.4
6.9　H.F. Winters: J. Chem. Phys. **63**, 3462 (1975)
6.10　S. Matsuo and M. Kiuchi: Jpn. J. Appl. Phys. **22**, 210 (1983)

6.11   T. Yara, H. Makita, A. Hatta, T. Ito, and A. Hiraki, Jpn. J. Appl. Phys. **34**, L312 (1995)

6.12   H. Makita, K. Nishimura, N. Jiang, A. Hatta, T. Ito, and A. Hiraki, Thin Solid Films, **281–282**, 279 (1996)

6.13   Y. Liou, A. Inspector, R. Weimer, and R. Messier, Appl. Phys. Lett. **55**, 631 (1989)

6.14   W.L. Hsu, J.M. Tung, E.A. Fuchs, K.F. McCarty, A. Joshi, and R. Nimmagadda, Appl. Phys. Lett. **55**, 2739 (1989)

6.15   K. Kobashi, K. Nishimura, Y. Kawate, and T. Horiuchi, Phys. Rev. B **38**, 4067 (1988)

6.16   S. Matsumoto, Y. Sato, M. Tsutsumi, and N. Setaka, J. Materials Sci. **17**, 3106 (1982)

6.17   M. Kamo, Y. Sato, S. Matsumoto, and N. Setaka, J. Cryst. Growth **62**, 642 (1983)

6.18   T.P. Ong and R.P.H. Chang, Appl. Phys. Lett. **55**, 2063 (1989)

6.19   T. Kawato, and K. Kondo, Jpn. J. Appl. Phys. **26**, 1429 (1987)

6.20   C.-P. Chang, D.L. Flamm, D.E. Ibbotson, and J.A. Mucha, J. Appl. Phys. **63**, 1744 (1988)

6.21   Y. Muranaka, H. Yamashita, and H. Miyadera, J. Vac. Sci. Technol. **A9**, 76 (1991)

6.22   J. Wei, H. Kawarada, J. Suzuki, and A. Hiraki, J. Cryst. Growth **99**, 1201 (1990)

6.23   M. Yuasa, O. Arakaki, J. S. Ma, H. Kawarada, and A. Hiraki, Diamond Rel. Mater. **1**, 168 (1992)

6.24   H. Komiyama, M. Ihara, and A. Yamaguchi, in Proceedings of the JNDF TX meeting (1993, JNDF, Tokyo), pp.16–21 (in Japanese)

6.25   K.A. Snail and C.M. Marks, Appl. Phys. Lett. **60**, 3135 (1992)

6.26   K. Mishuku, Y. Negishi, T. Yara, A. Hatta, T. Ito, and A. Hirkai, Advances in New Diamond Science and Technology, Proceedings of the 4th International Conference on New Diamond Science and Technology, ed. S. Saito, N. Fujimori, O. Fukunaga, M. Kamo, K. Kobashi, and M. Yoshikawa, MYU, Tokyo (1994) pp. 45–48

6.27   A. Hatta, H. Suzuki, K. Kadota, H. Makita, T. Ito, and A. Hiraki, Plasma Sources Sci. Technol. **5**, 235 (1996)

6.28   S. Samukawa, Jpn. J. Appl. Phys. **33**, 2133 (1994)

7. Other CVD Methods for Diamond Production

Norma J. Komplin and Robert H. Hauge

Chemistry Department, Rice University,
6100 Main Street, Houston, TX 77005, USA
e-mail: hauge@rice.edu

Springer Series in Materials Processing
Low-Pressure Synthetic Diamond Eds.: B. Dischler and C. Wild
© Springer-Verlag Berlin Heidelberg 1998

7.1 Introduction

Low-pressure synthesis of diamond films from a gas medium has been accomplished by many techniques, such as hot-filament chemical vapor deposition (CVD), microwave-plasma and radio-frequency (rf) plasma, direct current (DC) and alternating current (AC) plasma jets, combustion flame and oxyacetylene torches, ion beams, and so on. CVD techniques provide the capability to coat various shapes and produce many forms of diamond, such as fine powders, protective coatings, epitaxial and whisker growth, and coating of particles. Although there are many ways to produce diamond films and many applications, much needs to be accomplished before the full potential of diamond CVD is realized. In these methods, the feed gases are exposed to high temperatures and/or some high-energy activation to initiate the reactions necessary for CVD. Usually, the substrates are maintained at temperatures greater than 700°C for good-quality deposits. Hydrogen atoms are usually considered essential in facilitating deposition of diamond under graphite stable conditions, so the production of hydrogen radicals is key for most of the above processes [7.1].

The intent of this chapter is to provide a brief description of reports of novel methods that have been shown to produce diamond. Some of these methods do not require hydrogen. They may be divided into four activation methods that are labeled, respectively, laser activation, metal solvation, low-temperature thermal activation and chemical activation. With laser activation, both activation of the gas and laser ablation of solid surfaces have been shown to produce diamond. Metal solvation and reprecipitation of carbon as diamond has been demonstrated. Diamond has also been grown in hot-wall metal tube reactors, through which dilute mixtures of halocarbons and hydrogen are made to flow. Finally, chemical activation with thermally dissociated molecular chlorine of hydrocarbon and hydrogen mixtures has been shown to grow diamond down to temperatures as low as 200°C.

In each of the above methods, diamond growth has been clearly demonstrated. However, in all cases other than for chlorine activation, the chemical mechanisms that lead to diamond are not understood. The clear implication is that other chemical mechanisms do exist that can grow high-quality diamond. When they are better understood, one can expect new low-temperature diamond growth methods to be commercially viable.

7.2 Halogens in Diamond CVD

Atomic hydrogen has many roles in most diamond CVD, such as terminating the diamond lattice by forming C–H bonds, reacting with graphitic precursors, generating active radical sites at which hydrocarbons can adsorb by hydrogen abstraction, and assisting the incorporation of hydrocarbons into the lattice by further abstraction. It is also known that atomic hydrogen etches graphite faster than diamond [7.2]. Halogen atoms have bonding characteristics similar to those of hydrogen atoms, which makes them potential candidates to replace or partially replace hydrogen in diamond CVD.

A novel halogen-assisted CVD process was developed at Rice University by Patterson et al. [7.3]. This process uses halogens or halogen-containing gases that are heated to about 900°C with hydrocarbon gases, and react at atmospheric pressure to nucleate and grow diamond particles at substrate temperatures as low as 300°C. This halogen-assisted method has the potential to greatly influence diamond CVD technology. It promises many benefits, such as lower substrate temperatures, lower energy consumption, lower overall cost, and readiness for scale-up.

7.2.1 A Brief History

The halogen-assisted diamond CVD work at Rice University, and preceding halogen research, generated a worldwide interest in this process. Halogens and halogen-containing molecules have been introduced into rf-plasma, microwave-plasma and hot-filament reactors with the intent of lowering the substrate temperature and improving film quality. A brief review of these studies is presented.

Spitsyn and Deryagin were the first to report pyrolysis of halogens, such as CBr_4 and CI_4, in a process for depositing diamond on diamond at temperatures from 800°C to 1000°C and pressures of approximately 3×10^{-6} mm of mercury [7.4]. Rudder et al., at the Research Triangle Institute (RTI), utilized CF_4 in H_2 in a rf-plasma assisted CVD system to grow diamond on diamond, reporting that the nucleation density of diamond was increased with CF_4 in H_2 relative to CH_4 in H_2 with this technique [7.5]. Bai and coworkers used isotopic labeling and Raman spectroscopy to determine that chloromethanes produced diamond faster than methane in a hot-filament reactor [7.6]. Komplin et al. found that the addition of a few percent of HCl to a methane/hydrogen system in a hot-filament reactor enhanced diamond growth rates as much as tenfold at substrate temperatures near 670°C [7.7]. This enhancement was attributed to chlorine atoms involved in gas–surface reactions, as predicted from gas-phase chemistry. Chlorine atoms abstract hydrogen more readily than hydrogen atoms abstract hydrogen from the diamond surface, creating more open sites for diamond growth at these low substrate temperatures. Similarly, Wu and Hong studied the effect of chlorine addition on diamond growth in a CH_4/H_2 system in a hot-filament reactor [7.8]. Cl_2 molecules are mostly converted to HCl before arriving at the hot filament. Hong and coworkers previously reported on the growth characteristics of diamond films using CH_4, CH_2Cl_2, $CHCl_3$, and CCl_4 as carbon sources [7.9, 10]. Their results indicated that, in comparison to methane, the chlorocarbons generally yielded higher diamond growth rates, especially at lower substrate temperatures. Rego et al. obtained quantitative gas-phase species measurements using an *in situ* molecular beam spectrometer designed with a hot-filament reactor for a variety of C/Cl/H gas mixtures [7.11, 12]. Almost all chlorine was reduced to HCl, and the authors support the premise that chlorine atoms are involved in the gas–surface reactions that produce active growth sites on the diamond surface. Corat and coworkers demonstrated that a small amount of CF_4 added to the standard methane/hydrogen mixture resulted in diamond growth at lower temperatures with reasonable growth rates in a hot-filament reactor [7.13]. An activation energy of 11 kcal/mol was obtained for diamond growth with CF_4 addition for the temperature range of 250–850°C, which was about half the activation energy measured for a CH_4/H_2 mixture alone. A similar activation energy was obtained for the CCl_4/H_2 system [7.14]. The technique for low- temperature growth with the addition of CF_4 to the methane–hydrogen system was applied recently to diamond growth on tungsten carbide with 6% cobalt, to enhance adhesion of the diamond film to the tool bit [7.15]. The lower-temperature deposition improved the smoothness of the coating and minimized the thermal stress induced by the diamond CVD process, resulting in longer-lasting cutting tools. Schimdt et al. performed a study using CH_4, CHF_3, and CF_4 in H_2 in a hot-filament reactor, and determined that diamond could be grown at 400°C in the CHF_3/H_2 system with significant growth rates, but no diamond deposition was observed with the CF_4/H_2 system [7.16].

Kadono et al. reported diamond growth from CF_4/H_2 gas mixtures with microwave-plasma activation [7.17]. It was observed that the quality of the diamond films improved as the concentration of tetrafluoromethane was lowered, as in conventional methane–hydrogen systems. Maeda et al. examined the process of diamond deposition on silicon in a microwave plasma using CF_4 and CHF_3 as well as methane as the carbon source [7.18]. Their results agreed with those of Kadono et al., since quality diamond was only produced at CF_4 or CHF_3 concentrations of 1% or lower. Maeda and coworkers concluded that fluorine may aid in the removal of nondiamond phases above 850°C, but it terminates active sites for diamond growth at lower silicon substrate temperatures in their microwave-plasma system. Ramesham et al. also reported diamond thin-film growth using microwave-plasma assisted CVD and a gas mixture of CF_4/H_2 on silicon wafers at substrate temperatures between 800°C and 875°C [7.19]. Grannen and Chang have used fluorocarbon gases in a microwave-plasma system to deposit diamond on silicon carbide and tungsten carbide [7.20]. This builds upon work published in 1991 which employed CF_4 gas to grow diamond on nondiamond substrates in a microwave-plasma system [7.21].

Grannen and Chang observed that it was possible to use higher percentages of fluorocarbons to produce diamond than of corresponding hydrocarbons. Fox et al. observed that the activation energy for diamond growth using CF_4, CH_3F, or CH_4 in hydrogen was nearly the same in a microwave-plasma system, and ascertained that diamond may be grown from the same intermediate in these systems [7.22]. Kotaki et al. deposited diamond from an $Ar/CCl_4/H_2$ DC plasma jet and found that the deposited diamond film contained less amorphous carbon than when methane was the carbon source [7.23]. A similar observation was made in a Ar/H_2 DC plasma jet by adding dichlorobenzene which lead the authors to also conclude that chlorine abstraction plays an important role in activating the diamond surface, thereby giving rise to higher deposition rates and better removal or prevention of graphite and amorphous carbon [7.24].

A novel technique for the deposition of diamond by CVD and atomic-layer epitaxy (ALE) was developed by Hukka and coworkers [7.25, 26]. The radical precursors are produced when thermally generated atomic fluorine abstracts hydrogen from the precursor molecule. The best evidence of diamond deposition on diamond was using $CHCl_3$ as the precursor molecule. The $CHCl_3$ and H_2 were injected into a stream of atomic fluorine at pressures of 10^{-3} to 10^{-2} Torr. The growth rate was approximately 0.08 μm/hr. It was noted that metal fluoride deposits (NiF_2, MgF_2, AlF_3, CaF_2, and SiF_2) were observed on sections of some samples. Another novel method for diamond chemical vapor deposition was developed by Pan and coworkers [7.27]. Atomic chlorine generated thermally in a resistively heated graphite furnace at temperatures from 1300°C to 1500°C activated necessary diamond precursors. Chlorine abstraction of hydrogen from molecular hydrogen and the hydrocarbon present produced atomic hydrogen and the active carbon species. The quality of the deposited diamond depended on the substrate temperature and the H_2/Cl_2 mole ratios. In general, the best substrate temperature was approximately 150°C lower than that for a conventional hydrogen/hydrocarbon hot-filament system with similar growth rates. Similar work was also performed by Wu and Hong, in which diamond was deposited by injecting thermally decomposed Cl atoms into a methane/hydrogen hot-tube system at high flow velocities [7.28]. The novel processes will be detailed in forthcoming sections.

7.2.2 Hot-Wall Halogen-Assisted CVD

Halogen-assisted CVD of diamond films has received much attention because of the low substrate temperatures required for growth and the lack of energetic activation, except for chemical activation by reactions with fluorine or chlorine. Patterson et al. flowed mixtures of fluorine and hydrocarbons or hydrogen and halocarbons through a hot furnace at atmospheric pressure, and diamond growth was observed downstream form the hot zone at temperatures as low as 250°C [7.29]. All of the reactions were carried out in a simple flow tube system in which the reactant gases (the structures of which are comprised of carbon, hydrogen, oxygen, and halogen atoms) are passed through a resistively heated reaction chamber into which a suitable substrate has been placed.

The reactor was comprised of a tube of Monel 400 alloy for use with fluorine or SiO_2 with chlorine. Two separately controlled resistively heated furnaces with a maximum service temperature of 1100°C were used to heat the reactor. Water cooling was provided through a closed system at the ends of the reactor, just prior to the flanges. This kept the Teflon flanges near room temperature. The temperature gradient in the reactor typically ranged from 900°C to about 30°C.

The pretreated substrate was placed into the reactor before gas flow began. Diamond deposition times ranged from several hours to several days, depending on the amount of growth desired. The temperature gradient within the Monel reactor gave rise to a broad range of substrate temperatures during each reaction. The temperature usually ranged between 250°C and 750°C.

Conditions for a typical experiment in the fluorine-assisted CVD system were as follows. A typical gas pressure of 760 Torr was normally used, but reduced pressures as low as 100 Torr were also tested. The reactant gases were comprised of either elemental fluorine and a simple hydrocarbon such as methane or of hydrogen and a simple halocarbon. It was found that the addition of oxygen to the reactants in the form of pure O_2, air, H_2O, or CO_2, or bound to the carbon in the reactant molecule, such as in methanol and diethyl ether, aided the growth of diamond. The reactant gases were typically diluted with an inert gas such as helium or argon.

The diamond growth precursors are expected to be long-lived in order to survive all of the collisions in an atmospheric furnace, so the chemistry is probably quite different from that of other CVD techniques. In the reducing environment it is possible that metals are available in the diamond growth area through reduction of the various metal fluorides which are present, and are commonly observed along with the newly grown diamond. The main disadvantages are reactor failure due to corrosion in a hydrogen/fluorine environment, low deposition rates, metal halide contamination of the films, and non continuous films.

This halogen-assisted technique was verified by Wong and Wu when they successfully repeated experiments similar to the ones performed at Rice [7.30]. For example, diamond was successfully produced with the following deposition conditions: 1 standard cubic centimeter per minute (sccm) of CHF_3, 2 sccm of CH_4, 197 sccm of H_2, a central furnace temperature of 1000°C, and a deposition time of 29 h. Butler and coworkers at the Naval Research laboratories (NRL) also verified the experiments performed at Rice, using CH_3F in hydrogen at 800°C [7.31]. Problems related to corrosion of the reaction vessel in the fluorine system lead to replacment of the fluorine with chlorine. Diamond was then grown in an SiO_2 reactor using carbontetrachloride and chlorine in hydrogen, resulting in the major products of hydrogen chloride and diamond.

It has been shown that thermal halogen-assisted diamond CVD is capable of producing diamond films on nondiamond substrates at substrate temperatures ranging from 250°C to 650°C. The best carbon sources, such as dimethyl ether and methanol, also contained oxygen. Diamond growth at low substrate temperatures is not easily achieved with other CVD techniques. However, the introduction of fluorocarbons or chlorocarbons in other diamond CVD methods has allowed lower substrate temperatures to be used without compromising the quality of diamond produced. For our slow flow atmospheric pressure reactor, low-temperature growth cannot be rationalized by radical reaction chemistry.

7.2.3 Chlorine-Assisted Diamond CVD

It was first demonstrated by Hauge et al. that thermally dissociated chlorine will
activate molecular hydrogen by producing atomic hydrogen in quantities sufficient
to activate the growth of diamond through the atomic hydrogen – hydrocarbon free
radical process [7.32, 33, 34]. This has also been demonstrated subsequently by
other investigators [7.28]. The process has been modeled with a modified SPIN
code kinetics model [7.35].

The chlorine activation method is based on the fact that molecular chlorine is
dissociated to the atomic state at lower temperatures than molecular hydrogen. The
degree of dissociation is shown as a function of temperature in Fig. 7.1. The
reaction of atomic chlorine with molecular hydrogen is known to be rapid; thus a
steady-state equilibrium for the reaction [Cl + H$_2$ = H + HCl] is quickly
established, with an equilibrium constant near unity. In the gas mixture, the ratio of
atomic hydrogen to atomic chlorine follows the ratio of molecular hydrogen to
hydrogen chloride.

Pan et al. have demonstrated that diamond can be grown over a wide
temperature range, with chlorine activation of molecular hydrogen [7.34]. This
was accomplished in a reactor in which molecular chlorine was brought in through
a central graphite tube, which was resistively heated to near 1500°C. The hydrogen
and hydrocarbon were brought in through a concentric outer graphite tube and
mixed with the atomic chlorine immediately in front of the atomic chlorine orifice.

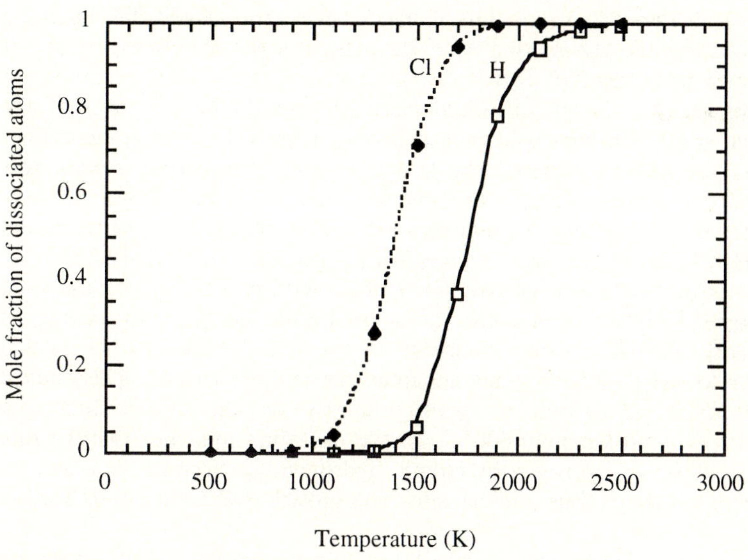

Fig. 7.1: Thermal dissociation of hydrogen and chlorine at 1 Torr.

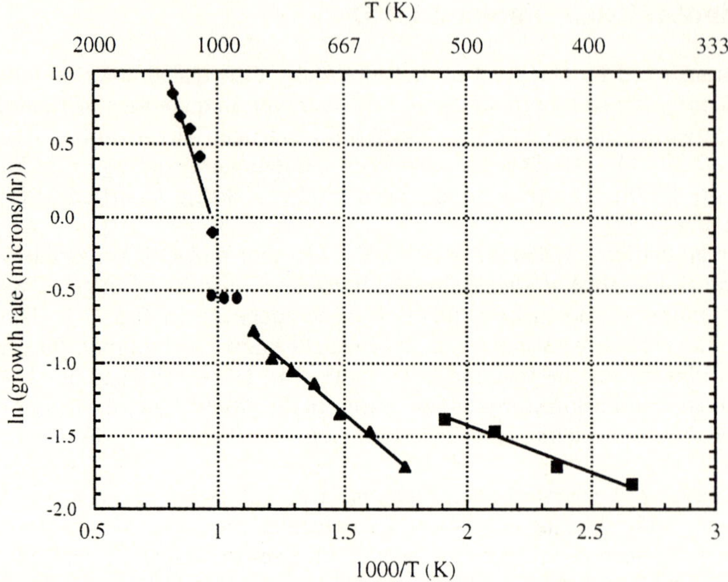

Fig. 7.2: Arrhenius plots of diamond growth rates as a function of substrate temperature.

It was found that diamond quality deteriorated (became more graphitic) if the H_2/Cl_2 ratio was lowered to values less than 30. At a ratio of ten, the diamond Raman peak was completely missing when growing at a substrate temperature of 760°C and a total pressure of 20 Torr.

It was subsequently found that diamond quality could be enhanced if the atomic chlorine gas stream was split into opposing streams directed at each other and parallel to the growth surface. The hydrogen and hydrocarbon stream was passed through the atomic chlorine streams and directed at the growth surface. This had the effect of spreading the diamond growth region over a larger area and lowering the H_2/Cl_2 ratio to 15, with retention of high-quality diamond [7.36].

Pan et al. carried out a quantitative study of growth on a (110) single-crystal diamond surface with the concentric chlorine heater design. An Arrhenius activation energy plot of the data is shown in Fig. 7.2. Three distinct growth regions appear to exist with constant apparent activation energies. The highest region above 700°C is thought to grow primarily via the atomic hydrogen mechanism. Below approximately 600°C, kinetic modeling indicates that the rate of surface abstraction of hydrogen by atomic hydrogen has become too small to produce sufficient surface free radical sites for growth via addition of methyl radicals.

Wu and Hong also produced diamond by injecting chlorine atoms formed by dissociating chlorine in a hot graphite tube and then injecting the Cl atoms into a methane/hydrogen mixture to initiate radical reactions in a hot-tube system at high flow velocities [7.28]. Typical deposition parameters are shown in Table 7.1.

Table 7.1: Typical deposition conditions in Wu and Hong's study [7.28].

CH$_4$ or C$_2$H$_2$ concentration	015–1.0%
Cl$_2$ concentration	4–8%
Hot-tube temperature	1500°C
Substrate temperature	400–750°C
Total flow rate	300–600 sccm
Reactor pressure	2–7.5 Torr

Diamond film quality was enhanced as the chlorine concentration increased from 4% to 8%, where it became saturated. It was noted that the film quality improved as the substrate temperature was reduced from 750°C to 600°C, but deteriorated from 600°C to 400°C. An Arrhenius plot of diamond film growth rates versus temperature revealed two activation energies, 3.6 kcal/mol for the region 600–750°C and 7.9 kcal/mol for 400–600°C. By comparison with Pan's studies, the growth rates are substantially lower and do not exhibit a low activation energy growth at very low temperatures. This may be due to the lower total pressures used by Wu and Hong.

Kinetic modeling has recently indicated that growth via production of surface free radical sites cannot explain most of the growth of diamond below 600°C [7.37]. Thus at the lower temperatures it seems likely that new chemical mechanisms with low activation energies must operate.

A possible mechanism that involves the direct insertion of atomic carbon into surface carbon–hydrogen bonds is currently under investigation. Kinetic modeling has indicated that with high atomic hydrogen concentrations, the surface concentration of atomic carbon increases as the temperature is lowered. This occurs because the atomic hydrogen surface concentration also increases with reduced temperature, since hydrogen loss, due to recombination with surface hydrogen, no longer occurs at significant rates. This mechanism is not unique to the chlorine activation method, since other methods that generate high atomic hydrogen concentrations at the growth surface will also grow diamond via direct insertion of atomic carbon into surface hydrogen–carbon bonds.

The abrupt increase in growth near 250°C is attributed to the adsorption of chlorine on the diamond surface. At these lower temperatures, the direct insertion of atomic carbon into the chlorine–carbon bond becomes a possible mechanism.

Important issues with respect to the gas dynamics and mixing of atomic chlorine and hydrogen gas streams should be investigated before large-scale reactors can be built. Ongoing studies are exploring the use of opposed gas streams, where the hydrogen–hydrocarbon stream is brought in through orifices in the substrate [7.37].

The disposal of the HCl product gas that is produced in the process must also be addressed before the method can be utilized in large-scale reactors. It currently appears feasible to convert the HCl back to hydrogen and chlorine with a dry electrolysis process analogous to that recently announced by Dupont [7.38]. With process optimization, this method appears to be capable of large-area coating, with high growth rates of diamond at the higher substrate temperatures, as well as large-area growth of diamond at substrate temperatures as low as 200°C.

7.3 Laser-Assisted Diamond Synthesis

The field of laser-assisted growth of diamond encompasses a broad variety of potential deposition methods. Lasers are capable of providing economical and technological benefits through reduced fabrication time and better quality growth. Laser chemical vapor deposition provides a clean energy source, the possibility for high growth rates, low substrate temperatures, selective deposition area and better surface integrity.

Chiefly, only single-laser systems have been used such as CO_2 lasers for photothermal processes or UV excimer lasers for photochemical and laser ablation processes. In the photothermal process, lasers can decompose gaseous molecules by pyrolysis. In laser pyrolysis, the gases are excited by the beam irradiation, and the substrate is heated to the desired temperature by controlling the power and the irradiation time of the beam. The excited gases then decompose on the hot surface. Most of the reported work on laser-assisted CVD of diamond involves the use of excimer lasers and the photodecompositon of gaseous molecules. In laser photolysis, the photons break the chemical bonds of the gaseous molecules and allow the product to be deposited on the substrate. One necessary factor is that the gases absorb the laser radiation. Recently, a novel process using a set of multiplexed lasers has been developed called the QQC process. The plasma formed by multiplexed lasers which range from the UV to the IR creates a unique set of crystal growth conditions. This has enabled diamond to be grown at atmospheric pressure, without hydrogen, without surface pretreatment and without substrate heating.

7.3.1 Infrared Laser-Assisted Growth

Infrared laser-assisted techniques use the laser to couple photon energy into the reactants and the substrate. This generated thermal energy leads to chemical reactions and physical processes that deposit films. Fedoseev et al. reported using a CO_2 laser to pyrolyze small particles of graphite or carbon black to form diamond powder [7.39, 40]. These results have been replicated by others using transmission electron microscopy (TEM), X-ray diffraction, and electron diffraction [7.41, 42]. The lack of Raman results implies that this data is inconclusive. Raman spectroscopy is approximately 50 times more sensitive to graphitic carbon than diamond, making it an excellent method for determining whether nondiamond phases are present. It is also sensitive to various possible noncrystalline phases of carbon that may be present, resulting in a qualitative view of the film generated. It has been discovered that electron diffraction does not differentiate between graphitic material and diamond, leading Kitahama et al. to retract their conclusion of diamond growth as confirmed by electron diffraction. In 1993, Molian and Waschek reported a CO_2 laser CVD process in which a gas mixture of 2% CH_4 in hydrogen was pyrolyzed to deposit diamond on a GaAs or Si substrate [7.43]. Raman spectroscopy and SEM analysis of the film revealed fine diamond crystallites.

7.3.2 Ultraviolet Laser-Assisted CVD

One of the problems associated with diamond film growth is the substrate temperature necessary for high-quality diamond films. In an attempt to lower the substrate temperature, a laser-assisted CVD process was devised by Kitahama et al., who reported the deposition of diamond on silicon wafers using 193 nm irradiation from a ArF laser to activate a 0.5%–2% C_2H_2 in H_2 mixture [7.44, 45]. It was hoped that the UV laser irradiation would produce chemically active species, since C_2H_2 has the highest absorption coefficient at 193 nm of all the light hydrocarbons [7.46]. Only scanning electron microscopy (SEM) and reflection electron diffraction were used to determine the presence of diamond. It was discovered that oriented graphite and polycrystalline diamond have similar diffraction patterns. The above deposits were re-investigated and found to be graphitic layers, as confirmed by Raman spectroscopy [7.47].

Goto et al. reported the deposition of diamond on Si using an ArF laser (193 nm) and a mixture of 0.1–1% CCl_4 in H_2 [7.48]. The presence of diamond was confirmed by Raman spectroscopy and reflection electron diffraction. To make this technique work, atomic hydrogen was provided from a microwave discharge or a hot filament. Hydrogen is transparent to 193 nm radiation. Celii et al. investigated Goto et al.'s experimental conditions, but concluded that laser irradiation induced a decrease in diamond growth relative to hot-filament deposition alone [7.49].

Laser-Excited CVD (LECVD)

Subramaniam and Rebello have grown diamond from a 95% CO – 5% CH_4 gas mixture excited at low power with a CO laser [7.50]. LECVD involves exciting the gas phase by optically pumping the vibrational modes of CO. A state of thermodynamic disequilibrium is maintained between the vibrational and translational modes of the reactant molecules by the IR laser irradiation. It is proposed that the laser energy absorbed by the CO is transferred to the methane. Subsequently, diamond may be deposited from the excited methane or from the resultant dissociated or reactive fragments (C_xH_y). Diamond has been grown on monocrystalline Si (001) substrates, pretreated by abrading with diamond grit, heated to a temperature of 600°C, and also on unheated substrates. This method does not require dilution with hydrogen. Samples were analyzed with a Raman microprobe and by SEM.

Laser Chemical Vapor Deposition (LCVD)

Rebel et al. used an ArF laser at 193 nm to dissociate a mixture of 0.7% CO in hydrogen to form diamond on heated and unheated monocrystalline (001) and (111) silicon substrates [7.51]. The laser beam was focused approximately 7 mm above the substrate. The CO is photodissociated by a multiphoton process in which at least two photons are required to break a CO bond. The fragments, C and O, are free to react with the hydrogen, resulting in diamond growth. Unlike the

LECVD process in which CH_4 was the precursor and CO was used as the medium to store significant energy in its vibrational modes, this process utilizes CO directly as the gas-phase precursor. This work was encouraging, since the Raman spectrum shows only the 1332 cm^{-1} peak characteristic of diamond and SEM shows a cluster of faceted crystals. The laser beam in these experiments impinged on the substrate surface, so transient substrate heating probably occurred. In an effort to grow diamond at low temperature, another experimental approach was designed. Rebello and Subramaniam reported diamond growth at room temperature on silicon (111) substrates by laser irradiation of CO/hydrogen mixtures at low pressures [7.52]. The laser beam could be directed on to the substrate surface with the gas flow parallel (configuration A and B) or normal (configuration C) to the substrate surface. These configurations are shown in Fig. 7.3. The ArF excimer laser was operated at a repetition rate of 20 Hz, with a pulse energy of about 50 mJ/pulse and a pulse duration on the order of 20–25 nsec. Typical experimental conditions were a CO volume flow rate of 1.1 standard cubic centimeters per minute (sccm), an hydrogen flow rate of 99 sccm, and a total pressure of 8 Torr. Generally, the duration of the experiment was two hours. Diamond growth appeared downstream from the focal point of the laser beam in the case of the gas flow parallel to the surface, as shown in configurations A and B. In the case of configuration C, where the gas flow was perpendicular to the substrate surface, the laser beam broke the symmetry. The diamond growth was observed in two locations on both sides of the focal beam, as shown in Fig. 7.3. Raman microprobe spectra demonstrated a sharp peak at 1332 cm^{-1}, characteristic of diamond, with no evidence of codeposition of nondiamond carbon. SEM images of the substrate after growth displayed irregularly faceted particles of the order of 5–10 µm. Energy-dispersion spectroscopy (EDS) of these particles showed the presence of carbon with trace amounts of oxygen. The oxygen is believed to come from trace amounts in the gas phase during growth and not from sample handling.

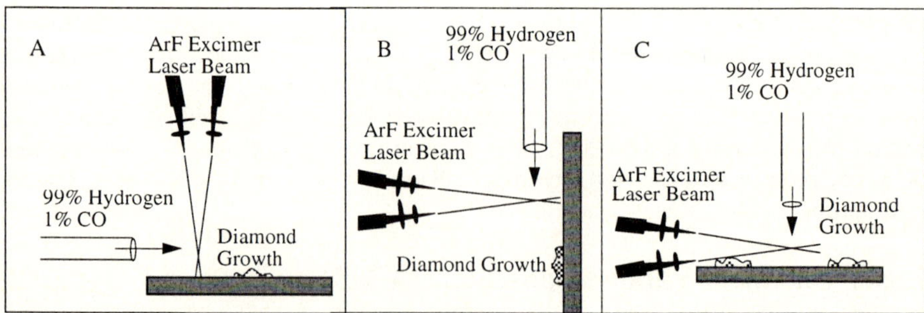

Fig. 7.3: The three different configurations used in LCVD experiments [7.51].

7.3.3 QQC Materials Deposition Process

QQC, Inc. has developed a novel diamond deposition process which does not require a vacuum, hydrogen, or surface pretreatment [7.53]. This process does not require substrate heating, and thereby offers great potential for coating aluminum, plastics, and other low-melting materials. Prior to this process, only single-laser systems had been used. CO_2 lasers were used for photothermal processes, and ultraviolet excimer lasers for photochemical processes and for laser ablation processes [7.54]. This process uses a combination of lasers, ranging from the ultraviolet to the infrared. The plasma formed by these lasers can create a wide variety of chemically reactive species as well as very hot electrons, which produces a unique crystal growth environment. The main carbon source for diamond growth was carbon dioxide. Deposition rates as high as 1 µm/sec have been reported.

QQC, Inc., a materials division of Turchan Technologies, Inc., of Dearborn, Michigan, has been involved in developing different methods for synthesizing very hard materials for surface enhancement of practically any substrate material, especially cutting tools. A breakthrough in diamond deposition occurred while working on a process to coat WC cutting tools. Interestingly, a carbon dioxide tank was accidentally hooked up instead of a nitrogen tank and the process parameters were fortitous enough to produce a new hard coating, which was subsequently analyzed as diamond.

The process utilizes three types of lasers; two ultraviolet excimer lasers, an infrared Nd : YAG laser, and an infrared CO_2 laser. Each laser emits a beam which is directed through an opening of a nozzle toward the substrate surface. The carbon dioxide carbon source and the nitrogen used as the shielding gas are directed in by nozzles.

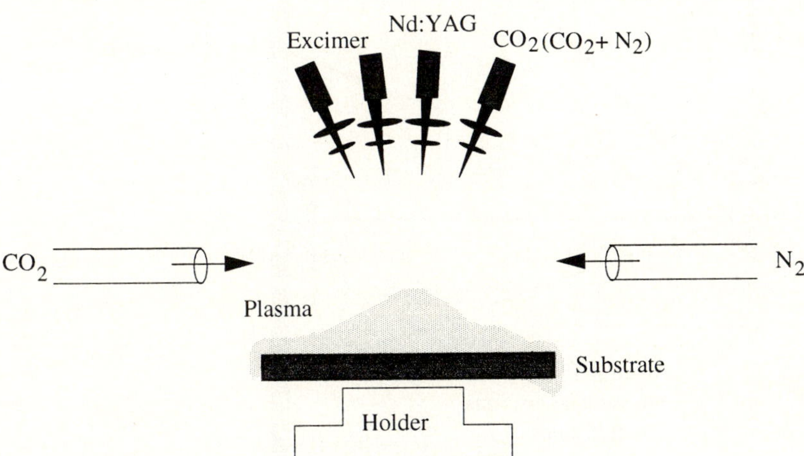

Fig. 7.4: A schematic of the QQC multiplexed laser process.

Figure 7.4 shows a schematic of the QQC materials cell. The coating cell is completely computer controlled, including workpiece loading and manipulation, laser pulsing and power modulation, gas flow and system process diagnostics. The process is carried out in the open air, with the CO_2 and N_2 gases shrouding the substrate. The duration of the deposition is typically 40 sec. The resultant coating would be 20–40 µm thick. The substrate temperature is below 50°C. The surface morphology of the coatings has been examined with SEM, and ranges from octahedron to cubic depending on the sample. The diamond coatings have a polycrystalline structure with typical grain sizes of 10–20 µm. The chemical composition has been analyzed using electron microprobe, X-ray photoelectron spectroscopy (XPS), and Auger electron microscopy. The films are mainly carbon, but may show some metal incorporation, depending on the substrate. There has been no adhesion problem, since the lasers interact with the solid and appear to grade the composition to diamond. The Raman spectra show a sharp peak at 1336 cm^{-1}, indicating some stress in the diamond films.

7.3.4 Pulsed Laser Deposition (PLD)

Pulsed laser deposition (PLD) has been used to create high-quality thin films of many complex materials [7.55]. A laser beam is focused incident to a rotating target, which is positioned normal to the substrate. The generated ablation plume contains energetic atoms, ions, and small molecules along this normal, resulting in the deposition of a film on the substrate. Typical target to substrate distances range from 2cm to 10 cm. A schematic of pulsed laser deposition is shown in Fig. 7.5.

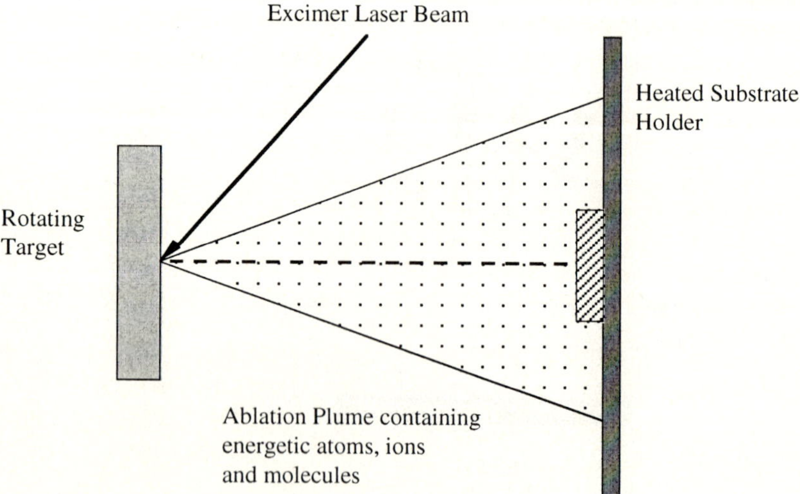

Fig. 7.5: A schematic of pulsed laser deposition.

Laser ablation of graphite has generally resulted in the deposition of amorphous diamond-like carbon (DLC) films [7.56]. Attempts at diamond deposition by PLD have been challenging unless a diamond target was utilized [7.57]. Some authors have reported the presence of diamond in their carbon films, but Raman spectroscopy has shown no evidence of diamond in these films. For instance, Rengan et al. found particles with diamond morphology in their diamond-like films grown at 860°C, using a pulsed XeCl (308 nm) laser synchronized with a hydrogen discharge produced with a capacitor [7.58]. However, diffraction and Raman spectroscopy analysis failed to prove that diamond was present. Bourdon et al. reported the growth of octahedral crystals up to 4 µm on fused silica substrates heated to 300°C in an hydrogen environment. Electron diffraction indicated the presence of diamond, but Raman did not show the characteristic diamond peak at 1332 cm^{-1}. Also, Seth et al. reported nucleating 0.01–0.15 µm diamond crystallites on self-biased (100) silicon substrates using a shorter laser wavelength (248 nm) combined with an argon rf plasma, but Raman did not show evidence of diamond in these films [7.59].

Recently, however, there has been evidence of diamond growth from ArF excimer laser ablation of nondiamond targets. Polo et al. succeeded in growing diamond crystals of the order of 1 µm on (100) silicon substrates heated to 450°C, using an ArF excimer (193 nm) laser with a fluence of 1.3×10^8 W/cm^2 focused on a pyrolytic graphite target at an angle of 45° [7.60]. This process was performed using 1 mbar of hydrogen as a buffer gas, which the authors have determined, enhances diamond nucleation [7.61]. Raman spectroscopy showed a sharp peak at 1332 cm^{-1}, which is characteristic of diamond. Polo et al. also discovered that no diamond growth occurred if a KrF (248 nm) laser was used instead of the ArF, even when all deposition parameters were held constant. The samples deposited by KrF laser ablation of graphite were amorphous sp^2 carbon, with two broad Raman peaks at 1340 and 1600 cm^{-1}. This indicates that the laser wavelength is a determining parameter in the growth of diamond by PLD from a graphite target. Since the laser power was held constant, the shorter-wavelength laser has a smaller ablation depth into the target and a higher photon energy. The resulting higher-energy carbon plume is probably necessary for diamond growth over amorphous carbon deposition. It was also noted by Molian, Janvrin and Molian that an ArF beam in their laser CVD process yielded better diamond growth on stainless steel substrates with less amorphous carbon deposited than the KrF laser-beam CVD [7.62].

Interestingly, in other reports of laser ablation of a carbon-based target leading to diamond formation, an ArF excimer laser beam has also been used for the deposition. Kozloz et al. have also shown that diamond can be grown from ablation of a nondiamond target composed of hard fullerenes [7.63]. The deposition was carried out using an ArF excimer laser in a vacuum chamber with a base pressure of 2×10^{-5} Torr. Freshly cleaved NaCl crystals were used as the substrate. Micro-Raman, electron energy loss, and electron diffraction measurements of the deposited films indicated microcrystalline diamond formation. The micro-Raman revealed a diamond peak at 1330 cm^{-1}, which would gradually broaden to the broad peaks characteristic of amorphous carbon after long exposure of the film to the exciting beam in the spectrometer. The Raman peak at 1330 cm^{-1} is characteristic of polycrystalline diamond, although it is broadened by the presence of graphitic carbon at 1350 cm^{-1} [7.64]. This leads the authors to propose that the amorphous carbon surrounds the diamond crystallites in the film. It is possible that deposition of the films in a hydrogen atmosphere may help to suppress some nondiamond film growth.

Lätsch and Hiraoka have claimed that excimer laser ablation of polymethylmethacrylate (PMMA) results in the formation of diamond crystallites on silicon substrates heated to 600°C [7.65]. Oxygen and hydrogen were added to the reaction chamber in order to help suppress graphitic deposits. This approach of employing organic and polymer targets for laser ablation was first employed by Athwal et al. [7.66]. They used Chrysene, PMMA, and other organic materials, but photoablation of these materials in vacuum did not result in diamond formation but in amorphous diamond-like carbon films. Lätsch and Hiraoka analyzed their deposited films with scanning electron microscopy (SEM) and micro-Raman. The SEM micrographs revealed a crystalline morphology of cubic blocks and octahedral crystals. The micro-Raman spectrum showed a broadened peak at 1330 cm^{-1} and a peak at 1597 cm^{-1}, revealing the presence of amorphous carbon and graphite along with the polycrystalline diamond.

The Thaler Method

A pulsed laser method was developed by Thaler to deposit diamond and other ultrahard coatings on substrates [7.67]. In one method, a plasma is generated by absorption of an intense impulse of laser radiation into a gas mixture at atmospheric pressure. This activated gas then mixes with a hydrocarbon precursor gas. The shock wave produced from the sudden absorption of laser irradiation transports ions, radicals, and molecular fragments rapidly on to a substrate, forming diamond or another ultrahard material. [7.68]. In a similar method, two distinct plasmas, one originating from the laser irradiation and the other from laser ablation of a solid substrate, are physically combined to initiate the chemistry to produce diamond. This process is referred to as laser–plasma coupling chemical vapor deposition (LPCCVD).

7.4 Hydrothermal Growth

Roy et al. have claimed that diamond can be synthesized in an intermediate-pressure regime. In 1992, a low-pressure solid-state source (LPSSS) method was utilized to grow diamond [7.69]. For example, mixtures of gold or silver with glassy carbon, water, and fine diamond seeds reacted at 800°C and 1 bar for 72 hours, precipitated diamond aggregates. The samples were analyzed by X-ray diffraction and SEM, and (when appropriate) TEM and microprobe Raman. Another system, reacting at 1.4 bar and 800°C for 48 hours, consisted of iodoform, water and diamond seeds in a gold tube [7.70]. Since diamond is not grown without the use of diamond seeds, some ambiguity exists in proving diamond growth. Szymanski et al. [7.71] also claimed hydrothermal growth, and Gogotzi et al. [7.72] claimed diamond synthesis via the reaction of SiC + H_2O, but a Raman peak at 1330 cm^{-1}, characteristic for diamond, may be from a ternary phase in the Si–O–C system [7.73]. Recently, evidence was presented to suggest that diamond is actually grown by the hydrothermal process using a metal–carbon–water system [7.74]. Spectroscopic, microscopic, and diffraction techniques were utilized both before and after growth experiments, enabling the new diamond grown to be distinguished from the diamond seeds added to nucleate growth, which indicated that small diamond aggregates were actually grown.

7.5 Summary

A large number of studies have shown that the addition of small amounts of fluorine and chlorine to reactors that utilize atomic hydrogen as the activator in diamond growth can have beneficial effects which range from improving diamond quality to increasing growth rates at low temperatures. In some cases, the presence of a halogen is thought to increase the production of atomic hydrogen, and at lower temperatures atomic chlorine becomes the primary generator of surface free radical sites.

Two of the methods discussed, hot-wall halogen-assisted diamond CVD and hydrothermal growth, do not grow diamond through a free radical process, which implies that other growth mechanisms do exist. When their respective mechanisms are well understood, it is possible that much less expensive methods for CVD growth of diamond will become available.

Chlorine activated diamond CVD (CACVD) and the QQC materials deposition process have demonstrated a continuous high diamond growth rate. The CACVD process avoids the high temperatures required to produce atomic hydrogen by direct dissociation, but adds the complication of handling and disposing of chlorine and hydrogen chloride. A good reactor design would deal with this problem: therefore, a well-designed large-scale CACVD reactor may offer an alternative to existing reactors in the future.

The QQC process has clearly demonstrated very high rates of diamond growth on tools, with the added advantage that the diamond is very adherent and the tool does not need to be independently heated. The method, however, is quite costly to implement, due to the use of four lasers. Thus high-volume production will be required to justify the initial cost and the maintenance of the lasers. The diamond produced also contains some inclusions of metal from the tool substrate.

The laser chemical vapor deposition (LCVD) method appears to offer very localized deposition of diamond at low temperatures, but only at low rates of growth. Pulsed laser deposition (PLD) has been extensively used to produce smooth hard films that contain both diamond and amorphous carbon. These films have found uses as cold cathode emitters and as wear-resistant coatings.

References

7.1   J.C. Angus and C.C. Hayman, *Science* **241**, 913 (1988)
7.2   N. Setaka, *J. Mater. Res.* **4**, 664 (1989)
7.3   D.E. Patterson, B.J. Bai, C.J. Chu, R.H. Hauge, and J.L. Margrave, in New Diamond Science and Technology, ed. R. Messier, J.T. Glass, J.E. Butler, and R. Roy, Materials Research Society, Pittsburgh, PA (1991), p. 433–8
7.4   B.V. Spitsyn and B.V. Deryagin, USSR Patent 399,134 (May 5, 1980)
7.5   R.A. Rudder, G.C. Hudson, J.B. Posthill, R.E. Thomas, and R.J. Markunas, Appl. Phys. Lett. **59**, 791 (1991)
7.6   B.J. Bai, C.J. Chu, D.E. Patterson, R.H. Hauge, and J.L. Margrave, J. Mater. Res. **8**, 233 (1993)
7.7   N.J. Komplin, B.J. Bai, C.J. Chu, J.L. Margrave, and R.H. Hauge, in Proceedings of the 3rd International Symposium on Diamond Materials, ed. J.P. Dismukes and K.V. Ravi, The Electrochemical Society Processing Series, Pennington, NJ (1993), p. 385
7.8   J.-J. Wu and F.C-N. Hong, J. Appl. Phys. **81**(8), 3647 (1997)
7.9   F.C-N. Hong, G.T. Lang, D. Chang, and S.C. Yu, in Applications of Diamond Films and Related Materials, Materials Science Monographs, ed. Y. Tzeng, M. Yoshikawa, M. Murakawa, and A. Feldman, Elsevier, Amsterdam (1991), p. 577–82
7.10   F.C. Hong, G.T. Lang, J.J. Wu, D. Chang, and J.C. Hsieh, Appl. Phys. Lett. **63**, 3149 (1993)
7.11   C.A. Rego, R.S. Tsang, P.W. May, M.N.R. Ashfold, and K.N. Rosser, J. Appl. Phys. **79**(9), 7264 (1996)
7.12   R.S. Tsang, C.A. Rego, P.W. May, J. Thumin, M.N.R. Ashfold, K.N. Rosser, C.M. Younes, and M.J. Holt, Diamond Rel. Mater. **5**, 359 (1996)
7.13   E.J. Corat, V.J. Trava-Airoldi, N.F. Leite, A.F.V. Pena, and V. Baranauskas, in Novel Forms of Carbon II, ed. R. Messier, J.T. Glass, J.E. Butler, and R. Roy, (Mater. Res. Soc. Symp. Proc. **349**, 1994), p. 439–444
7.14   E.J. Corat, R.C. Mendes de Barros, V.J. Trava-Airoldi, N.G. Ferreira, N.F. Leite, and K. Iha, to be published
7.15   V.J. Trava-Airoldi, B.N. Nobrega, E.J. Corat, E. Delbosco, N.F. Leite, and V. Baranauskas, Vacuum **46**, 5 (1995)
7.16   I. Schmidt, F. Hentschel, and C. Benndorf, Diamond Rel. Mater. **5**, 1318 (1996)
7.17   M. Kadono, T. Inoue, A. Midyadera, and S. Yamazaki, Appl. Phys. Lett. **61**, 772 (1992)
7.18   H. Maeda, M. Irie, T. Hino, K. Kusakabe, and S. Morooka, Diamond Rel. Mater. **3**, 1072 (1994)
7.19   R. Ramesham, R.F. Askew, and B.H. Loo, in Proceedings of the 3rd International Symposium on Diamond Materials, ed. J.P. Dismukes and K.V. Ravi, The Electrochemical Society Processing Series, Pennington, NJ (1993), p. 394
7.20   K.J. Grannen and R.P.H. Chang, J. Mater. Res. **9**, 2154 (1994)
7.21   K.J. Grannen, D.V. Tsu, R.J. Meilunas, and R.P.H. Chang, Appl. Phys. Lett. **59**, 745 (1991)
7.22   C.A. Fox, M.C. McMaster, W.L. Hsu, M.A. Kelly, and S.B. Hagstrom, Appl. Phys. Lett. **67**, 2379 (1995)
7.23   T. Kotaki, Y. Amada, K. Harada, and O. Matsumoto, Diamond Rel. Mater. **2**, 342 (1993)
7.24   T. Kotaki, N. Horii, H. Isono, and O. Matsumoto, J. Electrochem. Soc. **143**, 2003 (1996)
7.25   T.I. Hukka, R.E. Rawles and M.P.D'Evelyn, Thin Solid Films **225**, 212 (1993)

7.26   R. Gat, T.I. Hukka, R.E. Rawles, and M.P.D'Evelyn, "Progress toward atomic layer epitaxy of diamond: diamond films grown one layer at a time" in Proc., Ann. Tech. Conf. - Soc. Vac. Coaters, Albuquerque, NM (1993), p. 353

7.27   C.Y. Pan, C.J. Chu, J.L. Margrave, and R.H. Hauge, J. Electrochem. Soc. **141**, 3246 (1994)

7.28   J.-J. Wu and F.C.-N. Hong, J. Appl. Phys. **81**(8), 3652 (1997)

7.29   D.E. Patterson, C.J. Chu, B.J. Bai, Z.L. Xiao, N.J. Komplin, R.H. Hauge, and J.L. Margrave, paper presented at International Conference on Metallurgical Coatings and Thin Films, San Diego, CA, April 1991

7.30   M.S. Wong and C.H. Wu, Diamond Rel. Mater. **1**, 369 (1992)

7.31   M.S. Owens, E. Bier, D. Vestyck, and J.E. Butler, personal communication (1991)

7.32   R.H. Hauge and C. Pan, US Patent 5,589,231 (1996)

7.33   C. Pan, C.J. Chu, J.L. Margrave, and R.H. Hauge, J. Electrochem. Soc. **141**, 3246 (1994)

7.34   C. Pan, J.L. Margrave, and R.H. Hauge, in Novel Forms of Carbon II, ed. R. Messier, J.T. Glass, J.E. Butler, and R. Roy, Mater. Res. Soc. Symp. Proc. **349** (1994), p. 427

7.35   D.E. Kassmann and T.A. Badgwell, Diamond Rel. Mater. **5**, 895 (1996)

7.36   D. Satrapa, C. Pan, J.L. Margrave, and R.H. Hauge, in Proceeding of the International Conference on Metallurical Coatings and Thin Films, San Diego, CA, April 1994, sponsored by the American Vacuum Society

7.37   R.H. Hauge and T.A. Badgwell, personal communication

7.38   L.E. Manzer, Chem. Eng. News, p. 32 (June 30, 1997)

7.39   D.V. Fedoseev, I.G. Varshavskaya, A.V. Lavent'ev, and B.V. Deryagin, Powder Technol. (Switzerland) **44**, 125 (1985)

7.40   D.V. Fedoseev and B.V. Deryagin, in Proceedings of the 2nd International Conference on New Diamond Science and Materials, ed. R. Messier, J.T. Glass, J.E. Butler, and R. Roy, MRS Society, Pittsburg, PA (1991), p. 439

7.41   M. Alam, T. Debroy, R. Roy, and E. Breval, Carbon **27**, 289 (1989)

7.42   E. Breval, M. Alam, T. Debroy, and R. Roy, J. Mater. Sci. Lett. **9**, 1071 (1990)

7.43   P.A. Molian and A. Waschek, J. Mater. Sci. **28**, 1733 (1993)

7.44   K. Kitahama, K. Hirata, H. Nakamatsu, S. Kawai, N. Fujimori, T. Imai, H. Yoshino, and A. Doi, Appl. Phys. Lett. **49**, 634 (1986)

7.45   K. Kitahama, K. Hirata, H. Nakamatsu, S. Kawai, N. Fujimori, and T. Imai, in Photon, Beam and Plasma Stimulated Chemical Processes at Surfaces, ed. V. Donnelly, I. Herman, and M. Herose, Mater. Res. Soc. Symp. Proc. **75** (1987), pp. 309–15

7.46   T. Nakayama and K. Watanabe, J. Chem. Phys. **40**, 558 (1964)

7.47   K. Kitahama, Appl. Phys. Lett. **53**, 1812 (1988)

7.48   Y. Goto, T. Yagi, and H. Nagai, in Laser and Particle-Beam Modification of Chemical Processes on Surfaces, ed. A. Johnson, G. Loper, and T. Sigmon, Mater. Res. Soc. Symp. Proc. **129** (1989), pp. 213–217

7.49   F. Celii, H. Nelson, and P. Pehrsson, Appl. Phys. Lett. **62**, 2337 (1993)

7.50   J.H.D. Rebello, D.L. Straub, and V.V. Subramaniam, J. Appl. Phys. **73**(3), 1133 (1992)

7.51   J.H.D. Rebello, V.V. Subramaniam, and T.S. Sudarshan, Appl. Phys. Lett. **62**, 899 (1993)

7.52   J.H.D. Rebello and V.V. Subramaniam, J. Min. Met. Mater. Soc. **46**, 60 (1994)

7.53   P. Mistry, M.C. Turchan, S. Liu, G.O. Granse, T. Baurmann, and M.G. Shara, Innov. Mater. Res. **1**, 193 (1996)

7.54   J.G. Eden, Photochemical Vapor Depostion, John Wiley, New York (1992), pp. 5–46

7.55   D. Dijkkamp, T. Venkatesan, X.D. Wu, S. Shaheen, N. Jisrawi, Y.H. Min-Lee, W.L. McLean, and M. Croft, Appl. Phys. Lett. **51**, 619 (1987)

7.56   D.L. Pappas, K.L. Saenger, J. Bruley, W. Krakow, J.J. Cuomo, T. Gu, and R.W. Collins, J. Appl. Phys. **71**, 5675 (1992)

7.57   M.B. Guseva, V.G. Babaev, V.V. Khvostov, Z.Kh. Valioullova, A.Yu. Bregadze, A.N. Obraztsov, and A.E. Alexenko, Diamond Rel. Mater. **3**, 328 (1994)

7.58   A. Rengan, N. Biunno, and J. Narayan, Mater. Res. Soc. Symp. Proc. **162**, 162 (1990)

7.59   J. Seth, R. Padiyath, D.H. Rasmussen, and S.V. Babu, Appl. Phys. Lett. **63**, 473 (1993)

7.60   M.C. Polo, J. Cifre, G. Sanchez, R. Aguiar, M. Varela, and J. Esteve, Appl. Phys. Lett. **67**, 485 (1995)

7.61   M.C. Polo, J. Cifre, G. Sanchez, R. Aguiar, M. Varela, and J. Esteve, Diamond Rel. Mater. **4**, 780 (1995)

7.62   P.A. Molian, B. Janvrin, and A.M. Molian, J. Mater. Sci. **30**, 4751 (1995)

7.63   M.E. Kozlov, P. Fons, H.-A. Durand, K. Nozaki, M. Tokumoto, K. Yase, and N. Minami, J. Appl. Phys. **80**, 1182 (1996)

7.64   J. Wagner, C. Wild, and P. Koidl, Appl. Phys. Lett. **59**(7), 779 (1991)

7.65   S. Lätsch and H. Hiraoka, J. Min. Met. Mater. Soc. **46**, 64 (1994)

7.66   I.S. Athwal, A. Mele, and E.A. Ogryzlo, Diamond Rel. Mater. **1**, 731 (1992)

7.67   S.L. Thaler, US Patent 5,547,716 (August 20, 1996)

7.68   S.L. Thaler, US Patent 4,981,717 (January 1991)

7.69   R. Roy, H.S. Dewan, and P. Ravindranathan, J. Mater. Chem. **3**(6), 685 (1993)

7.70   R. Roy, D. Ravichandran, P. Ravindranathan, and A. Badzian, J. Mater. Chem. **11**(5), 1164 (1996)

7.71   A. Szymanski, E. Abgorowicz, A. Bakon, A Nidbalska, R. Salacinski, and J. Sentek, Diamond Rel. Mater. **4**, 234 (1995)

7.72   Y.G. Gogotzi, K.G. Nickel, and P. Kofstad, J. Mater. Chem. **5**, 2313 (1995)

7.73   R. Roy, D. Ravichandran, A. Badzian, and E. Breval, Diamond Rel. Mater. **5**, 973 (1996)

7.74   X.-Z. Zhao, R. Roy, K.A. Cherian, and A. Badzian, Nature **385**, 513 (1997)

8. Heteroepitaxy and Highly Oriented Diamond Deposition

Hiroshi Kawarada

School of Science and Engineering, Waseda University,
3-4-1 Ohkubo, Shinjuku-ku, Tokyo 169, Japan
e-mail: kawarada@mn.waseda.ac.jp

Springer Series in Materials Processing
Low-Pressure Synthetic Diamond   Eds.: B. Dischler and C. Wild
© Springer-Verlag Berlin Heidelberg 1998

8.1 Introduction

Single-crystal diamond films grown over a large area may open up a new field of solid-states physics and industry, because diamond has several advantageous properties such as high thermal conductivity, hardness and wide bandgap semiconducting characteristics. These advantages are most effectively obtained in the single crystalline phase. The highly oriented or heteroepitaxial growth of diamond is one of the most promising ways of obtaining single crystalline films. Of various substrates, heteroepitaxial growth on silicon (Si) or silicon carbide (SiC) has received much attention [8.1–8], because perfect and large single crystals can be provided in Si, and the development of large single-crystal SiC and heteroepitaxial SiC films on Si has advanced rapidly in recent years. However, highly oriented or heteroepitaxial diamond growth has also been realized on other materials, such as near noble metals, to exhibit the unique qualities of diamond with respect to nucleation and growth.

The growth of highly oriented diamond is the first step towards heteroepitaxy: however, the best substrate for diamond epitaxy cannot be identified at present. In general, due to the insufficient percentage of oriented particles at the initial growth stage, thick-film growth has been conventionally required to achieve relatively smooth highly oriented or heteroepitaxial films. The mechanisms of surface treatment, such as seeding and bias-enhanced nucleation (BEN) [8.9], need to be clarified. Although some particles are misoriented, less than one percent of oriented particles continue to grow during the selection process, in which severely misoriented particles drop out. This is one of the self-organizing features of diamond film growth, which can be controlled by the adoption of precise growth conditions.

In this chapter, on the basis of these characteristics of diamond growth, the following subjects have been selected. In Sect. 8.2, highly oriented or heteroepitaxial diamond growth on metals with a face-centered cubic structure, and semiconductors or insulators with a zinc blende or diamond structure, is surveyed to establish the characteristics of each process. The number of substrates is limited to six in this chapter. For the progress in other substrates, see [8.10]. In Sect. 8.3, nucleation and initial growth are discussed on the basis of BEN and subsequent island growth on β-SiC. In Sect. 8.4, selective growth and surface morphology changes are explained. The electrical properties of the highly oriented films are demonstrated in Sect. 8.5. Finally, heteroepitaxy and highly oriented diamond deposition are summarized in Sect. 8.6.

8.1.1 Deposition Methods

In principle, all of the methods explained in Chaps. 2–7 are applicable to highly oriented or heteroepitaxial growth of diamond. For oriented diamond nucleation, however, the quartz-tube-type microwave-plasma assisted chemical vapour deposition (MPCVD) developed by Kamo et al. [8.11] or the chamber-type method [8.12] are generally used, because of the presence of ions and the ability to control the bias between the substrate and the plasma at relatively high pressure (more than 20 Torr). The MPCVD systems are explained in detail in Chap. 2. Most oriented nucleation on the solid surface, – that is without diamond seeding – can be realized by BEN, in which positively charged particles are collected at the substrate due to the potential difference, which is larger than that of the self-biasing between the plasma and the substrate. The mechanism of BEN is discussed in Sect. 8.3.1. The DC-plasma assisted CVD (DC-plasma CVD) developed by Suzuki et al. [8.13] can also be used. It operates at a higher pressure than MPCVD, using the transition state between the arc and glow discharge. Most of the results of highly oriented or heteroepitaxial growth of diamonds in this chapter were obtained by MPCVD and DC-plasma CVD methods.

8.2   Heteroepitaxy and Highly Oriented Diamond on Various Substrates

The dominant crystallographic orientation relationships between diamond and non-diamond (ND) substrates are summarized by the following expressions:

- On (001) face-centered cubic, diamond and zinc blende structures:
  (001)diamond//(001)ND and $[\bar{1}10]$diamond//$[\bar{1}10]$ND.

- On (111) face-centered cubic, diamond and zinc blende structures:
  (111)diamond//(111)ND and $[\bar{1}10]$diamond//$[\bar{1}10]$ND.

- On (0001) hexagonal structures such as 2H, 4H and 6H phases:
  (111)diamond//(0001)ND and $[\bar{1}10]$diamond//$[\bar{1}\bar{1}20]$ND.

These relationships are valid even in highly mismatched systems such as diamond/SiC or diamond/Si.

8.2.1 Metal Substrates

Heteroepitaxial diamond growth has been reported on metals such as nickel (Ni) [8.14, 15], cobalt (Co) [8.16], platinum (Pt) [8.17] and iridium (Ir) [8.18]. These metals are classified in group VIII in the periodic table and have similar properties due to their similar valence structures. The parameters that might govern the epitaxial growth vary as the atomic number increases. These significant parameters are summarized in Table 8.1. The lattice constant and melting point of these metals increase as the atomic number increases. The lattice constant of Ni is the nearest to that of diamond. According to the binary phase diagram, the carbon solubility region exists in Ni and Co, but is scarce in Pt, and absent in Ir in the temperature range of CVD-diamond formation. Hence, the reactivity of these metals with carbon is expected to decrease in the order of Ni, Pt and Ir.

Table 8.1: Lattice constants, melting points, and carbon solubilities of various metals. The carbon solubilities are those at the temperature of CVD-diamond formation.

Material	Lattice constant (nm)	Melting point (K)	Carbon solubility (atom %)
Diamond	0.3567	~ 4000	–
Nickel	0.352	1728	5–6
Iridium	0.384	2720	~ 0
Platinum	0.392	2045	1.5

Monocrystalline bulk metals or heteroepitaxially grown metal films are used as substrates for the oriented growth of diamond. The surfaces of these metals, especially Ni or Pt, have been ultrasonically abraded in diamond powder-suspended ethanol or seeded with a coating of diamond powder. In some cases, after diamond seeding, the substrates are heated in an atomic hydrogen environment, using a hot filament to promote the reaction between diamond and metal. The surface damage of metal by a hot filament is much lower compared to that in a plasma.

Nickel

The oriented growth of diamond particles on Ni(111) and (001) substrates has been reported by Sato et al. [8.14], using a lower concentration of methane than normal to avoid the formation of a graphitic phase and the reaction between carbon and Ni. Since the dissolution of Ni into diamond retards diamond growth and degrades its crystallinity, a continuous film was not obtained. The suppression of the graphite phase by a nickel hydride or nickel–carbon–hydrogen (Ni–C–H) intermediate state has been proposed by Zhu et al. [8.15]. They applied the ternary eutectic phase of Ni–C–H to obtain oriented nucleation. A diamond-seeded Ni substrate was annealed at a high temperature (900–1200°C) in an atomic hydrogen atmosphere produced by a hot filament. In order to achieve a high degree of oriented diamond seeding on the Ni substrate, the random diamond seeds in the initial stage must react completely with the metal substrate. Using a real-time, *in-situ* laser reflectometry system to monitor changes in surface morphology, Sitar et al. [8.19] observed that, during the heating stage in atomic hydrogen ambient, the as-seeded surface in the initial stage was reconstructed to become smooth by the reaction between seeded diamond and Ni. This monitoring was verified by SEM observation. Up to 6 atom % of the carbon was dissolved in the Ni surface layer. They also observed that the formation of the Ni carbide phase occurred at the early stage of the process, before the diamond growth, and proposed that diamond does not nucleate directly on Ni, but rather on its carbide. This phase was explained as Ni_4C, which had been reported to have a Ni-rich face-centered cubic structure with a lattice constant close to that of diamond [8.20].

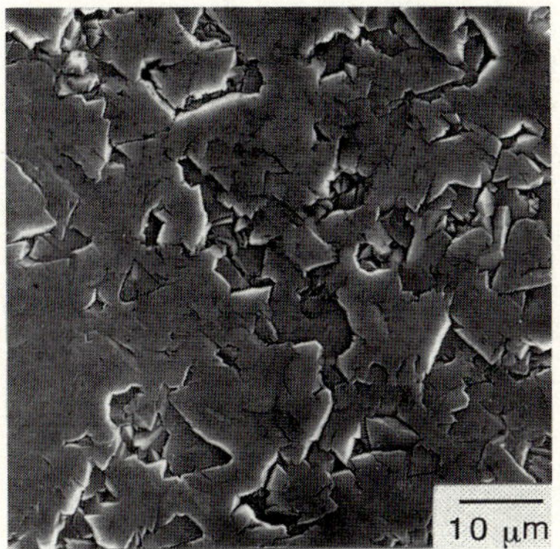

Fig. 8.1: An SEM image of epitaxially grown diamond on Pt(111) [8.17].

Platinum

In the case of Pt, where the solid solubility of carbon, according to the binary phase diagram, is less than 1.5 atom % at the process temperature of diamond growth, the interface reaction between carbon and metal is expected to be different from that of Ni. Tachibana et al. [8.17] have demonstrated the possibility of oriented and continuous diamond growth on a diamond-seeded Pt surface. Figure 8.1 shows a scanning electron microscope (SEM) image of a diamond film grown on Pt(111) [8.17]. Although the process is similar to that reported for a Ni substrate, not only hot filament, but also MPCVD can be used to heat the substrate in an atomic hydrogen atmosphere, indicating the stability of a Pt surface. On the basis of time-evolution observations of the surface morphology of the same area using SEM by interrupting the CVD process [8.21], and cross-sectional transmission electron microscope (TEM) observation [8.22], Tachibana et al. [8.21] have proposed the sequential growth model shown in Fig. 8.2. Diamond

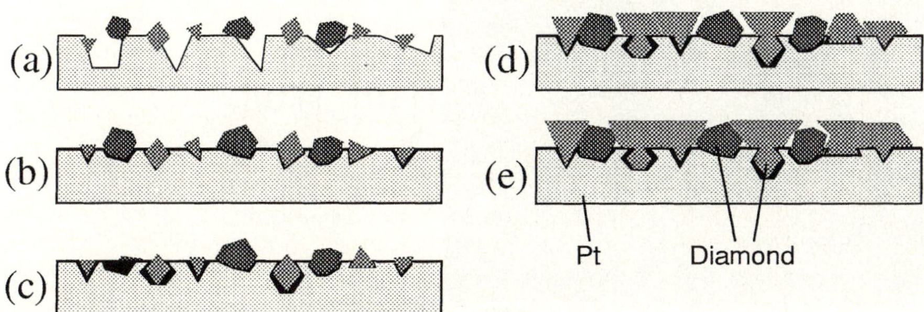

Fig. 8.2: A schematic model of diamond heteroepitaxy on Pt(111). When seeded diamond particles are small enough to move, they can take epitaxial alignment [8.21].

growth begins on the seeded diamond on a Pt surface (Fig. 8.2a). In the plasma environment, the surface grooves become shallow and diamond crystals dissolve on the molten surface of Pt and change their positions (Fig. 8.2b). Diamond crystals are partially aligned with the crystallographic orientation of the Pt (Fig. 8.2c). The <111> textured growth followed by the moulding stage produces coalesced (111) faces that grow over the non-oriented crystals (Figs. 8.2 c and d). Since the metal reaction with carbon can be ruled out in the present system, moulding of diamond directly onto the Pt surface is assumed to take place. The movement of fine diamond particles on metal surfaces during plasma deposition has been reported [8.23]. By rotation and/or translation, the small deposits in particle form can align epitaxially, provided that strong chemical bonds between the particle and the substrate are not expected [8.24]. On the molten state of the Pt surface, the particles change their positions to the energetically favourable sites.

Iridium

Heteroepitaxial diamond growth has been achieved on heteroepitaxial grown Ir thin films by Ohtsuka et al. [8.18]. In this method, the seeding process using diamond particles has not been used. The nucleation process is a kind of BEN performed by DC-plasma CVD [8.13] at 100 Torr, where the current due to positively charged ions impinging on the surface is thought to contribute to the diamond nucleation.

In spite of ion irradiation, the Ir surface retains a crystal structure after BEN, although the surface becomes roughened compared with the as-grown Ir surface. After BEN, oriented diamond particles with a density of 10^8 cm^{-2} have been grown at 200 Torr by DC-plasma CVD in the electron-assisted deposition mode; in other words, without ion irradiation in the same DC-plasma. The density of oriented particles is more than one order of magnitude higher than that on Ni or Pt with

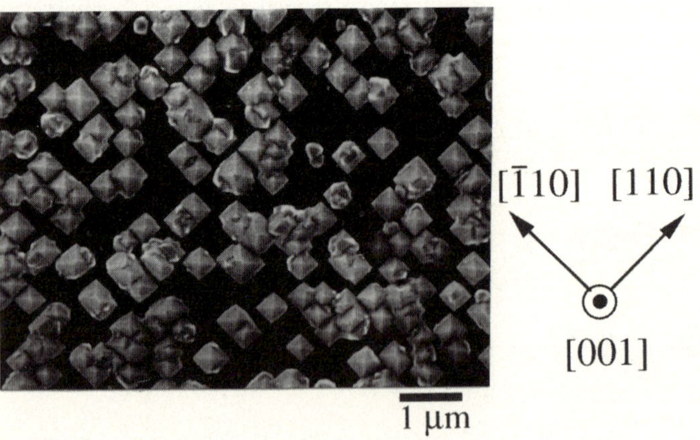

Fig. 8.3: An SEM image of epitaxially grown diamond particles on Ir (001) [8.18].

diamond seeds. While the initial growth stage has not yet been fully investigated, the degree of orientation of independent diamond particles is very high, as shown in Fig. 8.3 [8.18]. Continuous heteroepitaxial films have been also achieved for a film thickness of 2–3 µm [8.25]. Since the melting point of Ir is 700 K higher than that of Pt (see Table 8.1), and the eutectic phase is not present between carbon and Ir at the diamond deposition temperature, the possibility of the rearrangement of diamond in the molten phase of Ir decreases and the intermediate phase cannot be expected. Another type of oriented nucleation on this metal should be considered.

8.2.2 Substrates with Zinc Blende or Diamond Structure

Cubic Boron Nitride

Koizumi et al. [8.26, 27] successfully achieved heteroepitaxial growth of diamond on cubic boron nitride (c-BN), which has only a 1.3% lattice mismatch with diamond. Using DC-plasma CVD, diamond has been grown without surface treatment. A thin carbonaceous layer covers the boron-terminated {111} ({111}B) surfaces of c-BN prior to the formation of the epitaxial diamond islands, which indicates that the formation of a carbonaceous layer is closely related to the nucleation of diamond; this has been explained as the Stranski–Krastanov growth mode. A similar result was obtained on the {100} surfaces, where the particle density was 10^{10} cm^{-2} and the particles grew along the striations of the c-BN{001} surfaces. On the nitrogen-terminated {111} ({111}N) surfaces, however, the epitaxial nucleation is observed to be negligible. This has been considered to be due to the difference in adsorption energy of carbon atoms on the surfaces compared with that of hydrogen atoms [8.27]. According to the bond strength of a c-BN surface, a C–B bond (348 kJ/mol) is stronger than an H–B bond (320 kJ/mol), and a C–N bond (291 kJ/mol) is weaker than an H–N bond (391 kJ/mol). Hence, the chemisorption of carbon atoms on a (111)B surface can replace the surface hydrogen atoms, but that on a (111)N surface is not energetically favourable. This type of heteroepitaxial growth is closely related to interfacial chemical bonds.

Silicon Carbide and Silicon

From the phase diagram of carbon and Si, SiC is the only phase that exists and the transition phase cannot be expected. In spite of the large lattice mismatch (22% in the case of diamond/β-SiC), highly oriented growth of diamond has been achieved on β-SiC by Stoner and Glass [8.1] and on carburized Si surfaces by Jiang et al. [8.2]. The BEN process developed by Yugo et al. [8.9] is indispensable for heteroepitaxial nucleation. To explain the oriented nucleation in largely mismatched systems, 3-to-2 or 5-to-4 matching of lattice planes at diamond/Si or diamond/β-SiC interfaces, respectively, has been speculated (Table 8.2). To achieve relatively smooth highly oriented films, the growth of a film more than 50 µm thick is required [8.5], due to the insufficient percentage of epitaxially

Table 8.2: Matching of lattice planes at diamond/β-SiC and diamond/Si interfaces.

Interface	Lattice constant (nm)	Matching of lattice planes	Mismatch (%)
Diamond/β-SiC	0.3567/0.4360	5 : 4	− 2.2
Diamond/Si	0.3567/0.5431	3 : 2	1.5

oriented particles at the initial growth stage. The growth requires a selection process by which highly misoriented particles are removed [8.28, 29]. However, considerable thickness is required for such effective selection to take place.

Smoother and thinner (less than 5 μm in thickness) diamond films have been grown heteroepitaxially using high-quality β-SiC buffer layers on Si substrates [8.7, 30]. The process is comprised of three steps: (i) BEN on β-SiC grown on Si(001); (ii) [001] fast growth for the selection of epitaxially oriented particles; and (iii) smooth growth of the (001) surface. Control of the growth velocity ratio between the <100> and the <111> directions, as proposed by Wild et al. [8.28, 29], has been carried out. The particle density at the initial island growth stage (process (i)) is 10^{10}–10^{11} cm^{-2}, which is the highest amongst substrates for diamond heteroepitaxy [8.7]. This type of nucleation and initial growth is discussed later, in Sect. 8.3. After the 1 μm growth in process (ii), the particle density is decreased to 10^8 cm^{-2}, which is two orders of magnitude less than that of the island stage, indicating that less than 1% of the particles satisfying the epitaxial relationship have been selected. The tilting angle of the heteroepitaxial orientation has been improved to about 2° after the selection process, as explained in Sect. 8.4. The initially rough surfaces become smooth after process (iii), as shown in Fig. 8.4. The high surface energy or interfacial energy of diamond prevents the growth of high-index surfaces or, in some cases, of the high-angle grain boundary. This is a driving force for the selection and smoothing processes described above, illustrating the unique properties of diamond growth.

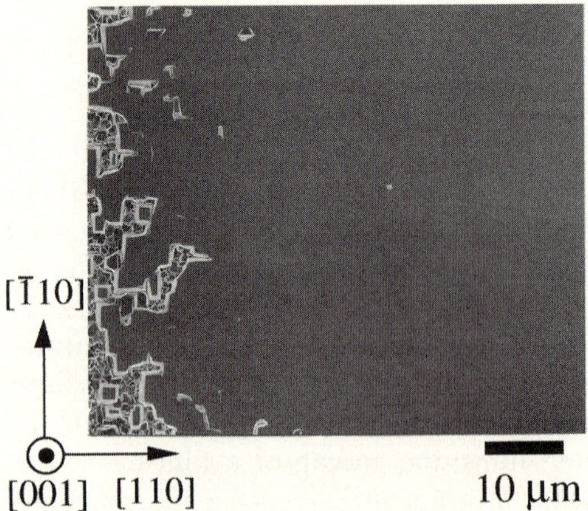

Fig. 8.4: An SEM image of epitaxially grown diamond on β-SiC(001). The left-hand side is the sample edge.

8.3 Nucleation and Initial Growth on SiC and Si

8.3.1 Bias-Enhanced Nucleation

The nucleation density of diamond on Si or SiC is very low – less than 10^5 cm^{-2} without substrate pretreatment. Scratching with diamond particles, however, cannot produce oriented growth in the case of Si or SiC. BEN was first realized by Yugo et al. [8.9]. For this technique, a DC-bias is applied between the substrate and the counter electrode contacting the plasma to control the potential difference between the plasma and the substrate. The applied bias voltage varies between 50 V and 300 V, depending on the particular apparatus or plasma conditions. Typical pressures are in the range 30–100 Torr. The reaction gas is CH_4, or another hydrocarbon gas diluted with H_2. The substrate temperatures are around 900°C. A nucleation density of 10^9–10^{11} cm^{-2} has been obtained on Si or SiC surfaces reproducibly.

The enhancement of diamond nucleation with negative substrate bias is considered to be due to the high flux of energetic hydrocarbon ions towards the substrate surface. The supersaturation for heteroepitaxial nucleation is effectively obtained by this ion flux, which is enhanced at lower pressures. However, ions with high energies can also destroy the substrate lattice, leading to a decrease in the percentage of heteroepitaxially nucleated particles. As a result of the substrate lattice damage, the percentage of heteroepitaxial particles decreases at pressures below 20 Torr. Heteroepitaxial nucleation of diamond requires a gentle acceleration of hydrocarbon ions under relatively high pressures in the 30–100 Torr range.

The mechanism of BEN on Si or SiC has been investigated by several groups, but no clear explanation of this complex process has yet been developed. The mechanisms can be divided into two concepts: surface effects, such as site and surface diffusion [8.31, 32]; and subplantation [8.33, 34]. In the former, the accelerated ions create special sites for heteroepitaxial nucleation and enhance the migration of species to gather at the site. In the latter, the acceleration energy is used to produce sp^3-bonding such, as in the formation of diamond-like carbon (sp^3-bonded a-C). The energy distribution of positively charged particles that induce nucleation for oriented growth is the key point in both nucleation mechanisms. Since the high collision frequency in the plasma prevents precise measurement of ion energy, the measured data are limited at present. Using a retarding probe on – 250 V biased substrates, Sattel et al. measured the ion energy distribution for their optimum condition for diamond nucleation to be 70–90 eV [8.34]. which is close to the optimum energy of the subplantation of carbon ions (90 eV) for the formation of sp^3-bonded a-C film. However, from the Monte Carlo simulation of the retardation of ion energy during a series of collisions in a sheath with an assumed thickness of 1 mm (a mean free path of 5 µm), the energies of ions reaching a negatively biased (– 200 V) substrate through the sheath become 10–20 eV on average, and 40–50 eV at most [8.35]. These values are much lower than the optimum energy for subplantation and correspond to the activation energy of adsorption and successive surface diffusion [8.34].

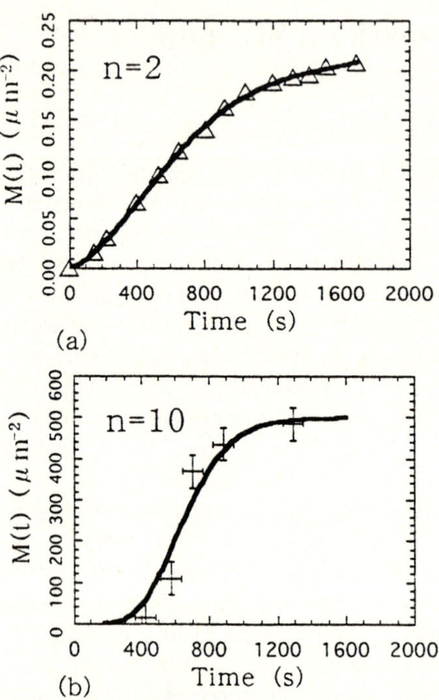

Fig. 8.5: Diamond particle densities $(M(t)/\mu\mathrm{m}^2)$ counted on Si (001) substrates as a function of CVD growth period and fitting curves in terms of nucleation kinetics. (a) $M(t)/\mu\mathrm{m}^2$ on scratched Si with diamond paste (open triangles) and fitted with critical cluster number $n = 2$ (solid line). (b) $M(t)/\mu\mathrm{m}^2$ on Si treated by BEN (crosses) and fitted with $n = 10$ (solid line) [8.36].

The enhancement of surface diffusion by BEN has been proposed by Jiang et al. [8.31] and analyzed by Tomellini [8.36]. If the accelerated ions adsorb on the surface and contribute to the enhancement of surface migration, the probability of reaching nucleation sites can be increased. Figures 8.5 a, b correspond to the particle density $(M(t)/\mu\mathrm{m}^2)$ obtained by counting on Si substrates scratched with diamond paste (open triangles in Fig. 8.5a) and treated by BEN (crosses in Fig. 8.5b), which were fitted in terms of the nucleation kinetics (solid lines) according to critical cluster numbers of $n = 2$ and $n = 10$, respectively [8.36]. $n = 2$ indicates a homoepitaxial nucleation process; for example, (001)–2×1:H surfaces where, dimer atoms form critical nuclei in two dimensions. In heteroepitaxial nucleation, the critical cluster volume, which depends on the interface or surface energies of the diamond/SiC or diamond/Si system, is generally larger than that of homoepitaxial nucleation, because the wetting angle is in principle 0° for homoepitaxial nucleation. Due to a larger critical cluster volume, a pronounced induction period was observed for BEN, as shown in Fig. 8.5b.

It is also possible that the bias treatment produces a geometrical change on the SiC or Si surface that lowers the activation energy of the nucleation process. In order to reduce the critical cluster number – that is, the critical nucleation volume – the geometrical factor plays an important role. Figure 8.6 shows a schematic representation of heterogeneous nucleation at a critical radius on flat and V-grooved surfaces, assuming that the wetting angle is 90°, as is usually observed in

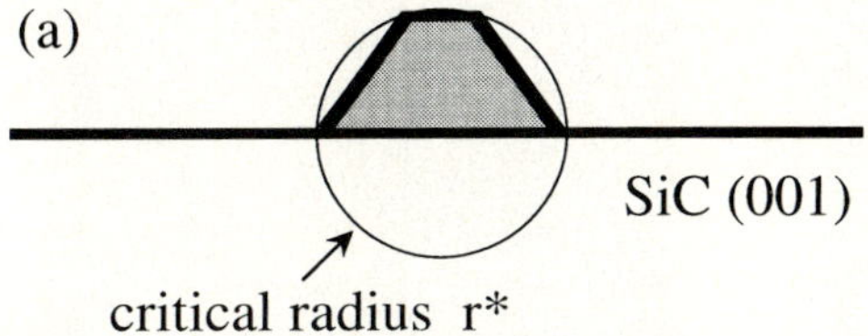

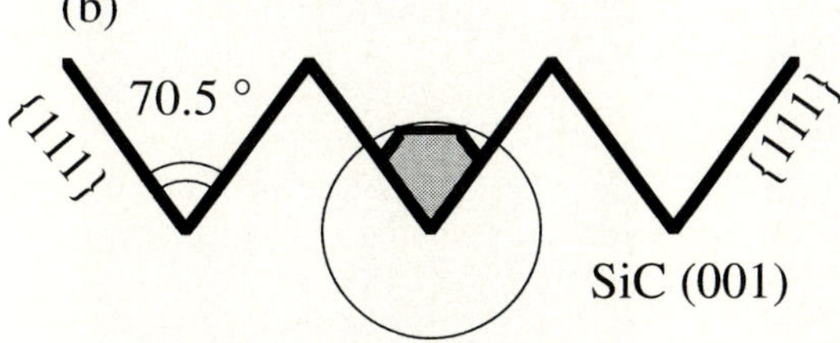

Fig. 8.6: A schematic representation of the critical nucleation volume on the flat and V-grooved surface, assuming the wetting angle to be 90°. The radius of the circles is the critical radius. **(a)** Flat surface. **(b)** V-grooved surface with groove angle = 70.5°, which corresponds to that of {111} facets on a (001) substrate. Hatched regions are critical nuclei.

diamond growth on an SiC or Si surface. Since the critical radius cannot be affected by the angle of the V-groove in the present geometry, the critical volume has been reduced as the groove angle decreases. Moreover, the activation energy of heteroepitaxial nucleation is reduced according to the critical volume. Therefore, nucleation from atomic-scale grooves or cracks can occur at very low supersaturation even when the wetting angle is relatively large. When combined with the enhanced surface diffusion model, this geometrical consideration provides special sites at which nucleation occurs preferentially compared with the normal surface of SiC or Si.

8.3.2 Initial Growth Observation on β-SiC

The effect of grooves on the initial stage of heteroepitaxial growth on β-SiC(001) has been explained in detail in [8.30]. Figure 8.7 shows SEM images and RHEED patterns of β-SiC(001) before and after BEN. After BEN for 2 min (Fig. 8.7b), [110] or [$\bar{1}$10] elongated protrusions (white stripes) and grooves (dark regions) were formed on β-SiC(001), and the RHEED pattern of β-SiC became spotty, indicating a grooved surface. These grooved surfaces are not observed on untreated β-SiC(001) (Fig. 8.7a) or on β-SiC(001) immersed only in H_2 plasma, indicating that the grooved structure is formed in the presence of CH_4 gas due to chemical etching by the plasma or deposition of β-SiC protrusions on β-SiC(001) substrate. After BEN for 5 min, diamonds are nucleated on the grooved structure,

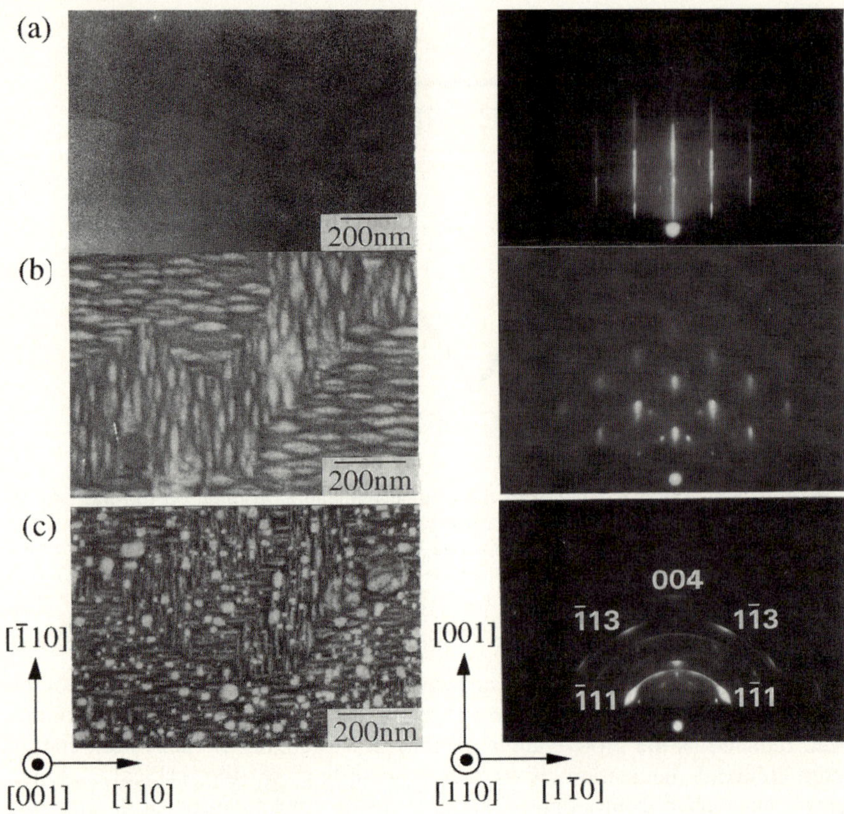

Fig. 8.7: SEM images (left) and RHEED patterns (right) of diamonds and β-SiC(001) surfaces before and after BEN. **(a)** Before BEN. **(b)** After BEN for 2 min. **(c)** After BEN for 5 min. [110] and [Ī10]-directed white stripes are β-SiC protrusions. Diamond particles are observed in **(c)**, and arcuate RHEED spots correspond to heteroepitaxially oriented diamond particles [8.30].

and an arcuate diffraction pattern of diamond with a tilting angle of 10° is observed (Fig. 8.7c). The density of the diamond particles is 10^{10}–10^{11} cm^{-2}. Diamond does not nucleate on β-SiC(001) without the grooved structure. The same finding has been reported by Maeda et al. [8.37] on carburized Si(001) subjected to BEN, with a similar type of grooved surface. Figure 8.8 shows the initial diamond growth for 5 min followed by BEN in Fig. 8.7c [8.30]. {111}-faceted diamond particles are elongated along the β-SiC stripes. Smaller diamond particles are observed mainly in the grooves of the β-SiC, which can provide nucleation sites under the low supersaturation.

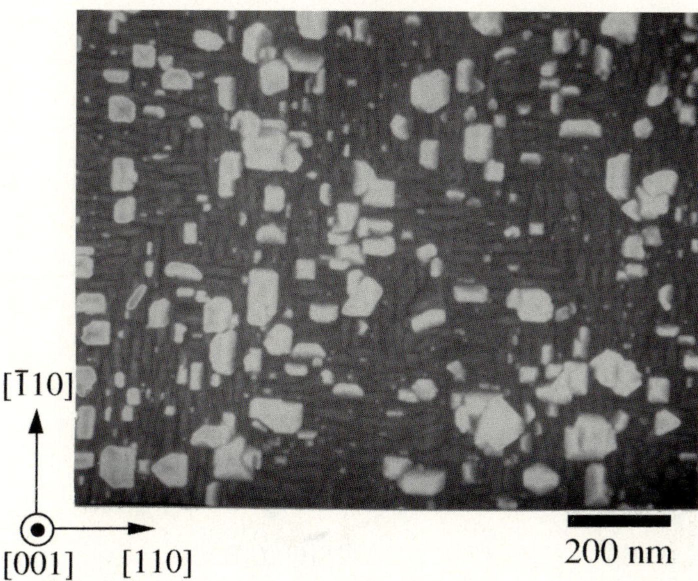

Fig. 8.8: SEM images of initial [001] fast growth of diamond on grooved β-SiC (001) surfaces after 5 min of BEN as in Fig. 8.7c. Rectangular diamond particles elongate along grooves [8.30].

The two kinds of domains, consisting of [110]- or [Ī10]-directed stripes, correspond to antiphase domains that exist in the β-SiC(001) surface. From the RHEED pattern with weak streaks directed to <111> [8.38] and geometrical considerations, the inclined surfaces of the grooves are thought to be composed of {111} facets [8.30]. On the condition that {111} facets are extrapolated to the bottom of the grooves to form atomic scale V-grooves, the bottom angle is 70.5° and the critical nucleation volume can be reduced by a factor of 2.5 compared with a flat surface. Figure 8.9 shows a schematic representation of the Volmer–Weber growth of diamond on β-SiC(001). Diamonds nucleate at the grooves formed on β-SiC(001) by the bias treatment (Fig. 8.9a). Diamonds grow mainly in the [110] or [Ī10] directions along the β-SiC grooves (Fig. 8.9b). It indicates that more reactive species are provided along the groove.

8.3.3 Interface Structure

In the case of diamond heteroepitaxy on an Si substrate, whether diamond nucleation occurs on Si directly or via a SiC phase simultaneously produced in BEN process is an important issue when considering the mechanism of heteroepitaxial nucleation. From surface chemical analyses such as X-ray

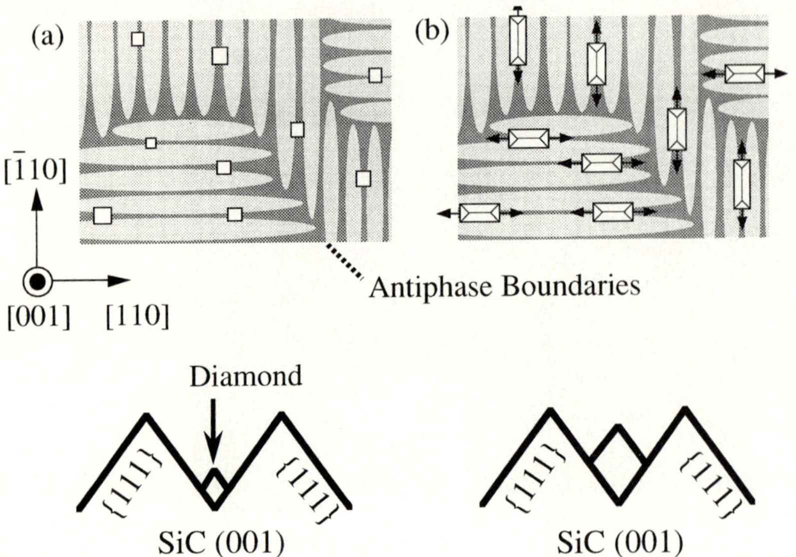

Fig. 8.9: A schematic representation of Volmer–Weber growth of diamond films on β-SiC(001). **(a)** Nucleation: diamonds nucleate in grooves (*dark area*) between β-SiC protrusions (*bright area*). **(b)** Diamonds grow along the grooves [8.30].

photoelectron spectroscopy (XPS), the surface Si 2p signal has been shifted from the binding energy of Si–Si to that of Si–C after BEN, indicating the formation of SiC in BEN process [8.37, 39]. To obtain crystallographic information regarding surface chemical bonding, Schaller et al. [8.40] used X-ray induced photoelectron diffraction (XPD) measurements, which are based on a real-space projection of photoelectron emission intensities reflecting the major crystalline directions. The diffractogram of the fitted C–C and Si–C components of the C1s photoemission signal after BEN showed four <111> and four <110> maxima, indicating that oriented nucleation of diamond occurred as oriented SiC was formed. They concluded that an oriented SiC is formed as an intermediate layer, on top of which oriented diamond starts to nucleate [8.40].

On the other hand, through a high-resolution TEM cross-sectional observation, Jiang et al. reported that the interface between diamond and Si is abrupt [8.41], and is composed of a periodic 3-to-2 registry of {111} atom planes of the diamond/Si interface (Table 8.2), where terminating lattice fringes for every third fringe of diamond are interpreted as individual 60° misfit dislocations. Because most of the interface was imaged as abrupt diamond/Si, they concluded that an SiC transition layer is not necessary for diamond heteroepitaxy. The proposed interface model using the near coincidence-site-lattice (CSL) concept, which is based on a 3-to-2 registry composed of three units of diamond and two units of Si, has only 1.5% mismatch [8.42]. However, a 5-to-4 registry of diamond/β-SiC with −2.2%

mismatch (Table 8.2) may have lower interfacial energy because of lower density of misfit dislocation. The above TEM observations have been carried out on continuous interfaces of diamond films and Si, where the starting point of diamond nucleation and growth cannot be identified. Using lattice imaging, Stammler et al. [8.43] and Wurzinger et al. [8.44] observed an SiC inter-mediated layer between small heteroepitaxial diamond particles and Si (001). Although there have been only few observations of island interfaces at present, important information regarding the initial growth stage can be obtained by studying the interfaces between the diamond nuclei and the substrate.

Since the nucleation density is 10^{11} cm^{-2} at most, the averaged information obtained by surface or interface analysis is mainly from regions which are unsuitable for heteroepitaxial nucleation. A modified or new method that can enhance the heteroepitaxial nucleation density by a factor of ten or more would be the most convenient way to allow the investigation of diamond heteroepitaxial nucleation.

8.4 Selective Growth and Surface Morphology

8.4.1 Texture Growth

Texture growth can be understood on the basis of the evolutionary selection of specific crystallite orientations as modelled by Van der Drift [8.45]. In this concept, the direction of fastest growth determines the film texture. This concept can be applied to explain the growth of CVD diamond. At the initial stage of the film growth, the nuclei grow larger to become independent particles. As their diameter reaches the average distance between the nucleation sites, their growth begins to be influenced by each other. The subsequent film growth is dominated by competitive growth between differently oriented growths. With increasing film thickness, more and more grains are overgrown and buried by the better oriented grains adjacent to them. Only crystals with their direction of fastest growth perpendicular to the surface will survive, and the less oriented ones (the majority) will be buried in the films.

$$\alpha = \frac{V_{100}}{V_{111}} \cdot \sqrt{3}$$

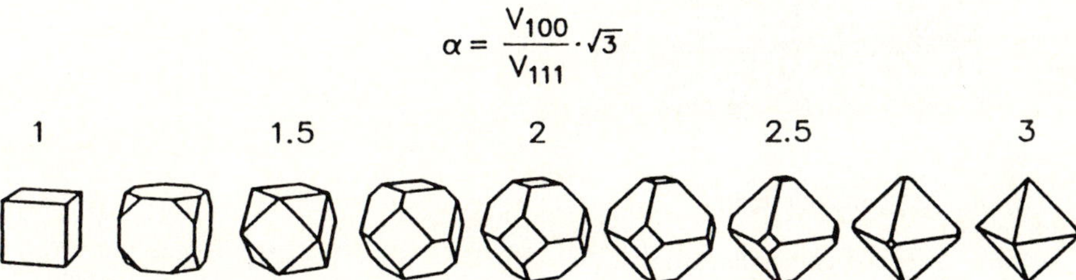

Fig. 8.10: Idiomorphic crystal shapes for different values of the growth parameter α. The arrow indicates the direction of fastest growth [8.29].

The crystal shape is determined by the relative growth rate on the {100} and {111} faces. Wild et al. [8.29] introduced a growth parameter $\alpha = \sqrt{3}\, V_{100}/V_{111}$, where V_{100} and V_{111} are the growth rates on the {100} and {111} faces, respectively. According to this parameter, various crystal shapes are obtained, as shown in Fig. 8.10. The growth temperature [8.46], impurities such as nitrogen [8.47] and/or the ratio of carbon to oxygen [8.25] influence the growth rate and its activation energy [8.48] on the (111) and (001) surfaces, suggesting a difference in the growth mechanism of diamond on these surfaces. The adsorption probability of growth precursors such as CH_3, CH_2 and C_2H_2 depends on the surface structures of diamond. While, on a (111)–1×1:H surface, the adsorption of CH_3 is more favourable than that of C_2H_2 [8.49], the adsorption of C_2H_2 is preferred on atomic steps formed on the dimer-row edge (Sb steps) on (001)–2×1:H. The difference in adsorption and migration under various experimental conditions affects the growth velocity on each surface.

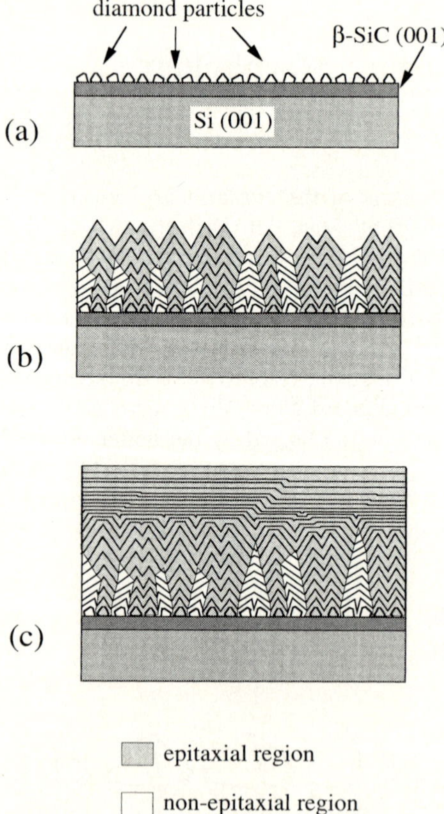

Fig. 8.11: A schematic of the growth of heteroepitaxial diamond on β-SiC(001). **(a)** BEN. **(b)** [001] fast growth with $\alpha \approx 3.0$. **(c)** Smooth growth of the (001) surface with $\alpha < 3.0$ (after [8.29, 30]).

8.4.2 Selective Growth for Improved Orientation

On the basis of two-dimensional simulation, Wild et al. applied the α parameter to control the selective growth of a particular orientation [8.29]. This idea can be explained, along with the diamond heteroepitaxial growth sequence on β-SiC(001), as shown schematically in Fig. 8.11. In order to obtain [001]-oriented diamond films after BEN (Fig. 8.11a), the direction of fastest growth must be [001]; that is, α must be close to or above 3.0. The alignment of the direction of fastest growth leads to an alignment of [001] directions perpendicular to the substrate surface (Fig. 8.11b). At this stage, the surface has a rough structure, with pyramid-like shapes composed of {111} facets. Subsequently, the α-parameter is decreased to below 3.0. As a result, (001) facets appear on the tips of these pyramids and grow larger rapidly, resulting in a film surface consisting of more or less parallel (001) facets. By this technique, a smooth surface can be realized (Fig. 8.11c).

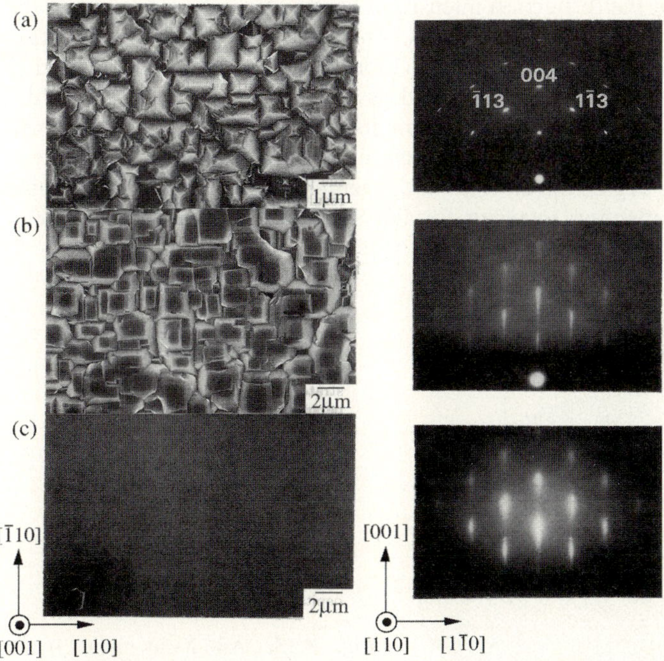

Fig. 8.12: SEM images (left) and RHEED patterns (right) of the surface of diamond films grown on β-SiC(001) after BEN. **(a)** [001] fast growth (1 μm thickness). **(b)** [001] fast growth (1 μm thickness) followed by smooth growth of (001). **(c)** [001] fast growth (1 μm) followed by smooth growth of (001) (4 μm thickness) [8.30].

Figure 8.12 shows SEM images and RHEED patterns of the typical growth stages of continuous diamond films on β-SiC(001) [8.30]. After BEN and the [001] fast growth to about 1 μm in thickness, textured films with only azimuthally oriented diamond particles are formed, and a spotty diamond bulk pattern is observed by RHEED (Fig. 8.12a). The width of the diamond 004 spot is reduced to approximately 2°. Furthermore, the [001] fast growth leads to a decrease of the particle density from 10^{10}–10^{11} cm^{-2} to 1×10^8 cm^{-2}. After the subsequent smoothing growth with $\alpha < 3.0$, expanded and smoothed (001) facets of azimuthally ordered particles are obtained (Figs. 8.12b, c). After the completion of the (001) smooth growth, grain boundaries are hardly observed on the film surface on an SEM image (Fig. 8.12c). Smooth diamond films (4 μm thick) with a streaky diamond RHEED pattern have been formed.

The angular spread (polar and azimuthal deviation) of crystals can be improved in thicker heteroepitaxial films. The 111 X-ray reflection intensities of heteroepitaxial diamond on β-SiC are stereographically projected in Fig. 8.13 [8.8]. Diamond crystals are aligned in both the polar and the azimuthal directions. The angular spread of the diamond crystal as given by the full width at half maximum (FWHM) of the diffraction intensity is about 1.5°, which is close to the angular resolution limit in the pole figures. From the fastest growth direction model shown above, only the polar (tilt) deviation decreases, while a reduction in the azimuthal deviations cannot be expected as a result of the increase in film thickness [8.50]. However, the azimuthal deviation obtained after the film growth is clearly lower than that obtained after the initial growth stage. Hence, it appears that there is also some degree of selection in the azimuthal direction. To determine the FWHM of the reflection intensity for the small polar deviation precisely,

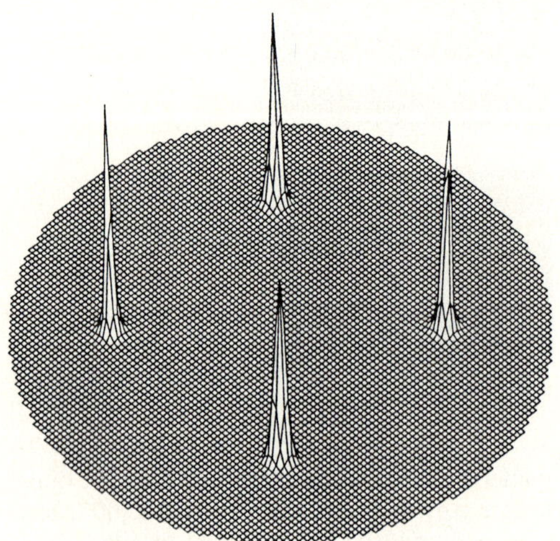

Fig. 8.13: An X-ray pole figure (polar angle χ, $0° < \chi < 80°$) of the 111 reflection of heteroepitaxial diamond on β-SiC(001). FWHM ≈ 1.5° [8.8].

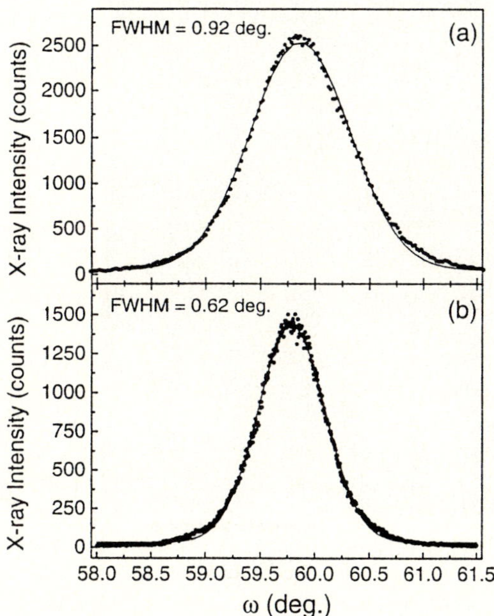

Fig. 8.14: Rocking curves of X-ray reflection profiles taken in ω-scan mode of heteroepitaxial diamond films after the (001) smoothing growth. **(a)** 20 μm in thickness. **(b)** 300 μm in thickness. For both samples, the reflections were 004 and the reflection profiles (*dots*) were fitted to Gaussian profiles (*solid lines*) [8.8].

rocking curve measurements are necessary. In thick diamond heteroepitaxial films on β-SiC with 20 μm and 300 μm thicknesses, the FWHM of the rocking curves (ω-scan) of the diamond 004 reflections was found to be 0.92° and 0.62°, respectively (Fig. 8.14) [8.8]. The improvement in polar deviation in thicker films is effective up to at least 500 μm.

Since the tilt deviation becomes less than 1° in the heteroepitaxial films, the effect of the inclined substrate can be investigated. A Nomarski optical micrograph of heteroepitaxial diamond grown on 4° off β-SiC(001) (inclined in the [$\bar{1}10$] direction) is shown in Fig. 8.15. Stripes are observed in the [110] direction, homogeneously distributed on areas larger than 5×5 mm^2. They are considered to be macro-steps (bunched steps) with (001) terraces, which form as a result of the inclination of the SiC surface from (001) [8.51]. Atomic force microscope measurements revealed that the step height is 40–60 nm and the step distance is 600–900 nm. The resultant vicinal angle is 3–4°, which reflects the substrate inclination of 4°. This indicates that the selective growth is effective even if the dominant orientation is inclined by several degrees from the perpendicular direction.

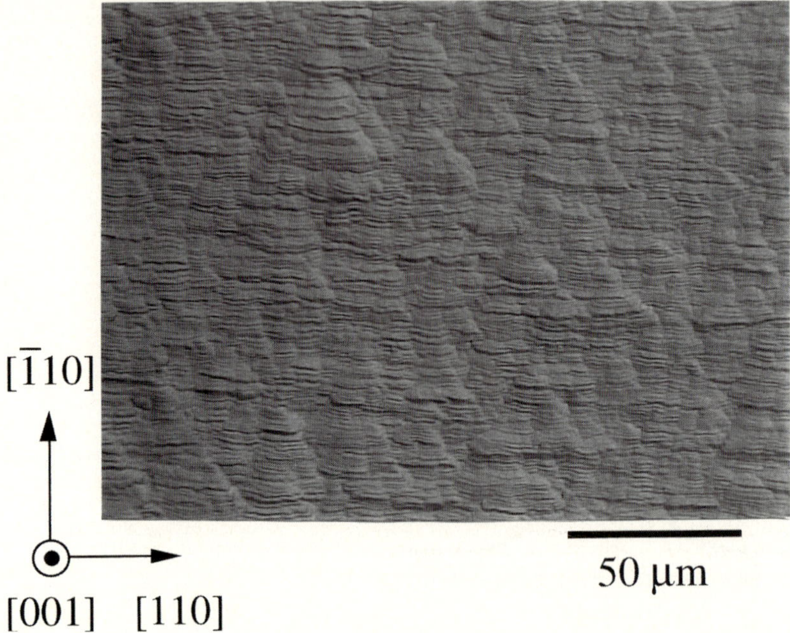

Fig. 8.15: A Nomarski optical microscopic image of heteroepitaxial diamond film (20 µm in thickness) on β-SiC, epitaxially grown on 4° off Si(001).

8.5 Electrical Properties of Heteroepitaxial Diamonds

Due to the low density, the reduced angle of the grain boundaries and the surface flatness, the electrical properties of heteroepitaxial layers have been expected to become comparable to those of homoepitaxial layers. Stoner et al. have shown that the hole mobility of boron-doped heteroepitaxial diamonds is 280 cm^2/Vs at room temperature [8.52], which is superior to that of polycrystalline samples and 1/3–1/4 of that of homoepitaxial layers with the same dopant concentration. The surface smoothness of heteroepitaxial layers has been also improved by boron doping [8.5], indicating the enhancement of surface migration of growth species by impurity doping [8.53]. The hole mobility may be further increased as a result of recent developments in heteroepitaxial growth.

As an electrical application of diamond, several types of p-channel field effect transistors (FETs) have been considered (see Chap. 17). Since these devices use only the surface regions of films, embedded grain boundaries are not detrimental to FET operation. Using the surface p-type conductive layer of undoped epitaxial films [8.54, 55], metal semiconductor FETs (MESFETs) [8.56, 57] have been fabricated on the relatively smooth surface of heteroepitaxial diamond on β-SiC(001) [8.51]. The I_{DS}–V_{DS} characteristics of a Cu gate MESFET operated in the

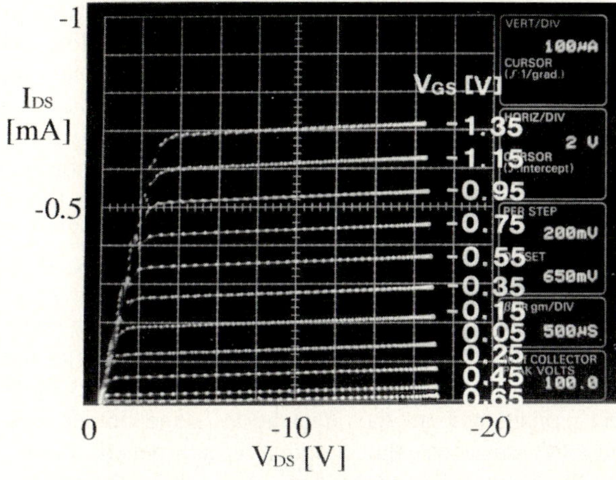

Fig. 8.16: I_{DS}–V_{DS} characteristics for various gate voltages, V_{GS}, of a Cu gate p-channel MESFET on heteroepitaxial diamond grown on β-SiC (001). The gate length and width are 5 µm and 60 µm, respectively. The transconductance is 7.4 mS/mm (from [8.51]).

depletion mode are shown in Fig. 8.16. Both enhancement- and depletion-mode operations have been realized by controlling the threshold voltage, which strongly depends on the metal electronegativity of the gate material [8.55]. Transconductances of 7–8 mS/mm, which are comparable to those obtained on homoepitaxial films and those of n-channel Si MOSFETs with the same gate length, have been obtained with MESFETs on heteroepitaxial films. The results indicate that the surface hole density (10^{13} cm^{-2}) and the surface hole mobility (30–40 cm^2/Vs) measured on the undoped and hydrogen-terminated surfaces of the homoepitaxial layer [8.54] are maintained even on the heteroepitaxial surface. The crystallinity of the surface region and the surface smoothness of the heteroepitaxial layers are sufficient for the fabrication of a surface channel device.

8.6 Conclusions

The characteristics of heteroepitaxial growth of diamond on various substrates can be summarized as follows.

8.6.1 Heteroepitaxial Nucleation and Initial Growth on Metal, SiC and Si

On metals, the nucleation density is generally less than 10^8 cm^{-2}. On a Pt surface, the alignment of diamond at the initial heating stage can be expected during the reconstruction of the metal surface layer in the molten phase. Since grain growth occurs at the process temperature in these metals, dynamic atom movements align small diamond particles in thermodynamically stable positions, leading to orientational growth. Lattice mismatch is not a decisive factor in obtaining crystallographic orientation with the substrate in the present stage of diamond oriented growth. The erosion of diamond during growth is suppressed as the atomic number of the metal substrate increases. A better degree of orientation has been obtained in a substrate material that has a higher melting point and a lower solubility of carbon.

On Si or SiC substrates, heteroepitaxial nucleation has been obtained by BEN, and the density is 10^9–10^{11} cm^{-2}, which is at least two orders of magnitude more than those achieved on metals. Enhancement of surface diffusion has been proposed and the size of critical nucleation has been estimated on the basis of the induction period of the diamond nucleation on an Si or SiC surface. Special geometrical factors must be considered at the nucleation stage under low supersaturation. On grooved SiC surfaces, the atomically sharpened valleys composed of {111} facets might be possible nucleation sites, where the critical size of nucleation is expected to be reduced by a factor of 2.5 compared with flat β-SiC(001) surfaces. The molten phase is not expected on Si or SiC surfaces. Special interface registries based on the CSL concept – that is, a 3-to-2 registry of atomic planes for diamond/Si and a 5-to-4 registry for diamond/β-SiC – are considered to overcome large lattice-mismatch, which reflects the surface chemical bonding, indicating a substantial difference in nucleation process in the case of a metal substrate. Although the nucleation density on SiC or Si is much higher than that on metals, significantly higher nucleation densities (in the order of 10^{12} cm^{-2}) are necessary to obtain atomic-level smoothness, as in the case of advanced semiconductor epilayers such as Si or GaAs.

8.6.2 Selective Growth

The fastest growth orientation can be applied to ensure the selection of better oriented particles. This is a kind of self-organizing feature of diamond film growth, which can be controlled by the adoption of precise growth conditions. The ratio of the growth velocities in the <100> and <111> directions (the α parameter) can be controlled by the temperature, impurities and reaction gases, reflecting the different growth mechanisms on {100} and {111} surfaces. Heteroepitaxial films with tilt angles of the diamond crystals below 1° have been obtained by applying appropriate selective growth processes.

References

8.1   B.R. Stoner and J.T. Glass, Appl. Phys. Lett. **60**, 698 (1992)
8.2   X. Jiang and C.-P. Klages, Diamond Rel. Mater. **2**, 1112 (1993)
8.3   S.D. Wolter, B.R. Stoner, J.T. Glass, P.J. Ellis, D.S. Buhaenko, E.E. Jenkins, and P. Southworth, Appl. Phys. Lett. **62**, 1215 (1993)
8.4   X. Jiang, C.-P. Klages, R. Zachai, M. Hartweg, and H.-J. Fusser, Appl. Phys. Lett. **62**, 3438 (1993)

8.5   B.R. Stoner, C. Kao, D.M. Malta, and R.C. Glass, Appl. Phys. Lett. **62**, 2347 (1993)
8.6   R. Kohl, C. Wild, N. Herres, P. Koidl, B.R. Stoner, and J.T. Glass, Appl. Phys. Lett. **63**, 1792 (1993)
8.7   H. Kawarada, T. Suesada, and H. Nagasawa, Appl. Phys. Lett. **66**, 583 (1995)
8.8   H. Kawarada, C. Wild, N. Herres, R. Locher, and P. Koidl, J. Appl. Phys. **81**, 3490 (1997)
8.9   S. Yugo, T. Kanai, T. Kimura, and T. Muito, Appl. Phys. Lett. **58**, 1036 (1991)
8.10 T. Suzuki and A. Argoitia, Phys. Status Solidi A **154**, 239 (1996)
8.11 M. Kamo, Y. Sato, S. Matsumoto, and N. Setaka, J. Cryst. Growth **62**, 642 (1983)
8.12 H. Kawarada, K.S. Mar, and A. Hiraki, Jpn. J. Appl. Phys. **26**, L1032 (1987)
8.13 K. Suzuki, A. Sawabe, H. Yasuda, and T. Inuzuka, Appl. Phys. Lett. **50**, 728 (1987)
8.14 Y. Sato, H. Fujita, T. Ando, T. Tanaka, and M. Kamo, Phil. Trans. R. Soc. Lond. A **342**, 225 (1993)
8.15 W. Zhu, P.C. Yang, and J.T. Glass, Appl. Phys. Lett. **63**, 1640 (1993)
8.16 W. Liu, D.A. Tucker, P. Yang, and J.T. Glass, J. Appl. Phys. **78**, 1291 (1995)
8.17 T. Tachibana, Y. Yokota, K. Nishimura, K. Miyata, K. Kobashi, and Y. Shintani, Diamond Rel. Mater. **5**, 197 (1996)
8.18 K. Ohtsuka, K. Suzuki, A. Sawabe, and T. Inuzuka, Jpn. J. Appl. Phys. **35**, L1072 (1996)
8.19 Z. Sitar, P.C. Yang, W. Liu, R. Schelsser, C.A. Wolden, and J.T. Prater, in Proceedings of the4th NIRIM International Symposium on Advanced Materials (1997), p. 7
8.20 H. Li, D. Pugh, J. Lees, and J.A. Bland, Nature **4791**, 865 (1961)
8.21 T. Tachibana, Y. Yokota, K. Miyata, K. Kobashi, and Y. Shintani, Diamond Rel. Mater. **6**, 266 (1997)
8.22 M. Tarutani, G. Zhou, Y. Takai, R. Shimizu, T. Tachibana, K. Kobashi, and Y. Shintani, Diamond Rel. Mater. **6**, 272 (1997)
8.23 S. Ojika, S. Yamashita, K. Kataoka, and T. Ishikura, Jpn. J. Appl. Phys. **32**, L1681 (1993)
8.24 H. Reiss, J. Appl. Phys. **39**, 5045 (1968)
8.25 K. Ohtsuka, S. Fukuda, K. Suzuki, and A. Sawabe, Jpn. J. Appl. Phys. **36**, L1214 (1997)
8.26 S. Koizumi, T. Murakami, T. Inuzuka and K. Suzuki, Appl. Phys. Lett. **57**, 563 (1990)
8.27 S. Koizumi and T. Inuzuka, Jpn. J. Appl. Phys. **32**, 3920 (1993)
8.28 C. Wild, P. Koidl, N. Herres, W. Müller-Sebert, and T. Eckermann, Electrochem. Soc. Proc. **91-8**, 224 (1991)
8.29 C. Wild, P. Koidl, W. Müller-Sebert, H. Walcher, R. Kohl, N. Herres, R. Locher, R. Samlenski, and R. Brenn, Diamond Rel. Mater. **2**, 158 (1993)
8.30 T. Suesada, N. Nakamura, H. Nagasawa, and H. Kawarada, Jpn. J. Appl. Phys. **34**, 4898 (1995)
8.31 X. Jiang, K. Schiffmann, and C.-P. Klages, Phys. Rev. B **50**, 8402 (1994)
8.32 W. Kulisch, L. Ackermann, and B. Sobisch, Phys. Status Solidi A **154**, 155 (1996)
8.33 J. Gerber, S. Sattel, K. Jung, H. Ehrhardt, and J. Robertson, Diamond Rel. Mater. **4**, 559 (1995)
8.34 S. Sattel, J. Gerber, and H. Ehrhardt, Phys. Status Solidi A **154**, 141 (1996)
8.35 S.P. McGinnis, M.A. Kelly, and S.B. Hagström, Electrochem. Soc. Proc. **95-4**, 73 (1995)
8.36 M. Tomellini, Ber. Bunsenges. Phys. Chem. **99**, 838 (1995)
8.37 H. Maeda, M. Irie, T. Hino, K. Kusakabe, and S. Morooka, J. Mater. Res. **10**, 158 (1995)
8.38 S. Aoyama and N. Shibata, J. Ceram. Soc. Jpn. **102**, 13 (1994) [in Japanese]
8.39 S. Yugo, T. Kanai, and T. Kimura, Diamond Rel. Mater. **1**, 388 (1992)
8.40 E. Schaller, O.M. Küttel, and L. Schlapbach, Phys. Status Solidi A **153**, 415 (1996)

8.41 X. Jiang and C.L. Jia, Appl. Phys. Lett. **67**, 1197 (1996)

8.42 C.L. Jia, K. Urban, and X. Jiang, Phys. Rev. B **52**, 5164 (1995)

8.43 M. Stammler, R. Stöckel, L. Ley, M. Albrecht, and H.P. Strunk, Diamond Rel. Mater. **6**, 747 (1997)

8.44 P. Wurzinger, N. Fuchs, P. Pongratz, M. Schreck, R. Heßmer, and B. Stritzker, Diamond Rel. Mater. **6**, 752 (1997)

8.45 A. Van der Drift, Philips Res. Rep. **22**, 267 (1967)

8.46 C. Wild, R. Kohl, N. Herres, W. Müller-Sebert, and P. Koidl, Diamond Rel. Mater. **3**, 373 (1994)

8.47 R. Locher, C. Wild, N. Herres, D. Behr, and P. Koidl, Appl. Phys. Lett. **65**, 34 (1994)

8.48 H. Maeda, K. Ohtsubo, M. Irie, N. Ohya, K. Kusakabe, and S. Morooka, J. Mater. Res. **10**, 3115 (1995)

8.49 D.R. Alfonso, S.H. Yang, and D.A. Drabold, Phys. Rev. B **50**, 15369 (1994)

8.50 M. Schreck and B. Stritzker, Phys. Status Solidi A **154**, 197 (1996)

8.51 H. Kawarada, C. Wild, N. Herres, and P. Koidl, Appl. Phys Lett. **72** (1998), April 13 issue

8.52 B.R. Stoner, personal communication

8.53 M. Hata, M. Tsuda, and S. Oikawa, Surf. Sci. **79/80**, 255 (1994)

8.54 K. Hayashi, S. Yamanaka, H. Okushi, and K. Kajimura, Appl. Phys. Lett. **68**, 376 (1996)

8.55 H. Kawarada, Surf. Sci. Rep. **26**, 207 (1996)

8.56 H. Kawarada, M. Aoki, and I. Itoh, Appl. Phys. Lett. **65**, 1563 (1994)

8.57 H. Kawarada, M. Itoh, and A. Hokazono, Jpn. J. Appl. Phys. A **35**, L1165 (1996)

Part III

Applications
of CVD-Diamond

9. Thermal Properties and Applications of CVD Diamond

Eckhard Wörner

Fraunhofer-Institut für Angewandte Festkörperphysik,
Tullastrasse 72, D-79108 Freiburg, Germany
e-mail: woerner@iaf.fhg.de

Springer Series in Materials Processing
Low-Pressure Synthetic Diamond Eds.: B. Dischler and C. Wild
© Springer-Verlag Berlin Heidelberg 1998

9.1 Introduction

One of many remarkable properties of diamond is its unsurpassed thermal conductivity. In contrast to metals, where conduction electrons are responsible for the high thermal conductivity, heat is conducted in electrical insulators by lattice vibrations. With a sound velocity of 17.5 km s^{-1} [9.1], diamond is the material with the highest Debye temperature (2220 K), exceeding that of most other insulating materials by an order of magnitude.

Figure 9.1 compares the thermal conductivity of nonmetallic crystals and copper with that of diamond. For all insulators, the temperature dependence of the thermal conductivity is similar. At low temperatures, the thermal conductivity is small due to phonon boundary scattering. At temperatures of about one-twentieth of the Debye temperature, it reaches a maximum, and for higher temperatures it decreases again. The only difference between diamond and the other insulating materials is a shift of the maximum to higher temperatures, thereby leading to the highest thermal conductivity of any material at room temperature.

High-power electronic and opto-electronic devices suffer severe cooling problems due to the production of large amounts of heat on a small area, thereby creating a tremendous heat flux (up to several kW/cm^2). For efficient cooling of these devices, it is essential to spread the narrow heat flux by placing a layer of high thermal conductivity between the device and the cooling system.

With a thermal conductivity of 20–25 W/cmK at room temperature [9.1–4], natural type IIa diamond is the material with the highest thermal conductivity, exceeding that of copper by a factor of five. Therefore, large-area CVD-diamond films have been proposed for many heat-transfer applications, although the first thermal conductivity measurements in the late 1980s were not very promising [9.5]. In recent years, however, the quality of CVD diamond has improved dramatically, and large-area CVD-diamond plates with thermal conductivities around 20 W/cmK have become available [9.6, 7]. Therefore, today, CVD

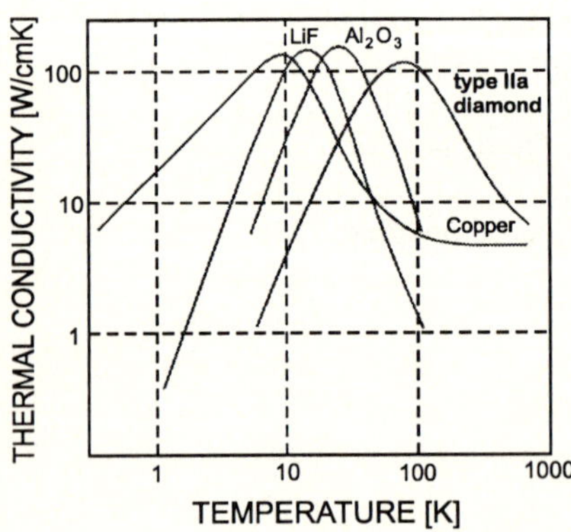

Fig. 9.1: The thermal conductivity of non-metallic crystals, copper and type IIa natural diamond [9.2].

diamond is used for various thermal management applications, such as submounts for integrated circuits, heat spreaders for high-power laser diodes or even as a substrate material for multi-chip modules (MCM) [9.8].

9.2 The Thermal Conductivity of Crystalline Insulators: the Klemens–Callaway Theory

With a band gap of 5.4 eV, diamond is classified as a wide-band-gap semiconductor. Therefore, even at temperatures as high as the Debye temperature (2220 K), virtually all of the heat is carried by phonons [9.9].

The theory of heat conduction due to lattice vibrations is known as the Klemens–Callaway Theory. On the basis of a relaxation time approximation for phonon scattering mechanisms and the Debye approximation for the phonon dispersion relation, the following expression is derived for the thermal conductivity of crystalline insulators [9.10–12]:

$$\kappa(T) = \frac{k_B}{2\pi^2 v}\left(\frac{k_B}{\hbar}T\right)^3 \int_0^{\Theta/T} \tau(x)\frac{x^4 e^x}{(e^x-1)^2}\,dx \qquad (9.1)$$

where k_B and $\hbar$ are the Boltzmann and Planck constants, ω is the phonon frequency, $\Theta = 2220$ K [9.13], the Debye temperature, $v = 1.32 \times 10^4$ m s^{-1} [9.14], the averaged phonon velocity, and x is defined as $\hbar\omega/k_B T$. Considering the various scattering mechanisms as independent, the total relaxation time τ is given by $1/\tau = \sum_i 1/\tau_i$, where τ_i is the relaxation time of an individual scattering mechanism (Sect. 9.3).

In (9.1), intrinsic normal scattering is not taken into account. Normal scattering is an important scattering mechanism for isotopically pure, nearly perfect crystals, but it is mostly neglected for crystals with a natural isotopic abundance (Sect. 9.3.5).

Equation (9.1) can be transformed into a more convenient form, which is very similar to $\kappa = 1/3Cvl$, obtained by applying the kinetic gas theory. While the kinetic gas theory assumes a constant specific heat C for all phonons, the Klemens–Callaway Theory takes the wavelength dependence of the specific heat (Fig. 9.2) into account. By transforming (9.1), the following expression is obtained:

$$\kappa(T) = \frac{1}{3}v\int_0^{\Theta/T} l(x)C(x)\,dx \qquad (9.2)$$

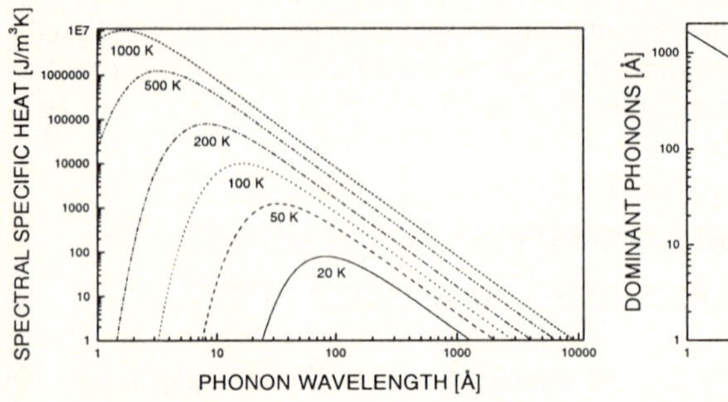

Fig. 9.2: The spectral specific heat as a function of temperature and phonon wavelength, showing a sharp peak (logarithmic scale) at $\lambda = h\nu/(3.83k_BT)$.

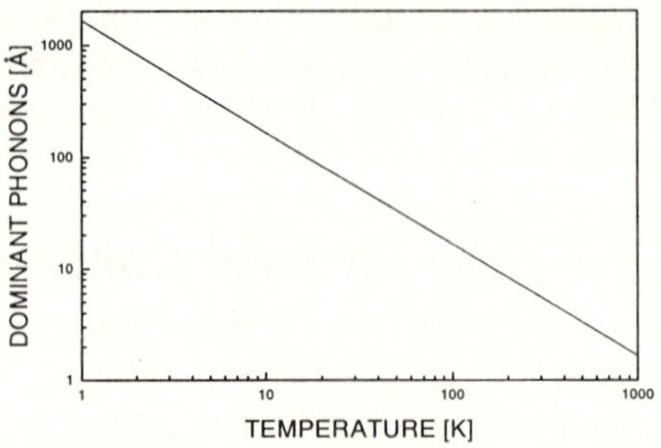

Fig. 9.3: The wavelength of the dominant phonons of diamond as a function of temperature.

where $C(x)$ is the spectral specific heat,

$$C(x) \equiv \frac{3k_B}{2\pi^2\nu^3}\left(\frac{k_BT}{\hbar}\right)^3\frac{x^4 e^x}{(e^x-1)^2} \tag{9.3}$$

and $l(x)= \nu\,\tau(x)$ is the frequency-dependent phonon mean free path.

In Fig. 9.2 the spectral specific heat is plotted as a function of temperature and phonon wavelength. It shows a sharp peak (logarithmic scale) at $\lambda = h\nu/(3.83k_BT)$. Phonons of this wavelength are called dominant phonons, since they conduct the majority of heat at a given temperature. Figure 9.3 shows the temperature dependence of the wavelength of the dominant phonons. As the temperature decreases, the dominant phonon wavelength increases. Therefore, changing the measuring temperature corresponds to probing the crystal with phonons of variable wavelength. Moreover, the phonon scattering cross-section and, thus the thermal resistance due to an individual scattering mechanism, becomes maximal if the wavelength equals the extension of the scattering center.

For this reason, temperature-dependent thermal conductivity measurements are a powerful tool with witch to assess the influence of structural defects.

9.3   Phonon Scattering Mechanisms in CVD Diamond

Due to its high Debye temperature, the room-temperature thermal conductivity of diamond is influenced by defects to a larger degree as compared to other materials.

Since the phonon wavelength distribution is a function of temperature (Fig. 9.2) and the scattering cross-section is a function of wavelength, different scattering mechanisms dominate the thermal resistivity at different temperatures. Figure 9.4 shows the temperature-dependent thermal conductivity as predicted by the Klemens–Callaway Theory.

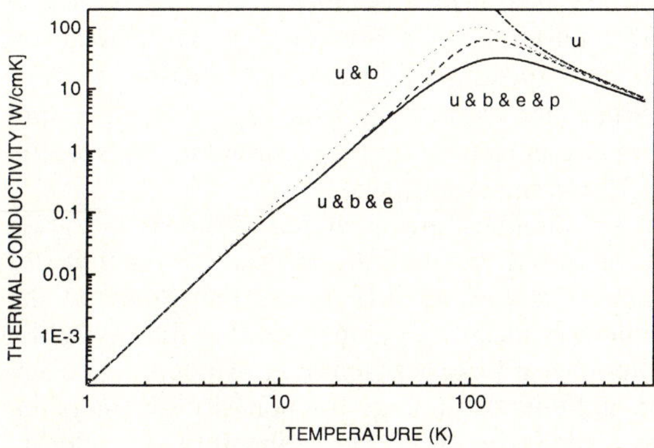

Fig. 9.4: The thermal conductivity of diamond according to the Klemens–Callaway Theory if only umklapp scattering is present *(u)*. The other curves are calculated with the successive addition of boundary scattering *(b)*, scattering at extended defects *(e)*, and point defects *(p)*, respectively.

If all scattering mechanisms except umklapp scattering *(u)* are neglected, the thermal conductivity diverges for $T \to 0$. However, for higher temperatures, the temperature dependence of the thermal conductivity can almost be described using umklapp scattering only. Boundary scattering prevents the thermal conductivity from diverging at low temperatures. Boundary scattering (Sect. 9.3.1) in combination with umklapp scattering *(u & b)* (Sect. 9.3.5) explains the high- and low-temperature range. For medium temperatures, defects such as extended defects *(e)* (Sect. 9.3.2) and point defects *(p)* (Sect. 9.3.3) influence the thermal conductivity (Fig. 9.4).

9.3.1 Scattering at Grain Boundaries and Sample Surfaces

From the viewpoint of heat conduction, the most important "defect" of CVD diamond is its polycrystallinity.

At low temperatures ($T < 20$ K) the wavelength of the dominant phonons is larger than 100 Å (Fig. 9.3), and thus larger than the extent of most crystal defects. Therefore all scattering mechanisms except diffuse boundary scattering are weak and the thermal conductivity is limited only by phonon scattering at boundaries. Actually, the thermal conductivity would diverge for $T \to 0$ without boundary scattering (Fig. 9.4).

Assuming boundary scattering to be diffuse for all phonons, the following expression for the boundary scattering relaxation time is obtained [9.15]:

$$1/\tau_B = \frac{v}{\alpha d}. \tag{9.4}$$

Here d is the sample size normal to the heat flow – in other words, the crystallite size for in-plane heat flow in CVD diamond – and α is a constant that depends on the sample geometry. For a circular or quadratic cylinder, one obtains $\alpha = 1$ or $\alpha = 1.12$, respectively [9.15]. Since the scattering time τ_B is temperature independent, low-temperature thermal conductivity is proportional to the specific heat, and therefore proportional to T^3 and the sample size.

To examine whether or not all phonons are scattered diffusely at grain boundaries, Morelli et al. [9.16] measured the in-plane thermal conductivity of CVD diamond down to temperatures as low as 0.15 K. As they lowered the temperature, they found that the phonon mean free path exceeded the crystallite size by a factor of ten. This implies that a large transmission of phonons occurs between grains at low temperature and thus that a large fraction of phonons is not scattered diffusely, but reflected in a mirror-like way. The probability of scattering or reflection at a grain boundary is dependent on the boundary roughness, on the scale of the phonon wavelength. Thus, depending on temperature, dominant phonons can be scattered or reflected at the same boundary. Therefore, Graebner [9.17] suggested rewriting the relaxation time of grain boundary scattering to take mirror-like phonon reflection into account.

Due to the textured nature of polycrystalline CVD diamond, phonons propagating perpendicular to the sample surface have fewer grain boundaries to cross than phonons propagating parallel to the surface. Therefore, the thermal properties of CVD diamond are far from isotropic (Sect. 9.3.4).

9.3.2 Scattering at Extended Defects

A remarkable difference in the thermal conductivity of hot-filament grown CVD diamond (HFCVD) and microwave-plasma assisted CVD-diamond films (MPACVD) was observed by Morelli et al. [9.18]. The HFCVD films showed a strong "dip" in the thermal conductivity between 15 K and 60 K, which was not observed or was very weak in MPACVD specimens. The "dip" indicates the existence of an additional defect in these films, which interacts strongly with the dominant phonons in this temperature range. The measured temperature dependence was explained either by resonant scattering from a defect or precipitate scattering from an extended defect [9.18], since both scattering mechanisms could account for the observed temperature dependence. To distinguish between these scattering mechanisms, Morelli et al. made measurements on natural diamond (type IIa) irradiated with fast neutrons. They observed, that due to irradiation, the thermal conductivity was reduced by 40% at room temperature and by a factor of 50 at 30 K (Fig. 9.5). This dramatic decrease was explained by the creation of extended defects about 15 Å in diameter. These defects also explain the characteristic "dip" in the thermal conductivity observed between 15 K and 60 K. While resonant scattering from vacancies could provide an alternative explanation for the characteristic "dip" in neutron-irradiated natural diamond as well as in HFCVD diamond, measurements on electron-irradiated diamond showed no such dip [9.9, 20]. Since, for electron-irradiated diamond,

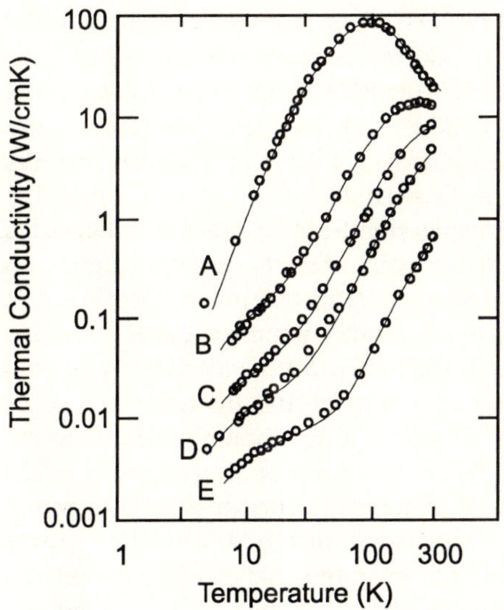

Fig. 9.5: The thermal conductivity of unirradiated natural type IIa diamond *(curve A)* and after irradiation with increasing fluences from 3×10^{16} *(curve B)* to 4.5×10^{18} neutrons cm^{-2} *(curve E)* [9.19]. The creation of the strong dip between 15 K and 60 K was attributed to scattering at extended defects.

only neutral vacancies and interstitials but no extended defects are created, Morelli concluded that the characteristic dip observed in irradiated type IIa diamond was due to scattering at extended defects.

Although this measurement does not prove that extended defects account for the observed dip in the thermal conductivity of CVD diamond, it is a strong hint, and therefore most authors use extended defect scattering to fit their temperature-dependent thermal conductivity data.

9.3.3 Point Defect Scattering and the Effect of the Natural Isotope Composition on the Thermal Conductivity

In general, point defects are defects with spatial dimensions that are small compared to the dominant phonon wavelength. Examples include impurity atoms that are incorporated substitutionally or on interstitial sites, vacancies and dislocations. However, different isotopes also act as point defects. Point defect scattering arises due to mass differences or differences in the strength of chemical bonds, thereby destroying the translation invariance of the Hamiltonian. Using second-order perturbation theory, Klemens [9.12] derived the following expression for the point defect relaxation time:

$$1/\tau_P = A_P x^4 T^4 \tag{9.5}$$

$$A_P = \frac{D_P a^3}{4\pi v^3} \left(\frac{k_B}{\hbar}\right)^4 \left(\frac{\Delta M}{M} + 2\gamma\alpha\right)^2. \tag{9.6}$$

Here, D_P is the point defect concentration, ΔM is the mass difference, and α is the volume difference between the impurity and the lattice atom; a^3 is the lattice volume (per atom) and γ is the Grüneisen constant of diamond ($\gamma = 1.1$). Point defect scattering influences the thermal conductivity over a wide temperature range (Fig. 9.4). Only at very low temperatures ($T < 4$ K) or for temperatures higher than 500 K does its impact on the thermal resistance vanish, since the dominant phonon wavelength (Fig. 9.2) becomes significantly larger (smaller) than the spatial dimensions of the defect. Thus, point defect scattering tends to influence the thermal conductivity most dramatically at its maximum (Fig. 9.4).

Not only defects such as vacancies or incorporated impurity atoms, but also different isotopes, act as point defects due to their mass difference ΔM. While the isotope effect on the thermal conductivity of single crystal diamond is widely discussed [9.1, 4, 21–24], there are only a few publications about this effect on the thermal conductivity of CVD diamond [9.25, 26]. Anthony et al. reported the first measurements on the thermal conductivity of isotopically enriched CVD diamond in 1991, but as they applied HFCVD to deposit diamond and the growth conditions were adjusted for efficient use of the rather expensive isotopically enriched methane gas, rather than for growing high-quality CVD diamond, no isotope effect on the thermal conductivity was observed [9.25]. In 1994, Graebner et al. [9.26] observed an enhanced thermal conductivity for isotopically enriched CVD-diamond films. They measured an in-plane thermal conductivity of $\kappa_\parallel = 21.8$, in contrast to 18 W/cmK for a sample without isotopic purification. Evaluating the temperature dependence of the thermal conductivity, they found that the scattering amplitude due to point defects was reduced drastically, indicating that the conductivity enhancement was caused by isotopic purification [9.26].

9.3.4 The Anisotropy of CVD Diamond

Due to its competitive growth, CVD diamond consists of roughly cone-shaped columnar structures [9.27] (Fig. 9.6). Close to the substrate, the crystallites are only some microns in diameter. With increasing thickness, the crystallite diameter increases, and reaches up to 100 μm for thicker samples. As a rule of thumb, the diameter is about one tenth of the sample thickness [9.7].

Due to this spatial anisotropy, a thermal anisotropy is expected; that is, the thermal conductivity κ should depend on whether the heat flow is perpendicular ($\kappa_\perp$) or parallel ($\kappa_\parallel$) to the sample surface. Furthermore, the local conductivity should depend on the vertical position in the diamond sample. Since phonons propagating perpendicular to the sample surface, in other words parallel to the grain structure, have to cross fewer grain boundaries, and since defects are expected to be concentrated at the boundaries [9.28, 29], the thermal conductivity should increase with increasing sample thickness, and $\kappa_\perp$ should be higher than $\kappa_\parallel$. Both effects were measured by Graebner et al. [9.30]. To investigate the thermal anisotropy, they applied a laser flash method to determine $\kappa_\perp$ and a heated bar

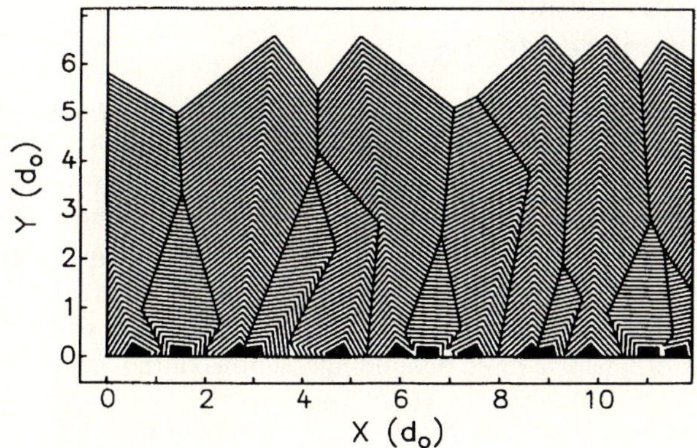

Fig. 9.6: Competitive growth simulation of a polycrystalline film [9.27].

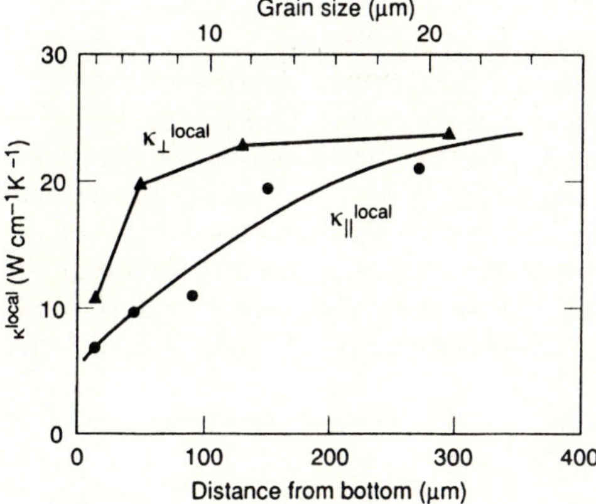

Fig. 9.7: The local room-temperature thermal conductivity of CVD diamond as a function of the height above the substrate and the grain size, measured parallel $(\kappa_\parallel)$ and perpendicular $(\kappa_\perp)$ to the diamond film [9.30].

technique to measure $\kappa_\parallel$. For a sample of inferior quality, the thermal conductivity perpendicular to the sample surface $\kappa_\perp$ was found to be more than 50% higher than the in-plane thermal conductivity $\kappa_\parallel$.

To study the height dependence of both $\kappa_\perp$ and $\kappa_\parallel$, a series of samples with different thicknesses, grown under identical conditions, was measured. A large through-the-thickness gradient for $\kappa_\perp$ and $\kappa_\parallel$ was found and the local thermal conductivities were derived. For $\kappa_\parallel$, the local value ranged from 7 W/cmK to at least 21 W/cmK at a height of 300 µm above the substrate [9.7] (Fig. 9.7). For $\kappa_\perp$, local values of up to 23.5 W/cmK were obtained.

By analyzing the temperature dependence, Graebner et al. [9.28] found that all defect-related scattering mechanisms decreased in strength with increasing sample thickness.

9.3.5 Phonon–Phonon Interaction: Umklapp and Normal Scattering

Since the phonon states in an hypothetical perfect harmonic crystal are stationary, no phonon–phonon interaction can take place, and thus any established phonon distribution would not relax to equilibrium, but would remain unchanged forever. Therefore, in the absence of defects, the thermal conductivity of an harmonic crystal would be infinite. However, since the potential energy of an atom in a real crystal lattice is not an exact quadratic function of displacement, interactions between phonons can occur. In first-order perturbation theory, there has to be a distinction between two types of intrinsic three phonon interactions of the form

$$q_1 + q_2 = q_3 + Q. \tag{9.7}$$

Those that conserve crystal momentum are called normal processes, for which $Q = 0$, and those that do not are called umklapp processes: they involve the addition of a nonzero reciprocal lattice vector Q. For normal processes, the total crystal momentum remains unchanged, so they do not directly give rise to thermal resistance, but rather simply redistribute the phonon distribution, thereby occupying states of higher energy. For these states, however, umklapp scattering or scattering at defects is more likely. Berman discussed the conditions under which normal scattering is expected to be important [9.31]. He found that normal scattering can be ignored for crystals containing a moderate amount of defects and several isotopes, but that it can be the dominant source of thermal resistance for very pure crystals with only one isotope. Umklapp scattering – that is, three phonon interaction involving a nonzero reciprocal lattice vector Q – leads directly to thermal resistance.

Since the calculation of the scattering rate due to umklapp processes is an unsolved theoretical problem, scattering rates that are introduced on an empirical basis are used to fit the thermal conductivity data. The following expression is generally used for umklapp scattering in diamond [9.4,32]:

$$1/\tau_u = A_u x^2 T^3 e^{-B_u/T}. \tag{9.8}$$

The parameters A_u and B_u were determined experimentally by Graebner et al. [9.32] using high-temperature thermal conductivity data of type IIa single-crystal diamond. They found that $A_u = 640/\text{sK}^3$ and $B_u = 470$ K.

Since the wavelength of the dominant phonons decreases as the temperature increases, defect scattering becomes negligible for higher temperatures. Therefore the only effect that limits the high-temperature thermal conductivity is umklapp scattering, leading to a thermal conductivity proportional to $1/T$.

To summarize, at low temperatures the thermal conductivity of CVD diamond is proportional to T^3 and the sample size. At medium temperatures, it reaches a maximum, and at high temperatures decreases as $1/T$. The thermal resistivity is dominated by grain-boundary scattering at low temperatures, followed by extended-defect and point-defect scattering at medium temperatures. At high temperatures $(T \approx \Theta_D/10$, depending on the defect state), intrinsic umklapp scattering is the main source of thermal resistance (Fig. 9.4).

9.4 Methods for the Measurement of the Thermal Conductivity or Diffusivity of Diamond

With the remarkable progress in growing CVD diamond, measuring its thermal conductivity has become a major field of interest. Thermal conductivity measurements not only provide important data for applications but, when carried out as a function of temperature, they are also a powerful method of characterizing the defect state of a crystal [9.9].

Thus many methods have been developed and applied for measuring the thermal conductivity or diffusivity of CVD diamond. While some offer remarkably high lateral resolution or require only very little sample preparation, others allow temperature-resolved measurements. The most interesting and widespread techniques will be discussed in the following subsections.

9.4.1 Steady-State Techniques

Basically, these techniques use a steady-state heat flux to establish a temperature gradient that is inversely proportional to the thermal conductivity. The temperature gradient is measured optically or by an array of thermocouples. A very common

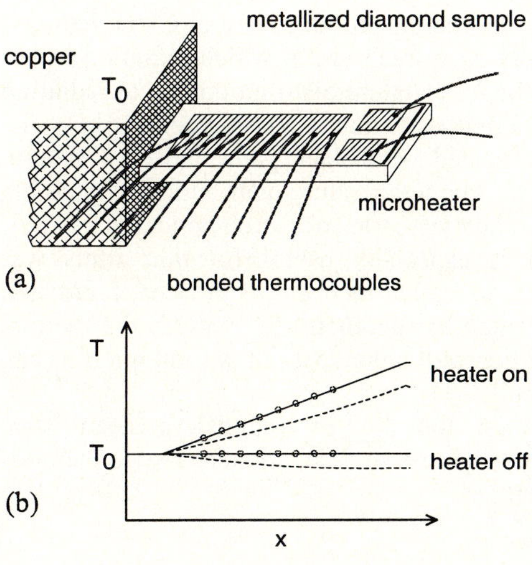

Fig. 9.8: (a) A schematic of the heated bar technique [9.6]. The diamond sample with a graphitic microheater at the free end and an array of eight bonded differential thermocouples is attached to a copper block, which determines the base temperature. **(b)** The temperature distribution in the case of large radiation losses (*dashed lines*) and in the case of no radiation losses (*solid lines*), with and without heat production in the microheater.

method, known as the *heated bar technique,* uses a bar-shaped geometry, in which one end of the sample is kept at a constant temperature and a heat flux is generated by a microheater at the other end (Fig. 9.8). By analyzing the resulting temperature gradient, the in-plain thermal conductivity is obtained.

Heating and cooling of the supporting copper block varies the base temperature of the measurement, thus allowing determination of the thermal conductivity as a function of temperature.

In this type of experiment, the influence of heat losses via other heat-transfer mechanisms, such as radiation or convection, is of crucial importance. While convection can easily be avoided by measuring in vacuum, radiation loss is most severe for long, thin plate-like samples with low thermal conductivity and high emissivity at elevated temperatures [9.33]. However, due to the high thermal conductivity of CVD diamond, radiation losses can often be neglected. Assuming a thermal conductivity of 5 W/cmK, 100 μm × 5 mm × 10 mm sample dimensions, and an emissivity of 0.2, only 0.5% of the total heat is lost due to radiation at room temperature [9.33]. However, if radiation has to be taken into account, the emissivity of the sample is obtained by evaluating the temperature distribution without heat production in the microheater (Fig. 9.8), thus allowing correction for radiation loss [9.6].

If heat loss due to radiation can be ignored, the temperature gradient is proportional to the heat flux $\mathbf{j} = -\kappa \, \nabla T$ (the Fourier law), where κ is the thermal conductivity.

The advantages of the heated-bar method are its accuracy, which has been proved in a very recent "round robin" experiment [9.34], and the possibility of temperature-resolved measurements. However, the sample preparation is usually very time-consuming. For considerable reduction of the preparation effort, Wörner et al. use a graphitic microheater, prepared by laser scribing and ultrasonically bonded gold wires in combination with the metallized diamond surface as thermocouples [9.6].

Ono et al. were the first to measure the thermal conductivity of CVD diamond in 1986 [9.5]. In contrast to other steady-state methods, in which radiation losses are generally avoided, Ono et al. used the temperature distribution due to radiation loss to determine the thermal conductivity of thin CVD-diamond films. To enhance radiation loss, the films were coated with black paint and held at both ends by heated supports, at about 150°C. The temperature distribution, containing direct information about the thermal conductivity, was measured optically.

Another steady-state method that is especially useful for thin films was developed by Jansen et al. [9.35]. They use micromechanical devices, fabricated by standard thin film technology, in heated-bar geometry to measure the thermal conductivity of CVD-diamond films. Films with a thickness of several microns can be measured between −195°C and 300°C.

Since CVD-diamond samples are far too thin to establish a steady-state temperature gradient perpendicular to the sample surface, all steady-state methods measure the in-plane thermal conductivity $\kappa_{\parallel}$.

9.4.2 Techniques with Periodic Heat Flow

Techniques with periodic heat flow analyze the propagation of a thermal wave, generated by a localized periodic heat source. By measuring the temperature transient, the thermal diffusivity D is obtained. It is related to the thermal conductivity κ by $D = \kappa/C$ where C is the heat capacity per unit volume. The advantages of these techniques are that, at sufficient high heating frequencies, heat loss due to radiation is negligible [9.36] and, generally, only a little sample preparation is required.

A periodic method applied by Hatta et al. uses large-area optical heating through a movable mask. The resulting heat wave parallel to the sample surface is measured by a fixed thermocouple mounted directly on to the sample [9.37–39]. A similar method has been applied by Kosky et al. They use a 5 W GaAlAs laser diode array to heat the bar-shaped diamond sample at its free end. The thermal wave propagation is measured optically by using an IR microscope [9.40]. Both methods measure the in-plane thermal diffusivity $D_\parallel$.

A method for measuring an average of $\kappa_\parallel$ and $\kappa_\perp$ near the sample surface was developed by Cahill et al in 1986. They use a small metal strip, evaporated on to the sample surface both as heater and as temperature sensor [9.41]. An alternating electrical current at frequency ω is applied to create a cylindrical heat wave, which changes the resistivity of the metal strip, due to its temperature-dependent electrical resistance. The voltage $U = IR$ then has a component at the frequency 3ω. By measuring the amplitude and phase of the resistance changes as a function of the excitation frequency, the thermal conductivity can be determined directly without having to determine the specific heat [9.42]. The method can be applied even at high temperatures, because errors due to radiation can be shown to be negligible [9.1,41].

The *mirage technique* (Fig. 9.9), a widespread periodic method with high lateral resolution, measures an average of $D_\parallel$ and $D_\perp$ near the sample surface. It uses a modulated, focused laser beam at frequencies of several 100 Hz to launch a

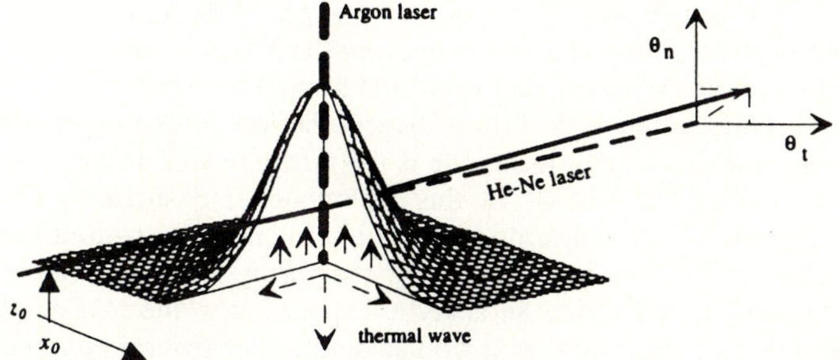

Fig. 9.9: The principle of the mirage effect: a thermal wave is generated by a modulated heating beam. The resulting temperature variation is measured using an He–Ne probe beam [9.43].

thermal wave at the surface of the sample. A second laser is deflected by the corresponding periodic changes in the refractive index (mirage effect) of the air above the sample. The resulting vector deflection of the probe beam is analyzed in phase and position. A multi-parameter least-square fit between the theoretical calculations and the experimental data is applied to determine the thermal diffusivity [9.44]. The sample surface is scanned with the probe beam either at a small angle ("bouncing geometry") or parallel to the sample surface ("skimming geometry"), allowing measurements on samples with relatively rough surfaces. [9.43]. The mirage method can also be used for *in situ* thermal diffusivity measurements.

Infrared radiometry methods also use modulated laser radiation to generate a thermal wave, but its propagation is measured by infrared radiometry using an IR camera [9.45, 46] or an IR detector [9.47–49]. These techniques use lower modulation frequencies (of several Hz) and measure the in-plane thermal diffusivity $D_\parallel$.

9.4.3 Pulsed Techniques

Pulsed techniques measure the time-dependent spread of a heat wave after a short heating pulse. The temperature transients depend on the thermal conductivity and the specific heat of the sample. If the total heat flux is measured, both values can be obtained separately. Because of the extreme thermal diffusivity of diamond, lasers are usually applied to generate very short heating pulses.

Lu et al. used the *converging thermal wave* technique [9.50, 51] to measure the in-plane thermal diffusivity $D_\parallel$ of diamond. This technique, which was originally proposed by Cielo et al. in 1985 [9.52], uses a laser pulse to create an annular-shaped heating pattern at the sample surface. The temperature development in the center of the annulus is measured optically, using a fast HgCdTe detector. A similar technique was developed by Wörner et al. [9.53]. They apply a short high-voltage pulse at the graphitic rim resistance of the circular diamond plate to generate a converging thermal wave. The rim resistance is automatically formed during laser cutting. While it is a disadvantage that relatively large sample sizes are needed (Lu et al. used an annulus of 22 mm in diameter [9.50] and Wörner et al. [9.53] use samples with $\varnothing \geq 10$ mm), the major advantages of these techniques are the focusing effect of the annulus, thereby increasing the signal-to-noise ratio, the independence of the sample thickness, and very little sample preparation.

A more traditional, but – due to the high diffusivity and the low thickness of the CVD-diamond samples – rather difficult technique is the *flash method*. It was first applied by Parker et al. in 1960 [9.54]. In this technique, one surface of a thermally insulated sample is heated uniformly with a high-intensity short-duration light pulse. The thermal diffusivity is obtained by measuring the temperature transient at the rear surface and by fitting an analytic expression to the measured data. While Parker et al. used a flash tube with a pulse duration of several hundred milliseconds and a chromel–alumel thermocouple to measure the diffusivity of metal plates with a thickness of several millimeters, Graebner et al. [9.55, 56], who first applied this method to CVD diamond, had to use a Q-switched Nd : YAG

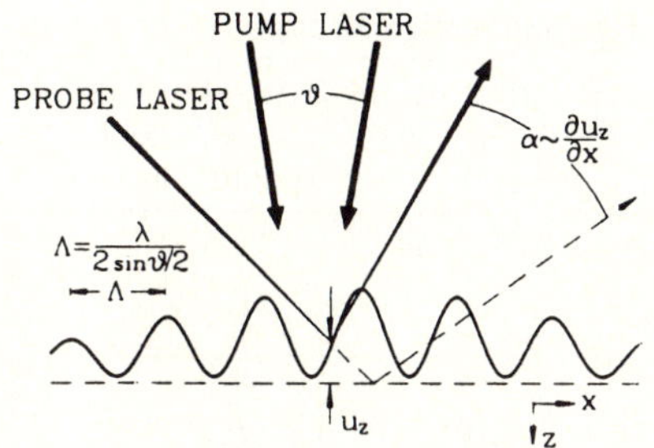

Fig. 9.10: Principle of the *transient thermal grating method*: A transient thermal grating (typical period 40 μm) is produced by two interfering laser beams. Its decay, measured by a third laser beam, is directly related to the in-plane thermal diffusivity [9.57].

laser with a pulse width of 8 ns in combination with a fast HgCdTe detector to measure the thermal diffusivity of CVD-diamond films. The flash method is the only technique so far to measure the thermal diffusivity $D_\perp$ normal to the sample surface, thus yielding information about the anisotropy (Sect. 9.3.4) of the thermal properties of CVD diamond.

A *transient thermal grating method* has been used by Käding et al. to measure the in-plane thermal diffusivity [9.57–60] (Fig. 9.10). The technique is based on surface displacement due to elastic thermal expansion. Two interfering pulsed laser beams create a thermal grating on the sample surface. The decay of this grating is measured by a third probe laser beam in reflecting geometry. The in-plane thermal diffusivity $D_\|$ is directly related to the decay time τ by $\tau \sim \Lambda^2 / D_\|$ [9.57].

The grating period Λ is usually several tens of microns, and the surface displacement typically about 1 Å. Since the technique measures the effective lateral thermal diffusivity averaged over the penetration depth, which in turn is correlated with the grating period Λ, a variation of the grating period allows to some degree depth-resolved analysis of the thermal diffusivity.

9.5 CVD Diamond for Thermal Management Applications

Due to the combination of excellent thermal conductivity ($\kappa_\|$ up to 20 W/cmK at 20°C) and high electrical resistivity ($>10^{12}$ ohm cm), CVD diamond is becoming increasingly attractive for various thermal management (TM) applications.

High-power electronic devices generate a large amount of heat over a small area, thereby creating a tremendous heat flux (up to several kW/cm^2). Since even the best cooling system cannot dissipate a heat flux of this magnitude, it is necessary to spread the heat flux before efficient cooling can take place.

As temperature reduction translates directly into improved reliability and performance, natural diamond has already been used for certain TM applications for a long time. However, since the cost of natural diamond increases damatically for larger dimensions, these applications were limited to small areas. With the

Table 9.1: Properties of electronic packaging materials and semiconductors.

Material	Thermal conductivity (W/cmK) at 20°C	Electrical insulator?	(ohm cm)	Thermal expansion coefficient (ppm/K) at 20°C
CVD diamond	10 ~ 20	Yes	$10^{12} \sim 10^{14}$	1.05
Copper	4.00	No	–	16.8
Silver	4.28	No	–	19.6
Beryllia, BeO	2.23	Yes	10^{14}	6.4
AlN	0.7 ~ 2.3	Yes	10^{13}	4.0
Silicon	1.5	Semi	–	2.62
GaAs	0.45	Semi	–	5.9

advent of CVD-diamond deposition techniques, large-area diamond plates with high thermal conductivity became available. The thermal and electrical properties of CVD diamond and other electronic packaging materials are summarized in Table 9.1.

From the viewpoint of performance (Table 9.1), no other material can compete with diamond. Diamond has the highest room-temperature thermal conductivity of all materials and, unlike other good heat conductors, such as silver and copper, diamond is an excellent electrical insulator. Another advantage is its coefficient of thermal expansion (CTE), which is close to that of silicon. This is an important feature, since thermal stresses induced during operation or mounting can enhance the thermal resistance between device and substrate – resulting in a higher operating temperature – or even destroy the device.

9.5.1 CVD Diamond for Integrated Circuit Packaging

Many reliability problems are related to the overheating of silicon chips. Near room temperature, the Arrhenius relationship between the temperature of an integrated circuit and its expected lifetime predicts that every reduction of 10 K will double the Mean Time Between Failure (MTBF) of the device [9.61].

Single-chip packages can now dissipate up to 20 W, and future generations are expected to operate at 50–100 W [9.8]. Since the heat production in electronic chips is far from homogeneous, diamond substrates can contribute to spreading the heat uniformly and thus preventing local overheating.

Boudreaux et al. [9.61] investigated the performance of CVD diamond as a substrate material for chip carriers by replacing the ground plate of a standard integrated circuit package with CVD diamond. Using a 5 W silicon test chip, they found that the thermal gradient was reduced by up to 50% when diamond was used instead of standard ceramic packages.

9.5.2 CVD Diamond for Multi-chip Modules (MCM)

The rapid changes in the electronics industry are driven by the attractiveness of faster and smaller electronic components. At the present time, the length of the electrical interconnections between packaged integrated circuits (IC's) limits the size and operational speed of electronics systems [9.62]. In order to increase the system speed, unpackaged dies mounted on the same substrate with high-density interconnections from die to die are increasingly being used in high-performance applications [9.8, 63]. However, in two-dimensional packaging, the maximum clock rate is determined by the time that a signal takes for corner-to-corner propagation. Thus the linear dimensions of a supercomputer operating at 1 GHz should not exceed 15 cm, since a signal propagating even at the speed of light travels only 30 cm in 1 ns. A method to overcome this problem is to pack as many chips as possible into a cube of dimensions $15 \times 15 \times 15 \, \text{cm}^3$ and thus use 3-D packaging, as was done in the construction of the CRAY 3 supercomputer.

There are two major difficulties to be coped with in 3-D packaging; namely, the cooling problem and the problem of vertical interconnections. In the case of CRAY 3, the $4 \times 4 \times \frac{1}{4} \, \text{in}^3$ modules were vertically held together by over 12 000 bare gold wires [9.63]. As there was enough free space between the modules, it was possible to use the interstices for spray cooling with liquid Flourient®. While this sophisticated approach is suitable for supercomputers such as CRAY 3, a simpler solution is needed for more ordinary systems. The interconnection problem could easily be overcome, since high-density arrays of vertical interconnections between adjacent boards exist, but unfortunately they suffer from the disadvantage of blocking the interboard space and they impede effective cooling. A solution to overcome both the interconnecting problem as well as the heat dissipation problem has been proposed by R. Eden. He suggests using a material of high thermal conductivity (1 mm thick CVD-diamond plates, $\kappa = 20 \, \text{W/cmK}$) to conduct the heat produced inside the configuration to the water-cooled edges (Fig. 9.11). In this approach, direct connection of the dies through holes drilled in the diamond substrate would be possible, thus ensuring short signal latency.

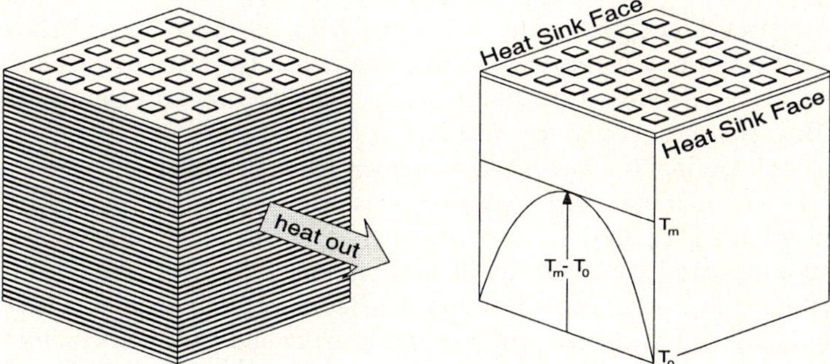

Fig. 9.11: CVD-diamond plates as an edge-cooled submount for multi-chip modules (MCM), proposed for 3-D supercomputers [9.63].

However, effective edge cooling is essential in this concept. To obtain this, Eden suggests making an extension of the diamond substrate area part of the heat exchanger itself. A 5 mm extension on two sides would be enough to create microchannels directly inside the diamond substrate by laser scribing. If such a 3-D MCM was cooled on two sides, leaving enough room for power supply and data exchange on the other sides, the temperature rise would be $T_m - T_0 = 31.25$ K for 20 kW heat production. If aluminum nitride would be used instead, the temperature rise would be as high as 367 K [9.63].

Although diamond is clearly the material of choice from the viewpoint of performance, cheap, large-area diamond substrates are needed. Economic studies have shown [9.64] that the cost of a metallized 1 mm thick 6 inch CVD-diamond wafer must be reduced to less than $10/cm^2.

9.5.3 Cooling of Laser Diodes and Laser Diode Arrays with CVD-Diamond Heat Spreaders

In the beginning when the efficiency of laser diodes was much lower, cooling with liquid nitrogen was required, and the first laser diodes, which operated at room temperature, used a diamond heat spreader [9.65].

Today, laser diode arrays have been demonstrated with continuous-wave output power levels of up to 100 W at room temperature [9.66]. These laser diode arrays were mounted on a CVD-diamond heat spreader, which reduced the thermal resistance by 50%. In addition to the reduction of the operational temperature, the temperature gradient over the laser diode can be reduced drastically by using a diamond heat spreader (Fig. 9.12). As a first commercial application, AT&T have used CVD-diamond heat spreaders to cool high-power semiconductor laser diodes, used in under-sea optical cables [9.67].

While large amounts of diamond are needed for integrated circuit packaging and MCM substrates, only a little material is required for the effective cooling of laser diodes and diode arrays, due to their small size. The rule of thumb is that the diamond substrate should be about five times the chip size, and its thickness about one-third of its linear extent [9.68]. Typical heat spreader sizes are $1 \times 1 \times 0.3$ mm^3 for laser diodes and $3 \times 10 \times 0.5$ mm^3 for laser diode arrays. Thus cutting a 2 inch CVD-diamond wafer yields 2000 heat sinks for laser diodes and 60 for laser diode arrays. It has to be stressed that the diamond substrate only spreads the heat flux; in other words, the heat has to be removed either by means of a large copper block (acting as a heat sink) or a microchannel cooler.

The advantage of diamond heat spreaders is illustrated by numerical simulations [9.69]. Figure 9.12 shows the results of two-dimensional simulations. The configurations considered are a laser diode array mounted directly on a silicon microchannel cooler (Fig. 9.12a) and the same configuration, but with a CVD-diamond heat spreader positioned between laser diode array and the microchannel cooler (Fig. 9.12b). For the calculation, a heat flux of 1000 W/cm^2 for the laser diode and a thermal conductivity of 20 W/cmK for the diamond layer were assumed.

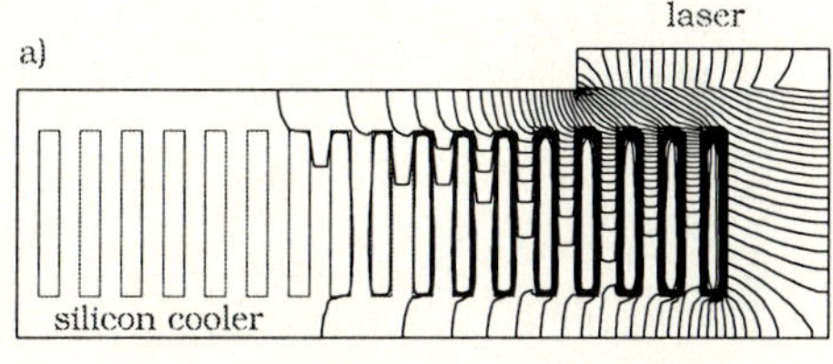

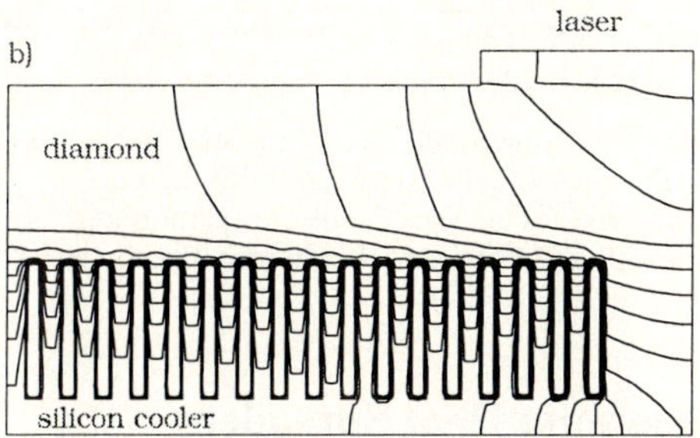

Fig. 9.12: The thermal simulation of a laser diode array (cross-section $100 \times 600 \ \mu m^2$) mounted on a silicon micro channel cooler (**a**) without a diamond heat spreader and (**b**) with a CVD-diamond heat spreader ($\kappa = 20$ W/cmK, cross-section 400 μm $\times$ 20 mm). Isothermal contours with $\Delta T = 1$ K are shown [9.69].

In Fig. 9.12, the temperature distribution is shown with isothermal increments of 1K. Without a diamond heat spreader the maximum temperature rise in the laser diode array (relative to the cooling water) amounts to $\Delta T = 45$ K. By adding a diamond heat spreader the heat flux is distributed over many more microchannels resulting in a significantly reduced temperature rise of $\Delta T = 18$ K. Even more important, the diamond heat spreader strongly influences the temperature gradient across the active zone in the laser diode. It is reduced from 17 K to below 3 K. Taking into account, that elevated operation temperatures strongly affect the lifetime and efficiency of semiconductor lasers and that there is a wavelength shift of several angstroms per Kelvin, the importance of diamond heat spreaders becomes obvious.

For the above simulation, width, height and thermal conductivity of the heat spreader were varied. In Fig. 9.13a the maximum temperature rise is shown as a function of thermal conductivity and height (width 20mm), in Figure 9.13b as a function of width and height ($\kappa = 20$ W/cmK).

Figure 9.13a shows that for any given thermal conductivity there is a specific optimal thickness. Increasing the heat spreader height beyond this value increases the thermal resistance. For CVD diamond ($\kappa = 20$ W/cmK), the optimal thickness for the configuration simulated is about 700 μm.

Another interesting question is how much diamond is actually needed for a certain temperature reduction. In Fig. 9.13b, the width and height of the diamond heat spreader are varied with fixed thermal conductivity ($\kappa = 20$ W/cmK), showing that the same temperature reduction can be obtained for various heat spreader configurations. This is an important point, since it is often cheaper to grow large area CVD-diamond films with moderate thickness than to grow thick films.

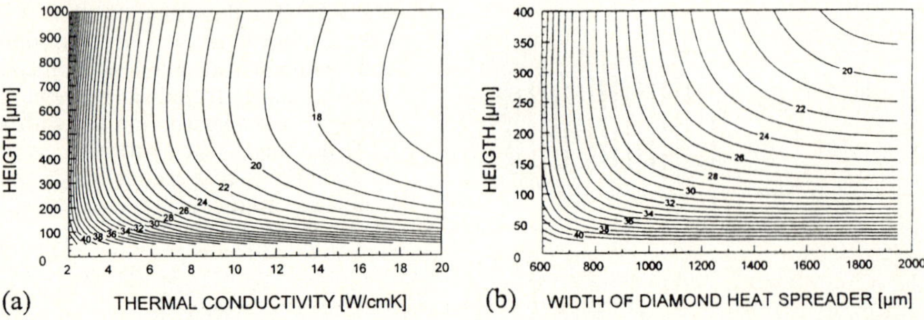

Fig. 9.13: (a) The maximum temperature rise of a laser diode array mounted on a silicon microchannel cooler as a function of the thermal conductivity and the height of the diamond plate. (b) The maximum temperature rise of a laser diode array mounted on a silicon microchannel cooler as a function of width and height of the diamond heat spreader ($\kappa = 20$ W/cmK).

9.5.4 The Production of CVD-Diamond Heat Spreaders

To produce heat spreaders, a CVD-diamond wafer has to be grown, polished, cut, and finally metallized. In recent years, a large variety of techniques for growing CVD-diamond films has been developed. For heat spreading applications, however, large-area high-quality material with a thermal conductivity of at least 12 W/cmK is needed. So far, only a few methods such as hot-filament thermal CVD, microwave-plasma assisted CVD and high-power DC-arcjet plasma CVD [9.70] have proved to be capable of growing large-area CVD diamond with a high thermal conductivity [9.6, 17, 70].

As-grown state-of-the-art CVD diamond has a rough polycrystalline surface, which has to be polished for most thermal and optical applications. Various techniques, ranging from thermochemical to laser polishing, have been developed [9.71]. A comparison of most important techniques is given by Bhushan et al. [9.72]. After polishing, the CVD-diamond plates have to be cut into the desired shape. In contrast to gemstones, which can be cut mechanically along certain well-defined relatively soft crystal orientations, CVD diamond is generally cut by using laser radiation [9.71, 73]. After polishing and cutting, the samples are metallized. A metallization layer is necessary for mounting electronic devices on to the heat spreader. As an adhesion layer, titanium is generally used, because of its carbide-forming property. To avoid alloy formation, a diffusion barrier layer (typically platinum) is used as a second layer, followed and capped by solder. In the past, durability problems have often arisen due to electromigration when soft solders such as eutectic PbSn or Indium were used [9.74]. These problems have been overcome by using hard solder, such as gold–tin eutectic alloy. Nowadays, diamond heat sinks are generally metallized with a multi-layer of Ti–Pt–Au, and captured with some microns of AuSn solder [9.75].

9.6 Summary

Diamond is the material with the highest room-temperature thermal conductivity, exceeding that of copper by a factor of five. Man-made large-area CVD-diamond films have been shown to reach thermal conductivities nearly as high as those of the purest type IIa natural diamonds. Therefore, these films are interesting for a large variety of applications in thermal management; some of which such as CVD-diamond heat spreaders for high-power laser diodes, are already well established. Other applications that take advantage of the extreme thermal conductivity are being pursued and will be realized as the cost of CVD diamond falls.

References

9.1   J.R. Olson, R.O. Pohl, J.W. Vandersande, A. Zoltan, T.R. Anthony, and W.F. Banholzer, Phys. Rev. B **47**, 14 850–6 (1993)
9.2   M. Seal, Phil. Trans. R. Soc. Lond. A **342**, 313–22 (1993)
9.3   R. Berman, in The Properties of Diamond, ed. J.E. Fields, Academic Press, London (1979)
9.4   D.G. Onn, A. Witek, Y.Z. Qiu, T.R. Anthony, and W.F. Banholzer, Phys. Rev. Lett. **68**, 2806–9 (1992)
9.5   A. Ono, T. Baba, H. Funamoto, and A. Nishikawa, Jpn. J. Appl. Phys. **25**, 808–10 (1986)
9.6   E. Wörner, C. Wild, W. Müller-Sebert, R. Locher, and P. Koidl, Diamond Rel. Mater. **5**, 688–92 (1996)
9.7   J.E. Graebner, S. Jin, G.W. Kammlott, J.A. Herb, and C.F. Gardinier, Appl. Phys. Lett. **60**, 1576–8 (1992)
9.8   G. Lu, in Applications of Diamond Films and Related Materials: Second International Conference, 1993, ed. M. Yoshikawa et al., MYU, Tokyo, pp. 269–74
9.9   D.T. Morelli, Chem. Phys. Carbon **24**, 45–107 (1994)
9.10 J. Callaway, Phys. Rev. **113**, 1046–51 (1959)
9.11 P. Carruthers, Rev. Mod. Phys. **33**, 92–138 (1961)
9.12 P.G. Klemens, in Solid State Physics, ed. F. Seits, and D. Turnbull, Academic Press, New York (1958)
9.13 J.E. Fields, in The Properties of Diamond, ed. J.E. Fields, Academic Press, London (1979)

9.14 J.W. Vandersande, Phys. Rev. B **13**, 4560–7 (1976)
9.15 H.B.G. Casimir, Physica **5**, 495 (1938)
9.16 D.T. Morelli, C. Uher, and C.J. Robinson, Appl. Phys. Lett. **62**, 1085–7 (1993)
9.17 J.E. Graebner, Diamond Films Technol. **3**, 77–130 (1993)
9.18 D.T. Morelli, T.M. Hartnett, and C.J. Robinson, Appl. Phys. Lett. **59**, 2112–4 (1991)
9.19 D.T. Morelli, T.A. Perry, and J.W. Farmer, Phys. Rev. B **47**, 131–9 (1993)
9.20 J.W. Vandersande, Phys. Rev. B **15**, 2355–62 (1977)
9.21 T.R. Anthony, W.F. Banholzer, J.F. Fleischer, L. Wei, P.K. Kuo, R.L. Thomas, and R.W. Pryor, Phys. Rev. B **42**, 1104–11 (1990)
9.22 T.R. Anthony, and W.F. Banholzer, Diamond Rel. Mater. **1**, 717–26 (1992)
9.23 R. Berman, in The Properties of Natural and Synthetic Diamond, ed. J.E. Fields, Academic Press, New York (1992)
9.24 R. Berman, Phys. Rev. B **45**, 5726–8 (1992)

9.25 T.R. Anthony, J.L. Fleischer, J.R. Olson, and D.G. Cahill, J. Appl. Phys. **69**, 8122–5 (1991)

9.26 J.E. Graebner, T.M. Hartnett, and R.P. Miller, Appl. Phys. Lett. **64**, 2549–51 (1994)

9.27 Ch. Wild, N. Herres, and P. Koidl, J. Appl. Phys. **68**, 973–8 (1990)

9.28 J.E. Graebner, J.A. Mucha, and F.A. Baiocchi, Diamond Rel. Mater. **5**, 682–7 (1996)

9.29 K.M. McNamara Rutledge, B.E. Scruggs, and K.K. Gleason, Appl. Phys. **77**, 1459–62 (1995)

9.30 J.E. Graebner, S. Jin, G.W. Kammlott, J.A. Herb, and C.F. Gardinier, Nature **359**, 401–3 (1992)

9.31 R. Berman, Thermal Conduction in Solids, Oxford University Press, Oxford (1976)

9.32 J.E. Graebner and J.A. Herb, Diamond Films Technol. **1**, 155–63 (1992)

9.33 J.E. Graebner, J.A. Mucha, L. Seibles, and G.W. Kammlott, J. Appl. Phys. **71**, 3143–6 (1992)

9.34 J.E. Graebner, to be published

9.35 E. Jansen and E. Obermeier, Phys. Status Solidi A **154**, 395–402 (1996)

9.36 M.B. Salamon, P.R. Garnier, B. Golding, and E. Buehler, J. Phys. Chem. Solids **35**, 851–9 (1974)

9.37 I. Hatta, Y. Sasuga, R. Kato, and A. Maesono, Rev. Sci. Instrum. **56**, 1643–7 (1985)

9.38 H.-B. Chae, Y.-J. Han, D.-J. Seong, J.-C. Kim, and Y.-J. Baik, J. Appl. Phys. **78**, 6849–51 (1995)

9.39 A. Maesono and R.P. Tye, Mat. Res. Soc. Symp. Proc. **416**, 91–9 (1996)

9.40 P.G. Kosky, Rev. Sci. Instrum. **64**, 1071–5 (1993)

9.41 D.G. Cahill and R.O. Pohl, Phys. Rev. B **35**, 4067–73 (1987)

9.42 D.G. Cahill, E. Fischer, T. Klitsner, E.T. Swartz, and R.O. Pohl, J. Vac. Sci. Technol. **A 7**, 1259–66 (1989)

9.43 K. Plamann and D. Fournier, Phys. Status Solidi A **154**, 351–69 (1996)

9.44 P.K. Kuo, L. Wei, R.L. Thomas, and R.W. Pryor, MRS Symp. Proc. **EA-19**, 119–22 (1989)

9.45 S. Albin, W.P. Winfree, and B. Scott Crews, J. Electrochem. Soc. **137**, 1973–6 (1990)

9.46 E.P. Visser, E.H. Versteegen, and W.J.P. van Enckevort, J. Appl. Phys. **71**, 3238–48 (1992)

9.47 A. Feldman, H.P.R. Frederikse, and S.J. Norton, Diamond Optics 3, 304–14 (1990)

9.48 H.P.R. Frederikse and X.T. Ying, Appl. Opt. **27**, 4672–5 (1988)

9.49 H.P.R. Frederikse, R.J. Fields, and A. Feldman, J. Appl. Phys. **72**, 2879–82 (1992)

9.50 G. Lu and W.T. Swann, Appl. Phys. Lett. **59**, 1556–8 (1991)

9.51 F. Enguehard, D. Boscher, A. Déom, and D. Balageas, Mater. Sci. Eng. **B 5**, 127–34 (1990)

9.52 P. Cielo, L.A. Utracki, and M. Lamontagne, Can. J. Phys. **64**, 1172–7 (1986)

9.53 E. Wörner, to be published

9.54 W.J. Parker, R.J. Jenkins, C.P. Butler, and G.L. Abbott, J. Appl. Phys. **32**, 1679–85 (1961)

9.55 J.E. Graebner, S. Jin, G.W. Kammlott, B. Bacon, L. Seibles, and W. Banholzer, J. Appl. Phys. **71**, 5353–6 (1992)

9.56 C.J.H. Wort, C.G. Sweeney, M.A. Cooper, G.A. Scarsbrook, and R.S. Sussmann, Diamond Rel. Mater. **3**, 1158–67 (1994)

9.57 O.W. Käding, E. Matthias, R. Zachai, H.-J. Füßer, and P. Münzinger, Diamond Rel. Mater. **2**, 1185–90 (1993)

9.58 O.W. Käding, M. Rösler, R. Zachai, H.-J. Füßer, and E. Matthias, Diamond Rel. Mater. **3**, 1178–82 (1994)

9.59 O.W. Käding, H. Skurk, A.A. Maznev, and E. Matthias, Appl. Phys. A **61**, 253–61 (1995)

9.60 J.E. Graebner, Rev. Sci. Instrum. **66**, 3903–6 (1995)

9.61 P.J. Boudreaux, in Applications of Diamond Films and Related Materials: Third International Conference, 1995, ed. A. Feldmann et al., NIST, Washington, pp. 603–10

9.62 A.P. Malshe, G.J. Glezen, G.J. Naseem, and W.D. Brown, Electrochem. Soc. Proc. **95**, 515–22 (1995)

9.63 R.C. Eden, Diamond Rel. Mater. **2**, 1051–8 (1993)

9.64 J.L. Davidson, W.D. Brown, and J.P. Dismukes, Proc. Electrochem. Soc **95-4**, 537–45 (1995)

9.65 I. Hayashi, M.B. Panish, P.W. Foy, and S. Sumski, Appl. Phys. Lett. **17**, 109–11 (1970)

9.66 M. Sakamoto, J.G. Endriz, and D.R. Scifres, Electron. Lett. **28**, 197–9 (1992)

9.67 C.T. Troy, Photonics Spectra **28**, 28–9 (1992)

9.68 M. Seal, Diamond Rel. Mater. **1**, 1075–81 (1992)

9.69 P. Koidl, C. Wild, E. Wörner, W. Müller-Sebert, M. Füner, and M. Jehle, Proceedings of the 1st International Diamond Symposium, 5–7 November 1996, Seoul, Korea (1996)

9.70 G. Lu, K.J. Gray, E.F. Borchelt, L.K. Bigelow, and J.E. Graebner, Diamond Rel. Mater. **2**, 1064–8 (1993)

9.71 A.P. Malshe, H.A. Naseem, W.D. Brown, and L.W. Schaper, in Applications of Diamond Films and Related Materials: Third International Conference, 1995, ed. A. Feldmann et al., NIST, Washington, pp. 611–8

9.72 B. Bhushan, V.V. Subramaniam, and B.K. Gupta, Diamond Films Technol. **4**, 71–97 (1994)

9.73 V.G. Ralchenko, S.M. Pimenov, T.V. Kononenko, K.G. Korotoushenko, A.A. Smolin, E.D. Obraztsova, and V.I. Konov, in Applications of Diamond Films and Related Materials: Third International Conference, 1995, ed. A. Feldmann et al., NIST, Washington, pp. 225–32

9.74 K. Mizuishi, J. Appl. Phys. **55**, 289–95 (1984)

9.75 A. Katz, C.H. Lee, and K.L. Tai, Mater. Chem. Phys. **37**, 303–28 (1994)

10. CVD Diamond for Optical Windows

Christoph Wild

Fraunhofer-Institut für Angewandte Festkörperphysik,
Tullastrasse 72, D-79108 Freiburg, Germany
e-mail: wild@iaf.fhg.de

Springer Series in Materials Processing
Low-Pressure Synthetic Diamond Eds.: B. Dischler and C. Wild
© Springer-Verlag Berlin Heidelberg 1998

10.1 Introduction

The development of high-power infrared (IR) lasers and the evolution of airborne IR-sensor systems have raised the demand for IR windows capable of withstanding high thermal and mechanical loads. In this context, diamond windows prepared by chemical vapor deposition (CVD) techniques have attracted considerable interest. The advent of CVD techniques for the manufacturing of large-area diamond windows has opened up the opportunity for a variety of new optical applications. These applications make use of the exceptional optical, thermal and mechanical properties of diamond, such as optical transparency from the UV to the far infrared, high thermal conductivity, resistance to thermal shock and high durability with respect to waterdrop impact and solid-particle abrasion. They benefit from the progress achieved on the synthesis of high-quality CVD-diamond wafers with properties approaching those of perfect type IIa natural diamond crystals.

The purpose of the present chapter is to point out the exciting potential of diamond optics and to assess the current status of CVD diamond as an IR window material.

10.2 CVD Diamond as an Infrared Material

The number of infrared materials, particularly for the 8–12 µm spectral range, is very limited. Furthermore, most of these materials suffer from certain physical limitations that make them unsuitable for window applications that involve severe environmental conditions, such as high power, high temperatures or aerodynamic load. Diamond, on the other hand, exhibits a combination of unusually favorable

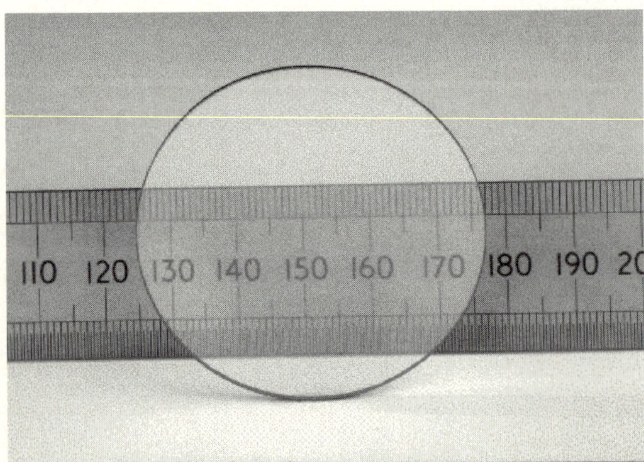

Fig. 10.1: A free-standing diamond window prepared by microwave-plasma CVD.

properties. Due to its large band gap (5.5 eV) and the lack of infrared active fundamental vibrational modes, diamond is optically transparent over a large wavelength range. Even at elevated temperatures, diamond remains transparent, since the large band gap does not allow the formation of free carriers. Amongst all solids, diamond possesses the largest thermal conductivity at room temperature and great mechanical strength. Until recently, there has been only little use of diamond in optics. This was simply due to the fact that natural diamonds or synthetic high-pressure diamonds could not provide the sizes and shapes required for optical components. This barrier was overcome by the development of CVD-diamond technology. Figure 10.1 shows a 2 inch diameter CVD diamond window prepared by microwave-plasma CVD.

10.2.1 Optical Properties of CVD Diamond

Infrared Absorption

Figure 10.2 depicts the transmission spectrum of a CVD-diamond film. The transmission range starts at 225 nm in accordance with the 5.5 eV indirect band gap. At longer wavelengths the only intrinsic absorption bands arise from two-phonon (1332–2664 cm^{-1}) and three-phonon (2665–3994 cm^{-1}) transitions. The maximum absorption coefficient in this spectral range amounts to 14 cm^{-1} at 2158 cm^{-1} [10.1]. Single-phonon transitions are located below 1332 cm^{-1}, the frequency of the triply degenerate zone-center mode. In principle, these transitions are forbidden by the cubic symmetry of a perfect crystal and the infrared selection rules. However, in imperfect crystals common defects such as impurities or

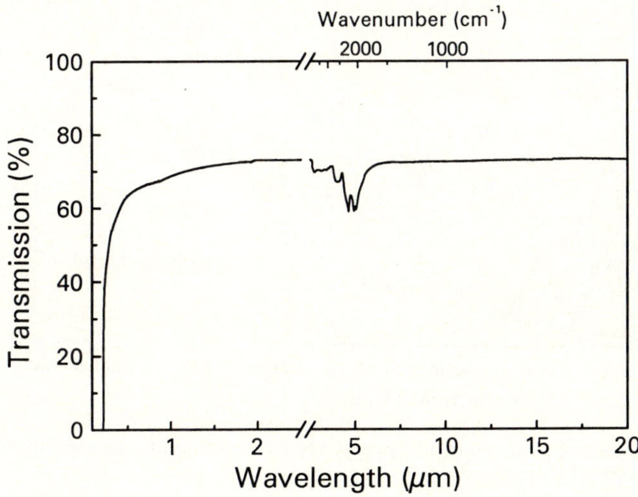

Fig. 10.2: The transmission spectrum of a high-quality CVD-diamond window ($d = 150$ μm).

192   C. Wild

dislocations can disturb the local symmetry, thus activating single-phonon absorption. Apart from the intrinsic absorption, the most frequently observed extrinsic absorption bands of CVD diamond are the CH-stretch vibrations between 2700 and 3000 cm^{-1}. The strength of these bands can be used to estimate the hydrogen content [10.2, 3]. In high-quality CVD-diamond films, these absorption bands are negligible.

The absorption around 10 μm is of particular interest, for CO_2-laser components and because many IR sensors operate within the 8–12 μm atmospheric window. A detailed analysis of diamond long-wave IR absorption has been performed by Thomas et al. [10.4]. They have developed an intrinsic infrared lattice vibration model based on an assembly of single oscillators with Morse interatomic potential. According to their conclusions, there is an intrinsic absorption floor in the 10 μm spectral range, which is caused by acoustic–acoustic two-phonon interactions. This two-phonon red wing can be approximated by [10.4, 6]:

$$\alpha \ (\mathrm{cm}^{-1}) = \frac{146\,\omega^5}{\omega_{max}^6} \quad (\text{for } \omega < \omega_{max}, \omega_{max} = 1143\,\mathrm{cm}^{-1})\ .$$

Figure 10.3 shows the absorption coefficient around 10 μm for two CVD-diamond samples and a type IIa natural diamond crystal. The smooth curve represents the theoretical absorption coefficient. Laser calorimetry data are shown as open symbols.

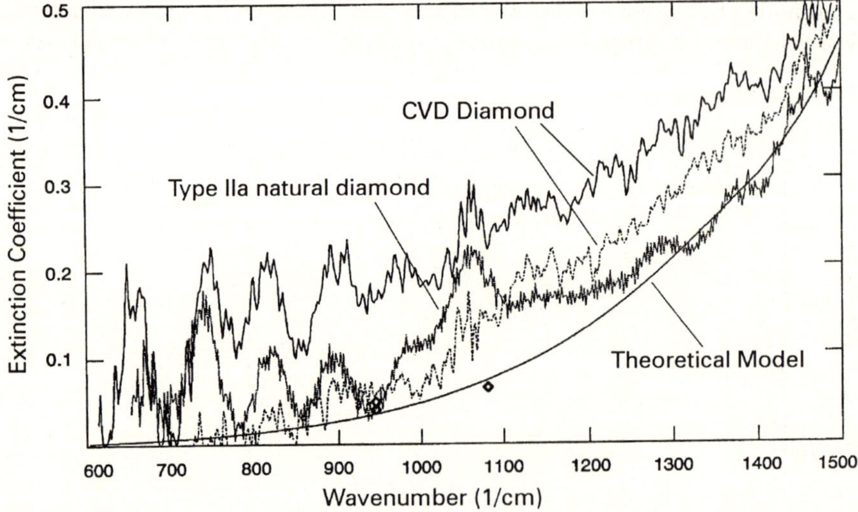

Fig. 10.3: Long-wave infrared transmission of natural type IIa diamond and two high-quality CVD-diamond samples [10.4, 5].

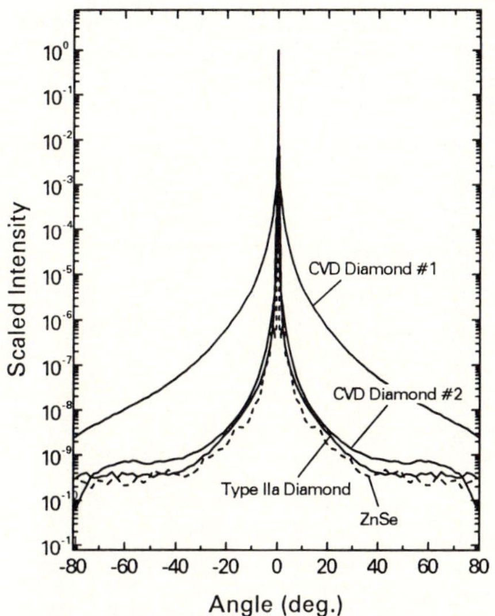

Fig. 10.4: Optical scattering of IR window materials at λ = 633 nm: BDTF function of two CVD diamond samples, a natural diamond crystal and a ZnSe window [10.15].

For high-power CO_2-laser applications, the absorption coefficient at 10.6 µm is of central importance. Typical values of type IIa diamond crystals are in the range 0.03–0.05 cm⁻¹, as determined by laser calorimetry [10.1, 7, 8, 9]. The lowest absorption coefficient reported for CVD diamond is comparable to these values [10.9]. In reference [10.10], measurements performed by different laboratories on three high-quality samples from various suppliers are summarized. The absorption coefficients at room temperature were found to be in the range 0.1–0.3 cm⁻¹ and to double upon heating to ~500°C.

Scattering Losses

The polycrystalline structure of CVD diamond and the concomitant grain boundaries raise the issue of optical scattering. To assess the optical quality of CVD diamond, the bidirectional transmission distribution function (BTDF) has been investigated by various groups [10.1, 10–14]. In Fig. 10.4, some of our own measurements are depicted [10.15]. The BTDF curves were measured with 633 nm He–Ne laser radiation. For transparent polished CVD-diamond windows, integrated scattering intensities (integrated from 2° to 80°) ranging from 0.79% to above 50% were observed. These findings indicate that optical scattering can be a major problem in CVD-diamond optics, and that the reduction of scatter requires a careful optimization of the grain structure.

Scattering in the infrared is usually an order of magnitude lower [10.12]. Values reported in the literature are 1% [10.1, 11] and 0.15% [10.13] (10.6 µm, CVD diamond, 2.5°–70°). No change in the optical scattering was found upon heating to 500°C [10.1, 10, 14]. Type IIa single crystals are roughly an order of magnitude lower in scatter than the best CVD-diamond samples [10.11, 14].

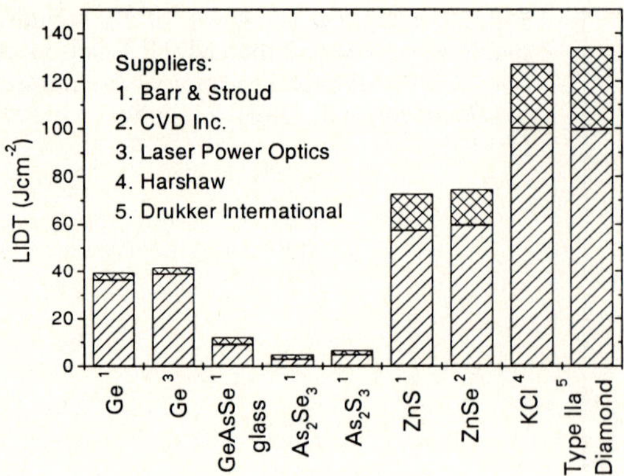

Fig. 10.5: The laser-induced damage threshold of various IR window materials at 10.6 μm [10.7].

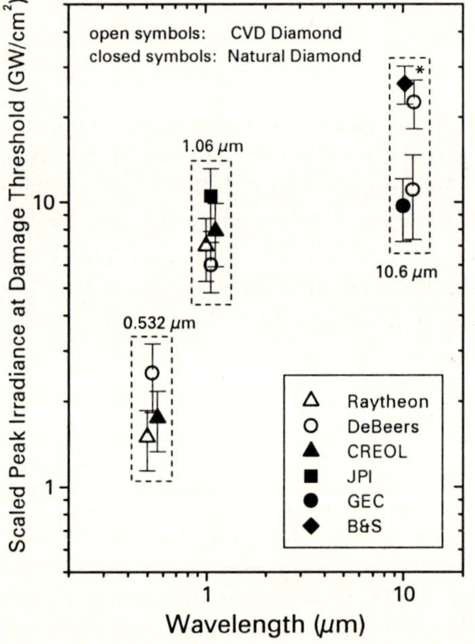

Fig. 10.6: The laser-induced damage thresholds of natural and CVD diamond: scaled peak irradiance at damage threshold vs. laser wavelength [10.9, 16]. According to Klein [10.16], the marked (*) data point overestimates the damage threshold by a factor of 2–3.

Laser-Induced Damage Thresholds

The ability of optical materials to withstand extremely high peak intensities is an important issue for high-power laser applications. In Fig. 10.5, the optical strengths as expressed by the laser-induced damage threshold (LIDT) of various CO_2-laser window materials are compared. Type IIa diamond exhibits one of the best LIDT values. LIDT measurements performed on diamond under various conditions have been reviewed by Klein [10.16] and Wood [10.17]. They pointed out that dielectric breakdown is the dominant damage mechanism for highly transparent diamond samples, and they demonstrated the applicability of the Bettis–House–Guenther (BHG) scaling law:

$$I_{p,th} = \frac{\text{const.}}{\text{spot size} \cdot \sqrt{\text{pulse width}}}$$

where $I_{p,th}$ refers to the peak irradiance at the damage threshold. After scaling to a spot size of 100 μm and a pulse duration of 1 ns, the data for natural and CVD diamond are in remarkable agreement (see Fig. 10.6) [10.16]. At peak intensities exceeding the LIDT, damage first occurs via a subsurface field-induced breakdown at the exit surface. This is a result of self-focusing and Fresnel reflections, leading to enhanced field strengths in the vicinity of the rear surface [10.16, 17].

10.2.2 Mechanical Properties

Mechanical Strength

The mechanical strength of window materials is of particular interest, since it determines the resistance to mechanical load and thermal stress. The strength of single-crystal diamond was found to be of the order of 2.8 GPa, which is 1/50 of the theoretical strength [10.18]. This discrepancy has been attributed to the presence of large bulk and surface flaws. Unfortunately, the measured strength of CVD-diamond wafers is still an order of magnitude lower than that of natural

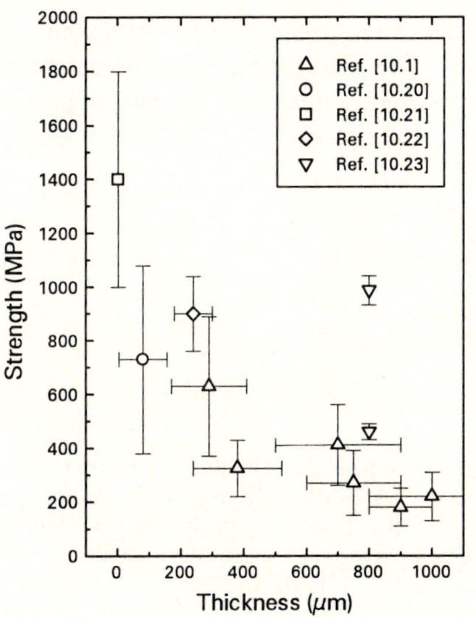

Fig. 10.7: The mechanical strength of CVD diamond vs. thickness.

diamond, reflecting the large amount of flaws, microcracks, and residual stresses [10.19] in this material. The results of various strength measurements are summarized in Fig. 10.7. A tendency for decreasing strength with increasing sample thickness is observed. This is because thin diamond membranes cannot have critical flaws as large as those of thicker disks [10.1]. In some cases, the strength values obtained with the nucleation side in tension were significantly higher than those obtained with the growth side in tension [10.1, 23]. The somewhat outstanding data point in Fig. 10.7 (987 MPa, 800 µm) was obtained using a three-point bending test, with the nucleation side in tension [10.23]. No decrease in the strength was found upon going to 1000°C [10.1, 10].

The mechanical reliability of brittle materials is usually expressed in terms of the Weibull modulus. This parameter reflects the degree of variability of the strength. The higher the Weibull modulus is, the higher is the probability of survival at stresses below the mean fracture strength. Typical values for ceramics are in the range 3–20. For CVD diamond, values between 6.5 [10.24] and 17.5 [10.23] have been reported.

Elastic Constants

In single-crystal diamond, the elastic constants are strongly anisotropic. For randomly oriented CVD diamond, averaging over all orientations predicts a modulus of 1140 GPa and a Poisson ratio of 0.069 [10.25]. These values are in good agreement with experimental findings [10.2, 25, 26, 27], indicating that the elastic properties of CVD diamond are comparable to those of single-crystal diamond.

Resistance to Rain and Sand Erosion

Resistance to rain and sand erosion is an important requirement for IR-transmitting windows and domes for airborne applications. The extreme hardness of diamond suggests that it should be the most durable IR-window material. Indeed, diamond has been found to exhibit the highest waterjet threshold velocity of any IR material

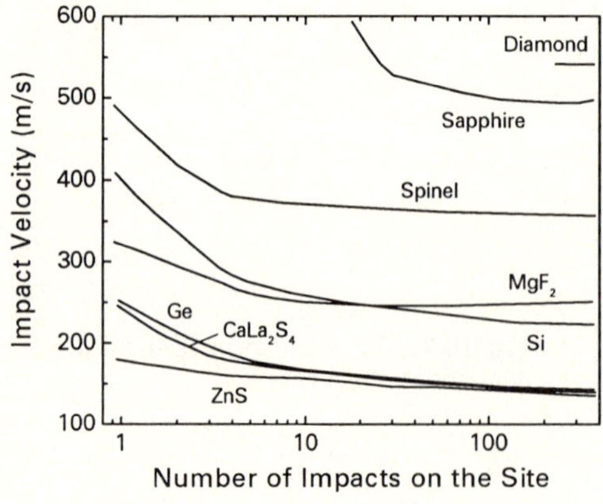

Fig. 10.8: Threshold velocity curves for various IR materials, indicating the boundary of the region within which circumferential cracking occurs [10.28].

(Fig. 10.8) [10.28]. The erosion resistance of CVD diamond has been investigated at the Cavendish Laboratory in Cambridge [10.28, 29]. Rain erosion was simulated with a high-velocity (80–600 m/s) waterjet from a 0.8 mm diameter nozzle. For natural diamond, the waterjet damage threshold velocity was found to be 515 ± 15 m/s [10.28, 29]. The primary damage mechanism is the formation of a ring crack at the edge of the loaded region, introduced by the Rayleigh surface wave. In the case of CVD diamond, a circumferential fracture was observed at slightly lower impact velocities (420 ± 60 m/s). The diameter of the ring crack was found to be 4.0 ± 0.2 mm, much larger than the diameter of the impacted zone. This is explained by bulk waves reflected at the back surface of the thin (0.5–1.0 mm) specimens. The reflected bulk waves reinforce the Rayleigh surface waves at larger radii.

Unfortunately, in addition to the circumferential fracture, many of the investigated CVD-diamond samples revealed the formation of fine central cracks at significantly lower velocities (200–350 m/s). This central damage mechanism was attributed to tensile stress that occured when release waves in the target propagating inwards from the jet edge meet on the axis [10.28].

In general, a large spread in the threshold velocities of various samples from different sources was observed, indicating that it may be possible to improve the erosion resistance through further optimization of the growth process [10.28].

Some sand erosion data was published by Jilbert et al. [10.29]. Tests were performed with 300–600 μm sand particles at velocities up to 250 m/s. While natural diamond is one of the most erosion resistant materials, CVD diamond is still considerably weaker.

10.2.3 Thermo-optical and Thermo-mechanical Properties

The performance of CVD-diamond IR windows at high temperatures or high optical power densities can be modelled provided that basic material parameters and the temperature dependence are known. In Fig. 10.9 the thermal conductivity, the thermal expansion coefficient, the thermal coefficient of the refractive index, and the variation of the optical thickness are plotted as functions of temperature. Figure 10.9a shows the thermal conductivity of a high-quality CVD-diamond sample [10.30]. The measurement technique is described in Chap. 9. Figure 10.9b shows the thermal expansion coefficient of high-quality and medium-quality CVD diamond [10.31]. The measurements were performed with a differential dilatometer using bar-shaped samples sized at $25 \times 5 \times 0.7$ mm^3. Prior to the measurements, the front faces of the samples were polished to obtain well-defined reference planes. The crosses in Fig. 10.9b refer to values recommended by Slack [10.32].

The variation of the optical thickness (Fig. 10.9c) was determined by laser interferometry at 633 nm, using a type IIa diamond crystal [10.31]. The data are in perfect agreement with that published by Patterson et al. [10.33]. Finally, in Fig. 10.9d the thermal coefficient of the refractive index of natural diamond is depicted. The low-temperature curve was taken from [10.34]. The two curves at

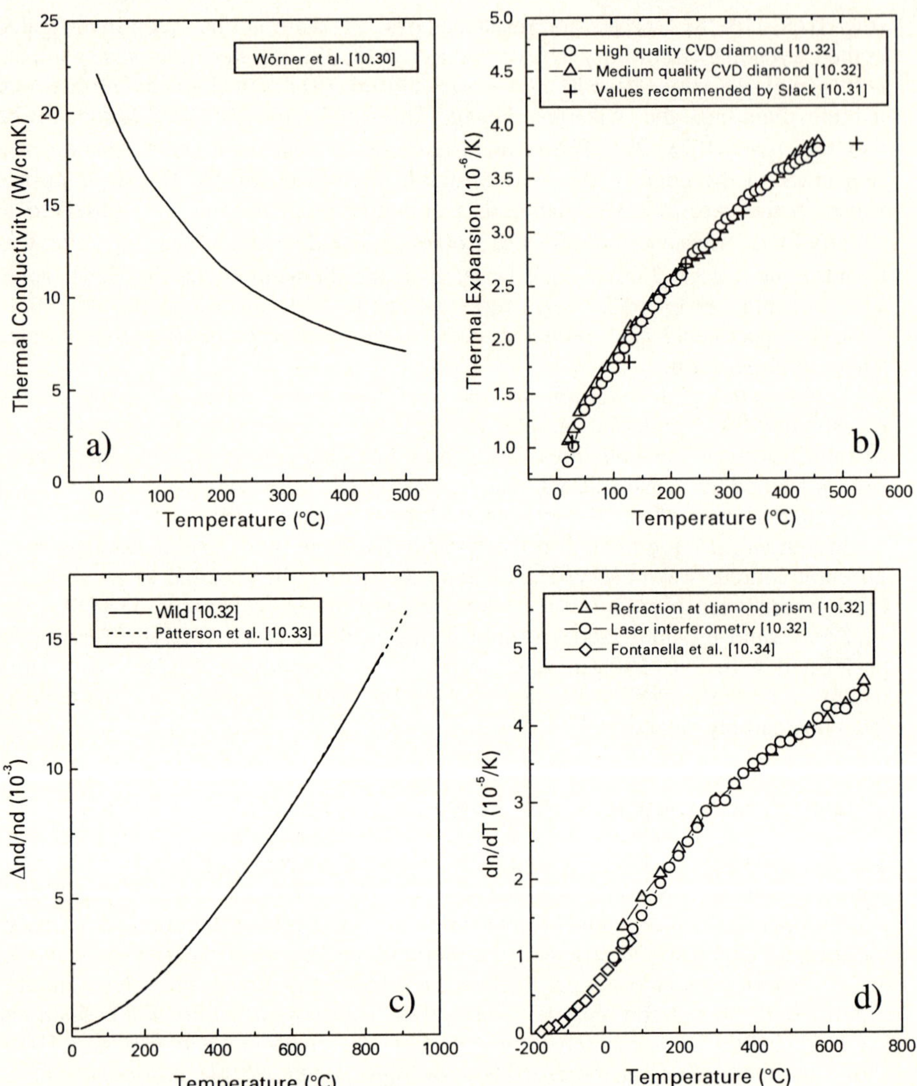

Fig. 10.9: The thermo-optical and thermo-mechanical properties of diamond.
(a) The thermal conductivity of a CVD diamond sample.
(b) The thermal expansion of two CVD diamond samples.
(c) The variation of the optical thickness *nd* for natural diamond.
(d) The thermal coefficient of the refractive index for natural diamond.

higher temperatures, determined by laser interferometry and by refraction through a diamond prism, are in excellent agreement. The data shown in Fig. 10.9 can be approximated by the following equations:

$$\kappa(T) = 5.62 + 16.295 \cdot \exp(-T/204.9) \qquad (T = 0, ..., 500°C)$$

$$\alpha(T) = 8.19E\text{-}7 + 1.107E\text{-}8 \cdot T - 1.48E\text{-}11 \cdot T^2 + 1.08E\text{-}14 \cdot T^3 \qquad (T = 20, ...,500°C)$$

$$dn/dT = 7.80E\text{-}6 + 9.80E\text{-}8 \cdot T - 9.84E\text{-}11 \cdot T^2 + 4.74E\text{-}14 \cdot T^3 \qquad (T = 20, ...,700°C)$$

$$(nd(T) - nd(25°C))/nd = -8.1E\text{-}5 + 3.02E\text{-}6 \cdot T + 2.91E\text{-}8 \cdot T^2$$
$$- 2.29E\text{-}11 \cdot T^3 + 9.52E\text{-}15 \cdot T^4 \qquad (T = 25, ...,800°C)$$

where the temperature T is in °C, the thermal conductivity κ is in W/cmK, the expansion coefficient α is in K^{-1}, dn/dT is in K^{-1} and $\Delta nd/nd$ is dimensionless.

10.3 The Performance of CVD-Diamond Windows

To date, the standard window material of high-power CO_2 lasers is ZnSe. The main advantage of ZnSe is its low absorption coefficient of 0.0005 cm^{-1} at 10.6 μm [10.35]. The power that can be transmitted through ZnSe windows is usually limited by absorption in the antireflection coatings. In comparison to ZnSe, diamond has a two orders of magnitude higher thermal conductivity and a much lower thermal expansion coefficient. This raises the possibility of dissipating the laser-beam induced heat to the window edges, where it can be removed by appropriate cooling techniques. The potential performance of edge-cooled CVD-diamond windows for high-power lasers was assessed by C.A. Klein [10.36]. Assuming liquid nitrogen cooling and an absorption coefficient of 10^{-4} cm^{-1}, he concluded that CVD diamond should be capable of handling power levels of the order of 750 kW.

To demonstrate the potential of CVD diamond as a high-power laser window material, an analysis based on the data presented in Sect. 10.2 will be given in the following subsections.

10.3.1 Heat Dissipation in Edge-Cooled Windows

For the following analysis, we assume a 2 inch diameter window through which a laser beam is transmitted. The laser beam is 30 mm in diameter with homogeneous intensity distribution ("top-hat distribution"). Part of the laser beam is absorbed in the window, resulting in a constant source of heat within the window. Neglecting thermal radiation and convective cooling, the temperature distribution of the window is given by

$$T(r) - T_0 = - \int_{r'=r_0}^{r} \frac{1}{\kappa(T) \cdot r'} \int_{r''=0}^{r} q(r'')dr''dr'$$

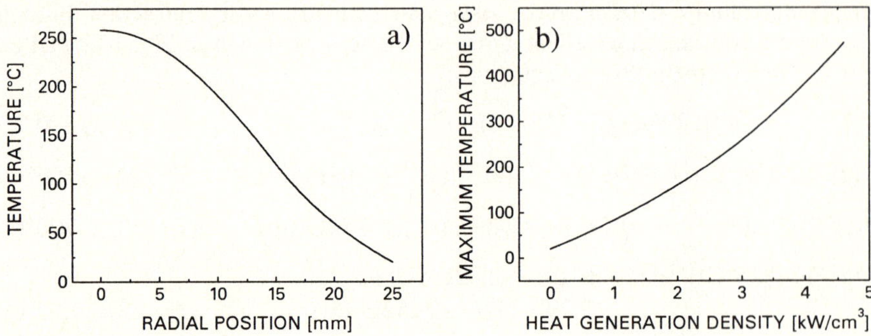

Fig. 10.10: The performance of edge-cooled CVD-diamond laser windows. **(a)** The radial temperature distribution for $3\,\text{kW/cm}^3$ heat generation density. **(b)** The maximum temperature versus heat generation density.

where T_o refers to the edge temperature, r_o to the window's radius, and $q(r)$ to the heat generation density. This integral can be solved numerically. The radial temperature distribution for a heat generation density of $3\,\text{kW/cm}^3$ is shown in Fig. 10.10a. Assuming an absorption coefficient of $0.1\,\text{cm}^{-1}$, this value corresponds to a laser-beam intensity of $210\,\text{kW}$. In Fig. 10.10b, the maximum temperature is plotted as a function of the heat generation density. Apparently, the power levels considered induce temperatures of the order of several hundred degrees centigrade and immense temperature gradients. This raises the issue of thermal lensing and thermal stress.

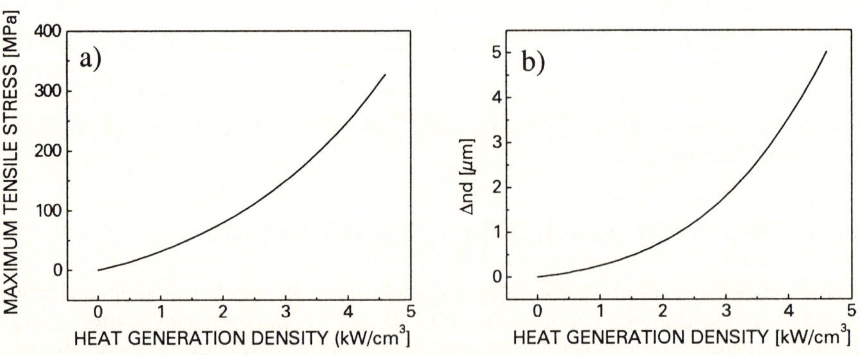

Fig. 10.11: The performance of edge cooled CVD diamond laser windows. **(a)** Maximum tensile stress. **(b)** Variation of the optical thickness nd (calculated for $d = 0.5$ mm) versus the heat generation density.

10.3.2 Thermal Lensing and Stress

On the basis of temperature distributions as shown in Fig. 10.10a, the thermal stress was calculated using thin-plate theory [10.37]. Figure 10.11a shows the maximum tensile stress within the diamond window as a function of the heat generation density. The values are not much lower than those for the mechanical strength (Fig. 10.7). Taking into account that a safety factor of four is usually applied, the thermal stress is certainly an issue that limits the power that can be handled by edge-cooled diamond windows. The optical path length difference across the window due to thermal gradients is shown in Fig. 10.11b. The resulting distortion of the transmitted wave front leads to defocusing and spherical aberration of the laser beam [10.36]. Usually, the maximum variation of the optical path length that can be tolerated is of the order of $\lambda/10$. At 10.6 µm, this value corresponds to a heat generation density of 2.4 kW/cm^3 (Fig. 10.11b).

10.3.3 Resistance to Thermal Shock

In contrast to high-power laser windows, where lateral temperature gradients dominate, thermal stress in missile domes is mainly due to thermal gradients through the thickness. The origin of these thermal gradients is the almost instantaneous rise in temperature after launch due to aerothermal heating. For most IR-window materials, brittle fracture induced by tensile thermal stress is the primary cause of failure [10.38]. The resistance to thermal shock is usually expressed in terms of the Hasselman figure of merit [10.39]:

$$R = \frac{\sigma_f (1-\nu)}{\alpha E}$$

where σ_f is the fracture strength, ν is the Poisson ratio, α is the thermal expansion coefficient, and E is the elastic modulus. After Klein [10.40], a more appropriate figure of merit for thermally thin missile domes is

$$R^* = \frac{\sigma_f^{5/3} (1-\nu)\kappa}{\alpha E} .$$

Here κ refers to the thermal conductivity. In thermally thin domes, the maximum temperature difference between the inner and the outer wall depends on the thickness of the dome and on its thermal conductivity. In turn, the minimum thickness is determined by the requirement that the dome must withstand mechanical stresses induced by aerodynamic pressure [10.40]. In Fig. 10.12, the figures of merit R and R^* are shown for various IR-transmitting materials. CVD diamond is one order of magnitude higher in R and three orders of magnitude higher in R^* than any other IR material.

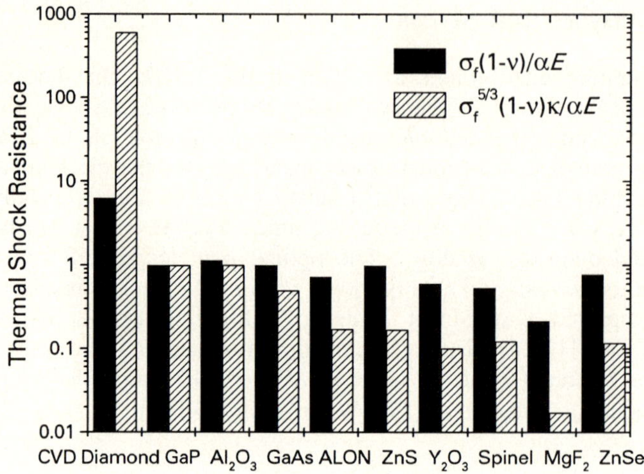

Fig. 10.12: Figures of merit for the thermal shock resistance of various IR-window materials.

10.3.4 Thickness Requirements

Due to the large mechanical strength, CVD-diamond windows can be made much thinner than those of other materials. The minimum thickness depends on the pressure difference Δp that the window or dome has to withstand. For flat windows, the minimum thickness as determined by pressure-induced fracture is given by

$$L_W = 0.554D\sqrt{\Delta p SF / \sigma_f}$$

where D is the window diameter, SF is the safety factor, and σ_f is the mechanical strength [10.36, 41]. Inserting $D = 50$ mm, $\Delta p = 1$ bar, $SF = 4$, and $\sigma_f = 500$ MPa, a minimum thickness L_W of 780 µm is obtained. Another aspect that must be considered in the case of laser windows is the pressure-induced bowing and the concomitant distortion of the laser resonator.

Missile domes must be able to withstand much higher pressure loads (up to 14 bar) [10.38]. After Klein [10.40], the minimum thickness is given by

$$L_W = 0.696R(\Delta p SF / \sigma_f)^{2/3}$$

where R is the dome radius. For $R = 35$ mm, $\Delta p = 10$ bar, $SF = 4$, and $\sigma_f = 500$ MPa, L_D amounts to 970 µm.

10.4 The Preparation of CVD-Diamond Windows

10.4.1 Diamond Deposition

The preparation of optical-quality CVD diamond has been demonstrated by various groups. In most cases, microwave-plasma CVD or DC-arcjet deposition has been applied [10.10]. Combustion flame synthesis has also been shown to be capable of producing optically transparent diamond [10.13], whereas the optical quality of hot-filament deposition diamond is inferior due to the incorporation of filament material [10.10]. Irrespective of the deposition techniques, optical-quality diamond requires deposition at low growth rates, typically below 5 µm/h [10.10, 11]. Silicon wafers or molybdenum disks are the most common substrates [10.13]. An important requirement for the economic manufacturing of CVD-diamond windows is homogeneous deposition on large areas. In this context, a microwave plasma deposition system developed at the author's institute [10.42] has found considerable interest. The system exhibits an ellipsoidal cavity which allows one to generate very intense, homogeneous and stable plasmas in close contact to the substrate.

10.4.2 Machining and Figuring

As-deposited CVD-diamond disks are usually very rough, non uniform in thickness, and slightly bowed. On the other hand, optical windows must fulfill strict specifications as far as flatness, parallelism, and surface roughness are concerned. Due to the extreme hardness, the figuring and polishing of CVD-diamond surfaces is a challenging task. Several techniques, including mechanical lapping with diamond abrasive, laser planarization [10.43], ion-beam or plasma etching [10.44, 45] and thermo-chemical polishing with hot iron disks, have been employed. It appears that mechanical lapping is the most useful and reliable technique for the polishing of large areas [10.46]. The surface roughness and flatness of CVD-diamond surfaces prepared at the Fraunhofer-Institut IAF are shown in Figs. 10.13 and 10.14. The rms roughness amounts to 20 Å and the deviation from an ideal plane of a 2 inch diameter diamond disk is below ±0.7 µm.

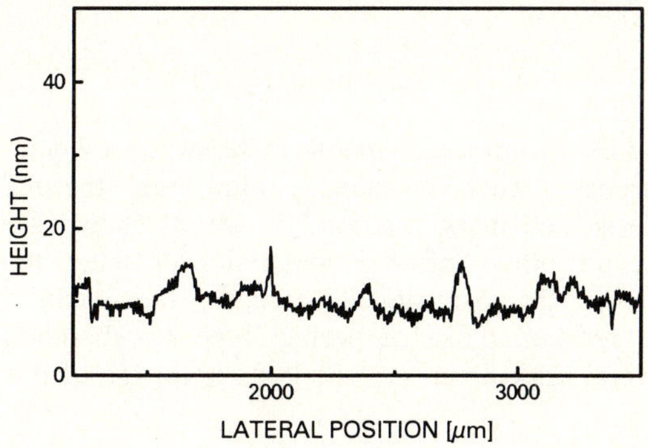

Fig. 10.13: The surface profile of a polished CVD-diamond window.

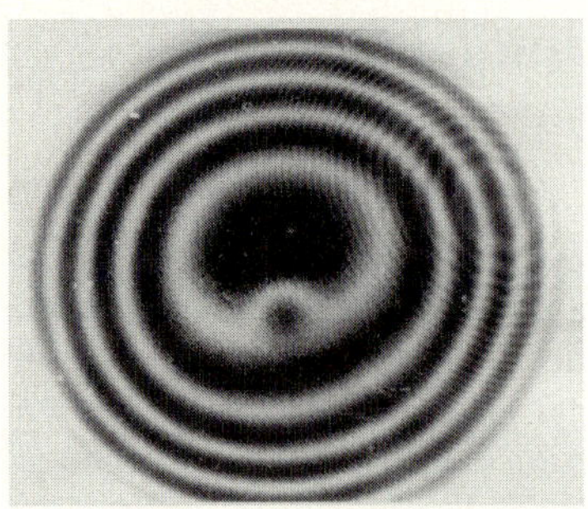

Fig. 10.14: The surface interference fringes of a CVD-diamond window (2 inch diameter, $\lambda = 633$ nm, still attached to the Si substrate).

10.4.3 Antireflection and Antioxidation Coatings

The transmittance of an uncoated diamond window amounts to 71.4% (at 10.6 µm). To eliminate reflection losses, the two surfaces must be coated with suitable antireflection (AR) coatings. In the case of single-layer coatings, the refractive index of the coating material should be close to 1.54 – the square root of the refractive index of diamond. Since diamond windows are supposed to operate under severe environmental conditions, AR coatings on diamond must fulfill specific demands, such as good adhesion, thermal and mechanical stability, and a well-matched thermal expansion coefficient.

Yttrium oxide (Y_2O_3, $n = 1.61$) films have been successfully applied as AR coatings on diamond [10.4, 47], resulting in transmittance values exceeding 90% at 10.6 µm. Another approach that provides optical impedance matching between diamond and air uses binary optical refractive surface structures. Those "moth-eye" structures have been reported by Harker and DeNatale [10.48].

Apart from antireflection coatings, diamond windows operating at high temperatures ($\geq 700°C$) need coatings that provide protection from oxidation. Klemm et al. [10.47] have demonstrated that diamond windows coated with Y_2O_3 can be heated to peak temperatures exceeding 1300°C without showing any oxidation damage. By comparison, for uncoated diamond windows a significant increase in light scattering due to surface etching was observed after 75 s at 800°C in air [10.49].

10.5 Summary

The superior properties of diamond as an infrared window material are evident. The broad-band optical transparency, wear resistance, ultra-high thermal conductivity, and thermal shock resistance make diamond a natural choice for many optical applications, where no other infrared material can meet the requirements. With the exception of the mechanical strength, the intrinsic properties of CVD diamond have reached those of perfect type IIa diamond crystals. However, there remain several impediments to the large-scale use of this material in optics. A major impediment is the machining and figuring of CVD-diamond windows, especially the control of the optical figure (the exact required geometry) of the workpiece and the machining of curved surfaces (domes). Affordability is another issue that must be considered. In any case, there are no fundamental barriers to hamper the optical application of diamond, and there is no doubt that diamond optics will become reality within the very near future.

References

10.1   D.C. Harris, SPIE **2286**, 218 (1994)

10.2   K.J. Gray, SPIE **1759**, 203 (1992)

10.3   J.M. Trombetta, J.T. Hoggins, P. Klocek, and T.A. McKenna, SPIE **1534** (1991)

10.4   M.E. Thomas, W.J. Tropf, and A. Szpak, Diamond Films and Techn. **5**, 159 (1995); M.E. Thomas, SPIE **2286**, 152 (1994)

10.5   M.E. Thomas and W.J. Tropf, SPIE **2286**, 144 (1994)

10.6   M.E. Thomas and W.J. Tropf, Johns Hopkins APL Technical Digest **14**, 16 (1993)

10.7   S.P. McGeoch, D.R. Gibson, and J.A. Savage, SPIE **1760**, 122 (1992)

10.8   K. Harris, G.L. Herrit, C.J. Johnson, S.P. Rummel, and D.J. Scatena, Appl. Opt. **30**, 5015 (1991)

10.9   M. Massart, P. Union, G.A. Scarsbrook, R.S. Sussmann, and P.F. Muys, SPIE **2714**, 177 (1996)

10.10 D.C. Harris, Naval Air Warfare Ctr. Weapons Division TP 8210 (1994)

10.11 D.C. Harris, in Applications of Diamond Films and Related Materials: Third International Conference, 1995, ed. A. Feldmann et al., NIST special publication **885**, 539 (1995)

10.12 C.F. Hickey, J. DeRosa, and K.A. Snail, SPIE **1760**, 154 (1992)

10.13 K.V. Ravi, SPIE **2286**, 174 (1994)

10.14 M.B. McIntosh, J.R. McNeely, R.E. Clausing, and W.B. Snyder, SPIE **1995**, 246 (1993)

10.15 R. Locher, to be published

10.16 C.A. Klein, SPIE **2428**, 517 (1995)

10.17 R.M. Wood, SPIE **2428**, 594 (1995)

10.18 J. Field (ed.), The Properties of Natural and Synthetic Diamond, Academic Press, London (1992)

10.19 A.B. Harker, D.G. Howitt, S.Chen, J.F. Flintoff, and M.R. James, SPIE **2286**, 254 (1994)

10.20 G.F. Cardinale and C.J. Robinson, J. Mater. Res. **7**, 1432 (1992)

10.21 Y. Aikawa and K. Baba, Jpn. J. Appl. Phys. **32**, 4680 (1993)

10.22 T.J. Valentine, A.J. Whitehead, R.S. Sussmann, C.J.H. Wort, and G.A. Scarsbrook, Diamond Rel. Mater. **3**, 1168 (1994)

10.23 C.J.H. Wort, J.R. Brandon, B.S.C Dorn, J.A. Savage, R.S. Sussmann, and A.J. Whitehead, in Applications of Diamond Films and Related Materials: Third International Conference, 1995, ed. A. Feldmann et al., NIST special publication **885**, 569 (1995)

10.24 J.M. Trombetta, J.T. Hoggins, P. Klocek, T.A. McKenna, L.P. Hehn, and J.J. Mecholsky, SPIE **1760**, 166 (1992)

10.25 C.A. Klein and G.F. Cardinale, Diamond Rel. Mater. **2**, 918 (1993)

10.26 M.P. D'Evelyn, E.B. Stokes, P.J. Codella, and B.E. Williams, in Applications of Diamond Films and Related Materials: Third International Conference, 1995, ed. A. Feldmann et al., NIST special publication **885**, 547 (1995)

10.27 R.S. Sussmann, J.R. Brandon, G.A. Scarsbrook, C.G. Sweeney, T.J. Valentine, A.J. Whitehead, and C.J.H. Wort, Diamond Rel. Mater. **3**, 303 (1994)

10.28 C.R. Seward, C.S. Pickles, R. Marrah, and J.E. Field, SPIE **1760**, 280 (1992); C.R. Seward, E.J. Coad, C.S.J. Pickles, and J.E. Field, SPIE **2286**, 285 (1994)

10.29 G.H. Jilbert, C.S.J. Pickles, E.J. Coad, and J.E. Field, in Applications of Diamond Films and Related Materials: Third International Conference, 1995, ed. A. Feldmann et al., NIST special publication **885**, 561 (1995)

10.30 E. Wörner, C. Wild, W. Müller-Sebert, R. Locher, and P. Koidl, Diamond Rel. Mater. **5**, 688 (1996)

10.31 C. Wild, to be published

10.32 G.A. Slack and S.F. Bartram, J. Appl. Phys. **46**, 89 (1975)

10.33 M.J. Patterson, J.L. Margrave, R.H. Hauge, Z. Ball, and R. Sauerbrey, Electrochem. Soc. Proc. **95-4**, 503 (1995)

10.34 J. Fontanella, R.L. Johnston, J.H. Colwell, and C. Andeen, Appl. Opt. **16**, 2949 (1977)

10.35 P. Klocek (ed.), Handbook of Infrared Optical Materials, Marcel Dekker, New York (1991)

10.36 C.A. Klein, SPIE **1624**, 475 (1992)

10.37 B. Boley and J. Weiner, Theory of Thermal Stresses, John Wiley, New York (1960)

10.38 C.A. Klein, Electrochem. Soc. Proc. **93-17**, 407 (1993)

10.39 D. Hasselman, Ceram. Bull. **49**, 1033 (1970)

10.40 C.A. Klein, SPIE **1760**, 338 (1992)

10.41 M. Sparks and M. Cottis, J. Appl. Phys. **44**, 787 (1973)

10.42 M. Füner, C. Wild, and P. Koidl, Appl. Phys. Lett., in print

10.43 R.D. Schaeffer, L. Chen, and W. Ho, SPIE **2703**, 265 (1996)

10.44 S. Holly and N. Koumvakalis, SPIE **2114**, 127 (1994)

10.45 I.P. Llewellyn, R.C. Chittick, and R.A. Heinecke, SPIE **2286**, 198 (1994)

10.46 A. Feldman and D.C. Harris, SPIE **2428**, 580 (1995)

10.47 K.A. Klemm, H.S. Patterson, L.F. Johnson, and M.B. Moran, SPIE **2286**, 347 (1994)

10.48 A.B. Harker and J.F. DeNatale, SPIE **1760**, 261 (1992)

10.49 C.E. Johnson, J.M. Bennett, and M.P. Nadler, SPIE **2286**, 247 (1994)

11. CVD Diamond for X-ray Windows and Lithography Mask Membranes

Peter K. Bachmann and Detlef U. Wiechert

Philips Research Laboratories Aachen,
Weisshausstrasse 2, D-52066 Aachen, Germany
e-mail: bachmann@pfa.philips.de

Springer Series in Materials Processing
Low-Pressure Synthetic Diamond Eds.: B. Dischler and C. Wild
© Springer-Verlag Berlin Heidelberg 1998

11.1 Introduction

Diamond possesses a number of interesting properties that make it an ideal candidate for a variety of X-ray applications. The most important ones in this context are high X-ray and visible light transmissivity, high mechanical stiffness and dimensional stability, high thermal conductivity, a low thermal expansion coefficient, high radiation stability, and high corrosion resistance. This unique combination of properties, along with the advent of CVD techniques that allow growth of diamond in the form of thin films on large (> 50 mm diameter) substrates, has led to the exploitation of diamond films as X-ray detector [11.1–7] and X-ray tube window materials [11.8], as high-resolution X-ray lithography (XRL) mask membranes [11.9–19], and as deep lithography (DXRL) mask supports for LIGA (German acronym for Lithographie-Galvanik-Abformung) microfabrication technology (see, e.g., [11.12, 20, 21]). The purpose of this chapter is to summarize the present state of the art of CVD-diamond films in the above-mentioned application areas, to outline advantages and drawbacks of CVD diamond, and to compare it with standard technology and competing materials.

11.2 CVD-Diamond Windows for X-ray Detectors

In energy- or wavelength-dispersive X-ray fluorescence analysis, element-specific X-rays excited by an electron beam or by irradiation with high-energy X-rays are analyzed by a semiconducting detector, usually a nitrogen-cooled lithium-drifted silicon single crystal. Detector vacuum windows are used to avoid contamination of the cold crystal and to protect it from the environment when breaking the vacuum of the analytic instrument. Up until approximately 1989, 8–12 μm thick beryllium foils with apertures of 5–10 mm were commonly used as window material. Such thick foils are required for vacuum-tight operation because of the morphology and the corrosion behavior of Be as a window material.

Unfortunately, the intensity of soft X-ray fluorescence radiation emitted from light elements such as boron, carbon, nitrogen, oxygen, or fluorine is substantially reduced when passing through such thick windows. Figure 11.1 illustrates the decreasing X-ray transmissivity of an 8 μm thick Be foil for radiation generated from light elements [11.3]. Thinner Be windows are difficult to manufacture. Leak-tight operation was a problem and corrosion of the window material upon exposure to humidity, vacuum oil, and so on often led to window damage.

Windowless detector operation was and is an alternative in order to increase sensitivity. However, this technique requires complex mechanical turret systems to displace and replace the window for measurements and sample loading, and safety interlocks are needed to prevent unintentional venting of the system with the window removed. Alternative materials did either not exist prior to 1988 or were, in terms of performance (transmissivity, permeability, and stability), inferior to beryllium.

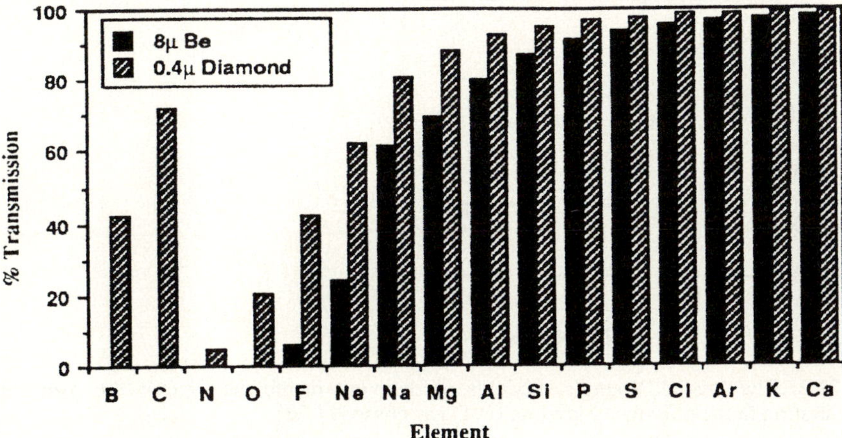

Fig. 11.1: The X-ray fluorescence radiation transmission data of an 8 μm beryllium and a 0.4 μm thick diamond window [11.5].

11.2.1 Properties and Requirements for X-ray Detector Windows

Thin diamond films are a logical choice as an alternative X-ray window material. However, because of the slightly higher atomic number, diamond windows need to be substantially thinner to match or surpass the X-ray transmissivity of Be. Fortunately, the manufacture of diamond membranes that are thin enough is feasible [11.1–7]. Figure 11.1 contains data for a 0.4 μm thick diamond window [11.5] and the increased detector signal, especially for the interesting elements boron, carbon, nitrogen, oxygen, and fluorine, is clearly visible. Such diamond films were initially found to be corrosion resistant and did not deteriorate upon exposure to humidity, oil, or corrosive vapors.

Diamond nucleation densities of more than 10^{10} nuclei/cm^3 and operation of the deposition system under clean-room conditions are required to reproducibly obtain pin-hole free membranes with the required thickness of only 0.3–0.4 μm. The commonly used pretreatment by polishing with diamond paste gives sufficiently high nucleation densities: however, particles, scratches and contamination may cause pin hole formation. Pre-deposition of a nucleation layer from a high carbon concentration CVD gas phase seems to be an applicable alternative, especially when low-pressure DC glow discharge plasma CVD is used as the deposition method [11.2].

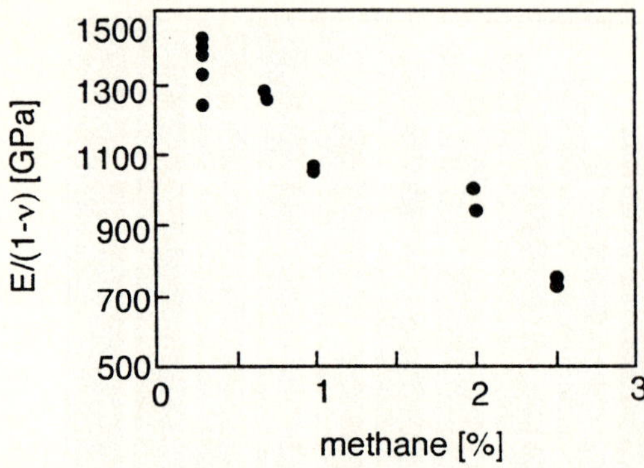

Fig. 11.2: The biaxial Young´s modulus of diamond membranes versus the methane concentration in the microwave-plasma CVD gas phase [11.22].

11.2.2 The Mechanical Stability of Thin Diamond Windows

The theoretical burst differential pressure of diamond membranes in a drum-like configuration (bulge test), as derived by Windischmann et al. [11.22], is given by

$$P_{burst} = 4.9\,((1-\nu)/E)^{1/2}\,\sigma_{UTS}^{3/2}\,t/a$$

where σ_{UTS} is the ultimate tensile strength (UTS) of the membrane, $E/(1-\nu)$ is the biaxial modulus of diamond, t is the membrane thickness, and a is the aperture radius. The biaxial modulus and UTS of diamond membranes depend strongly on the deposition conditions. Figure 11.2 depicts the dependence of the biaxial Young's modulus on the methane concentration in the CVD gas phase for 2–4 µm thick diamond membranes grown at 7–13 Torr at approximately 800°C from a 500 W methane/hydrogen-microwave plasma [11.22].

It is not surprising that the phase purity of the membrane material affects the mechanical properties of the films. Consequently, the ultimate tensile strength of diamond also depends on the C concentration in the CVD gas phase, as shown in Fig. 11.3 [11.22].

With an average value of 2.5 GPa obtained for > 100 diamond membranes and using a diamond biaxial modulus of approximately 1300 GPa, the burst pressure (in atmospheres) of an unsupported membrane can be simplified to [11.23]

$$P_{burst}\,(atm) = t\,(\mu m)\,/\,2a\,(cm).$$

Thus, the maximum differential pressure sustained by a 0.5 µm thick 0.5 cm diameter free-standing window is approximately 1 atm, and even good-quality membranes are not strong enough to tolerate loads of more than 1 atm for a window thickness of less than 0.5 µm at window diameters of 5 mm or more. Therefore, X-ray windows based on ultra-thin diamond films need to be supported by an (etched) substrate material (usually silicon) grid structure with an X-ray transparency of ≥70%.

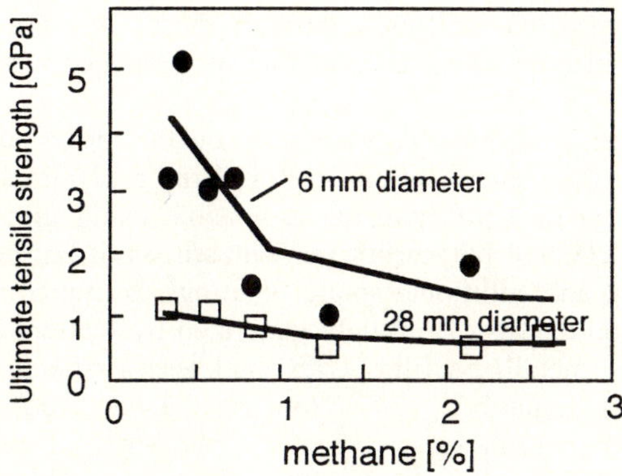

Fig. 11.3: The ultimate tensile strength (bulge test data) of diamond membranes of 6 mm and 28 mm diameter versus methane concentration in the microwave-plasma CVD gas phase [11.22].

The commercial manufacture of such windows was first announced by Crystallume Inc., CA, USA, in 1988 [11.1–3]. Figure 11.4 depicts a diamond window for X-ray fluorescence analysis along with the specifications given by the manufacturer.

11.2.3 Problems and Alternative Window Materials

Despite its interesting features and the impressive specifications given in Fig. 11.4, diamond X-ray detector windows never really penetrated the market. Initial tests at Tracor Northern, Inc., USA, were very successful: however, long-term

Specifications
film thickness:	$< 0.5\ \mu$m
aperture diameter:	6 mm
grid transmission:	70%
pressure rating:	1.25 atm
pressure cycling:	> 5000 (1 atm)
He leakage rate:	$< 10^{-9}$ Torr l/sec

Fig. 11.4: An energy-dispersive X-ray fluorescence (EDXRF) diamond window (Crystallume, Inc., USA [11.3]).

testing revealed that water vapor sometimes diffuses through these ultra-thin diamond films (most likely through pin holes) and reproducible performance was difficult to obtain [11.24].

Standard 8 µm Be windows (often additionally coated for better corrosion resistance and leak-tightness) are still the material of choice for X-ray analysis of *heavier* elements. Nowadays, detection of *light* elements is feasible using thin, dense polymer foils, such as polypropylene or polycarbonate, with silicon or boron nitride honeycomb support structures, and additional sputter coatings to improve leak-tightness and to protect the crystal detector from light generated by scattered and secondary electrons. Meanwhile, metallized films with thicknesses of 0.3–0.6 µm are available from a number of vendors (see, e.g., `http://www.kevex.com`, `http://www.philips.com/axr`, or `http://moxtek.com/windows/winselg.htm`). Ultra-thin boron nitride films (< 0.5 µm) on a boron nitride support structure turned out to be impermeable to water vapor, leak-tight, and, due to low X-ray absorption, well-suited for light-element analysis [11.24]. Thus, windows based on ultra-thin diamond films are unlikely to penetrate the analytic X-ray instrumentation market in the foreseeable future.

11.3 CVD-Diamond Windows for X-ray Tubes

Radiation generated in an X-ray tube exits the evacuated body of the tube through a vacuum window with an aperture of 15–20 mm diameter. This window needs to be highly X-ray transparent (Fig. 11.5) and, as for detector windows, beryllium is – despite its toxicity – still the material of choice. For tubes, however, support structures for the window material are not tolerable, because they reduce the X-ray

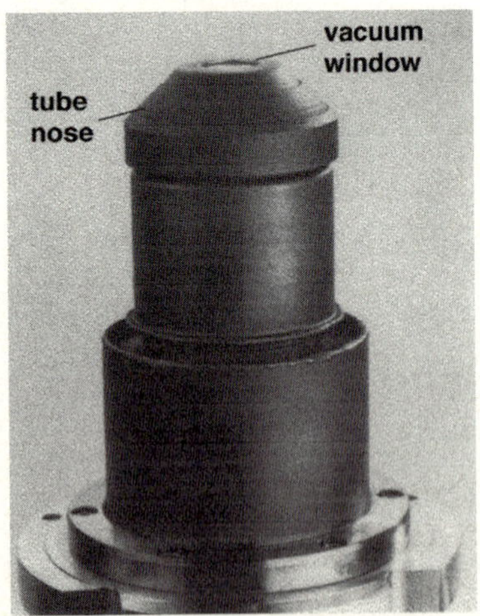

Fig. 11.5: The body, nose, and vacuum window of the PHILIPS high-intensity "Super Sharp End Window" X-ray tube.

intensity considerably and, in addition, cause unwanted secondary radiation. Such windows are usually unsupported and, therefore, need to be much thicker than for X-ray detectors. Beryllium windows of 75–125 µm thickness are common. Diamond windows as a nonpoisonous alternative to beryllium were investigated by Bachmann et al. [11.8].

11.3.1 Requirements for Diamond X-ray Tube Windows

In order to perform similarly to 75 µm thick Be, the diamond window thickness must not exceed 10 µm. The window aperture of modern tubes (Fig. 11.5) is approximately 20 mm in diameter and defines the size of the required free-standing diamond foil. Be windows are usually brazed to the tube housing. Such a metal connection that withstands elevated temperatures of several hundred degrees centigrade during tube bake-out, evacuation, and sealing of the tube is also required for the diamond alternative. The vacuum window needs to withstand differential pressures of more than 1 atm and X-ray absorption needs to be low. High radiation hardness is needed in order to ensure long-term tube performance.

11.3.2 Preparation and Properties
of Diamond X-ray Tube Windows

In order to fabricate free-standing diamond membranes, silicon substrates are usually polished with diamond paste to foster diamond nucleation. Subsequently, a diamond layer of the desired thickness is grown by any of the CVD techniques available (see Chaps. 2–7). Bachmann et al. used microwave-plasma CVD for their tube window investigations [11.8]. After film growth, part of the silicon substrate is removed by wet etching using, for example, HF/HNO_3. Then the diamond membrane is brazed directly to the tube housing or to a carrier ring that acts as a window frame.

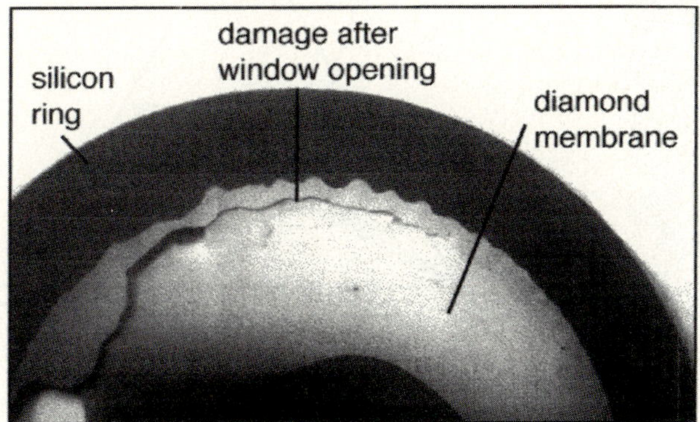

Fig. 11.6: A damaged 10 µm thick diamond membrane grown from acetone/oxygen mixtures. All membranes of similar thickness grown near the center of the C/H/O diagram suffer damage upon window opening due to mechanical weakness [11.8].

Because of the higher growth rates achievable in the center of the C/H/O diagram of diamond CVD [11.25, 26], deposition from C/H/O mixtures with fairly high oxygen levels is desirable. Unfortunately, such windows have turned out to be mechanically much less stable than windows of identical thickness grown from simple C/H mixtures. Despite high phase purity and good crystallinity, windows less than 20 μm thick, grown from C/H/O-mixtures with high oxygen levels (e.g., from acetone/oxygen, methane/oxygen, acetone/CO_2, or methane/CO_2) tend to break under the mechanical (stress) load transferred to the membrane upon removal of the substrate material. Figure 11.6 illustrates this surprising result. SEM studies reveal (Fig. 11.7) that the surface morphology of films grown from high oxygen concentrations in the gas phase is rougher and that re-entry corners occur more often than for films grown without or only little oxygen present in the CVD gas phase. TEM studies [11.27] indicate that the presence of oxygen

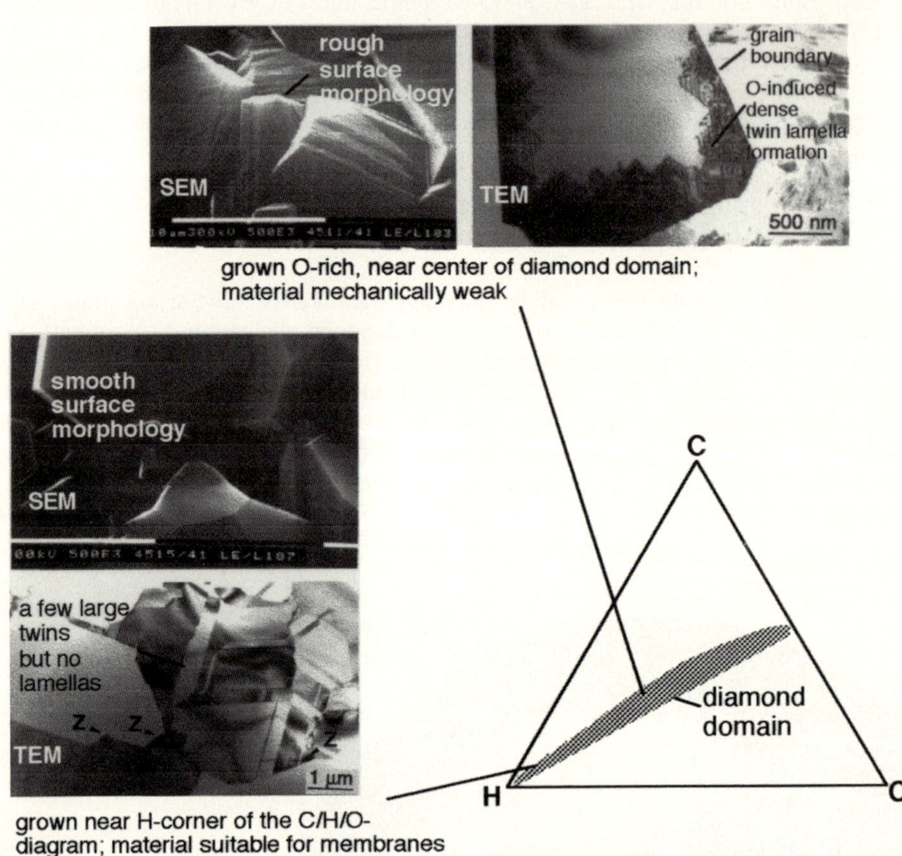

Fig. 11.7: The SEM surface morphology and TEM grain structure of diamond membranes grown at different positions in the C/H/O diagram of diamond CVD [11.8, 25, 26]. Oxygen-induced twinning [11.27] may be the reason for the mechanical weakness of diamond films grown near the center of the diagram.

Fig. 11.8: An 8 µm thick diamond X-ray tube, vacuum-mounted in a tube end cap. This C/H-grown window is leak-tight, has an aperture of 20 mm, and can withstand differential pressures of > 1 bar.

in the CVD gas phase leads to enhanced twin formation (Fig. 11.7). Oxygen-induced twinning and decoration of such twins with nondiamond carbon phases may explain the mechanical weakness as well as the limited thermal conductivity of such films [11.28, 29]. Bachmann et al. concluded that the growth of diamond membranes for X-ray tube windows (thickness < 10 µm; apertures approximately 1–2 cm) is only feasible if the amount of oxygen in the CVD gas phase is kept low.

Figure 11.8 depicts a C/H-grown diamond window mounted to the end cap of the type of X-ray tube shown in Fig. 11.5. Brazing at 600–900°C is feasible using Ti-activated silver/copper braze. The window shown in Fig. 11.8 has a thickness of only 8 µm and an unsupported aperture of 2 cm. The window is helium leak-tight ($> 10^{-9}$ Torr l /sec) and tube evacuation at a rate of 1 bar/sec is feasible.

However, despite excellent inherent mechanical properties, a 10 µm thick diamond membrane is inevitably more fragile than a 100 µm thick beryllium disk. Handling is more difficult and additional safety precautions are necessary to avoid damage. In addition, the fabrication of diamond membranes – that is, the use of a substrate, substrate pretreatment, diamond growth, masking, etching, and so on – is considerably more complicated, less reproducible, and, thus, more expensive than simply cutting a beryllium disk from a commercially available sheet of beryllium. Beryllium will, therefore, for the foreseeable future, remain the material of choice as an X-ray tube window material.

11.4 CVD-Diamond X-ray Lithography Membranes

Soft radiation X-ray lithography (XRL) is expected to be the replication technique that will allow mass fabrication of gigabit dynamic random access memory chips

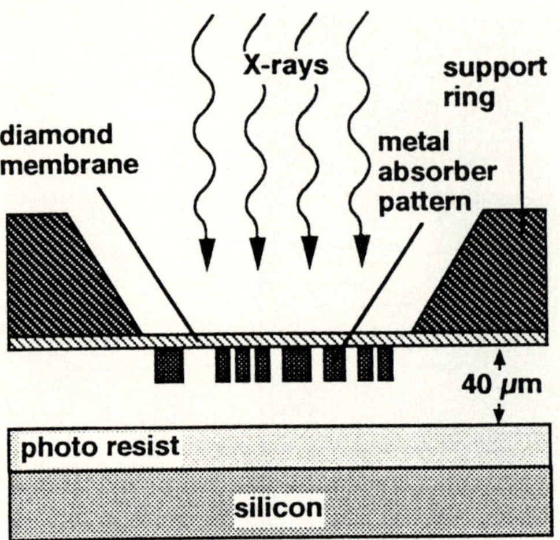

Fig. 11.9: A schematic of proximity X-ray lithography.

(DRAMs) with feature sizes of less than 100 nm. LIGA technology, based on deep XRL (DXRL) using hard X-rays, is expected to open up new roads in the mass production of micromechanical elements and microsystems.

The status of X-ray lithography, the various support membrane/absorber material combinations investigated by different groups (SiN/Ta [11.30], pSi/W [11.31], SiN/Au [11.32], SiC:H/W, or Au [11.33]), and the role of diamond as a possible mask support material are excellently reviewed by Ravet and Rousseaux [11.12]. Details on the still very much dominating deep UV optical high-resolution lithography can be found in [11.34]. LIGA is extensively covered in [11.20, 21]. Therefore, this section only briefly summarizes the present state-of-the-art of X-ray lithography.

11.4.1 Requirements for Diamond Lithography Membranes

In XRL and DXRL, absorbing metal (W, Ta, or Au) patterns are deposited on to a suitable, X-ray transparent, thin support material. The pattern is transferred to a photoresist by illumination with soft- (XRL) or hard (DXRL) (synchrotron) X-rays. Due to the lack of suitable X-ray optics, the absorbing features need to be of the same size as in the final product (Fig. 11.9).

During mask manufacture, the pattern is transferred to a photoresist on the membrane material by nanometer-resolution e-beam lithography. After resist development, the metal absorber pattern is electro-deposited directly onto the membrane material. Due to the lack of X-ray optics, the mask has to be placed very close to the surface of the photoresist layers in the subsequent replication steps (proximity lithography). This requires mechanically stable mask supports. Mask preparation and handling (positioning, re-positioning, irradiation, etc.) define the following requirements for XRL membranes [11.12]:

- high X-ray transparency to achieve high contrast

- high stiffness and moderate *tensile* stress to avoid mechanical pattern distortion

- high fracture strength to avoid damage during positioning (proximity XRL) and to support the thick absorber structures (hard X-rays) used in LIGA DXRL

- low surface roughness (only a few nm) for maximum resolution

- high optical transparency (> 50% at 632.8 nm) for optical mask alignment in commercial X-ray stepper systems

- high thermal conductivity (especially for high-flux DXRL) to avoid temperature-induced distortion

- high radiation stability to avoid damage and distortion.

Many materials (Si, SiN, SiC, BN, and amorphous carbon (aC)) are considered as XRL and DXRL membranes. On the basis of its unique combination of favorable properties, the use of CVD-diamond film as a mask support was first investigated in detail by Windischmann et al. [11.9, 22] and later by a number of groups in the USA, Japan, and Europe [11.5–7, 10–19, 35–41].

11.4.2 Preparation and Properties of Diamond Lithography Membranes

Similar to the procedure described for X-ray vacuum windows, lithography membranes are fabricated by first depositing a diamond layer on to a silicon substrate, usually by hot-filament (Chap. 5) or by microwave-plasma CVD (Chap. 2). Subsequently, an aperture is etched in the silicon substrate to form the free-standing membrane. Substrate pretreatment with diamond paste, to ensure high nucleation densities and growth conditions that foster high secondary nucleation and growth of fine-grain polycrystalline diamond layers, is chosen in order to maximize the optical transparency of the XRL diamond membranes at 630 nm, the operating wavelength of most commercial positioning systems in X-ray steppers. The thickness, grain size, surface roughness, and phase purity of the material need to be optimized to achieve a mechanically stable membrane with minimum thickness and maximum transparency. The deposition temperature and the C/H/O ratio in the CVD gas phase determine the stress state and stress level of the membrane material. Films that are under compressive stress are absolutely useless in X-ray lithography. High tensile stress levels may result in difficulties and damage during and after deposition of the metal absorber pattern. Ideal XRL membranes show a tensile stress of 50–200 Mpa. Further details about stress control of diamond membranes can be found in [11.37–40].

The membrane shown in Fig. 11.10 exhibits a tensile stress of 200 MPa and was grown from an 800 W microwave plasma at 12 Torr, using a mixture of acetone and hydrogen at approximately 850°C. For optimized, 1.0–1.5 μm thick

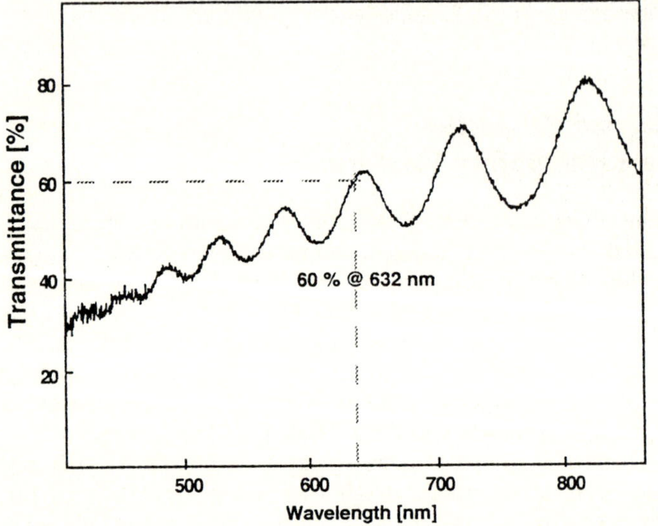

Fig. 11.10: The optical transmission spectrum (*bottom*) of a 1.3 μm thick diamond XRL membrane (*top*).

XRL membranes with apertures of 40–70 mm, transmission values of 50–65% are feasible [11.13, 35]. A corresponding spectrum is given in Fig. 11.10. Surface roughness values (rms) of 15–20 nm on the growth and 3–5 nm on the substrate side of the diamond membrane were measured by atomic force microscopy. The size of the diamond crystals is approximately 30 nm, and the films show a 1332 cm^{-1} diamond Raman signature peak widths of 10–15 cm^{-1} [11.13, 35].

Ravet et al. [11.13] were able to demonstrate that such membranes are smooth enough for patterning with a gold grating of 40 nm line width (Fig. 11.11, top). The mask alignment signal of gold marks on optimized as-grown membranes was found to be strong enough for optical positioning in a commercial stepper (Fig. 11.11, bottom). These authors were also able to demonstrate that diamond membranes are suitable for proximity replication with mask-to-wafer distances of ≤ 40 µm. They conclude that, with alignment accuracies of 30 nm, optimized diamond membranes perform in a comparable way to SiC membranes. Polishing of the films, however, might improve the transparency of diamond membranes even further, although it will make membrane manufacture more expensive [11.13].

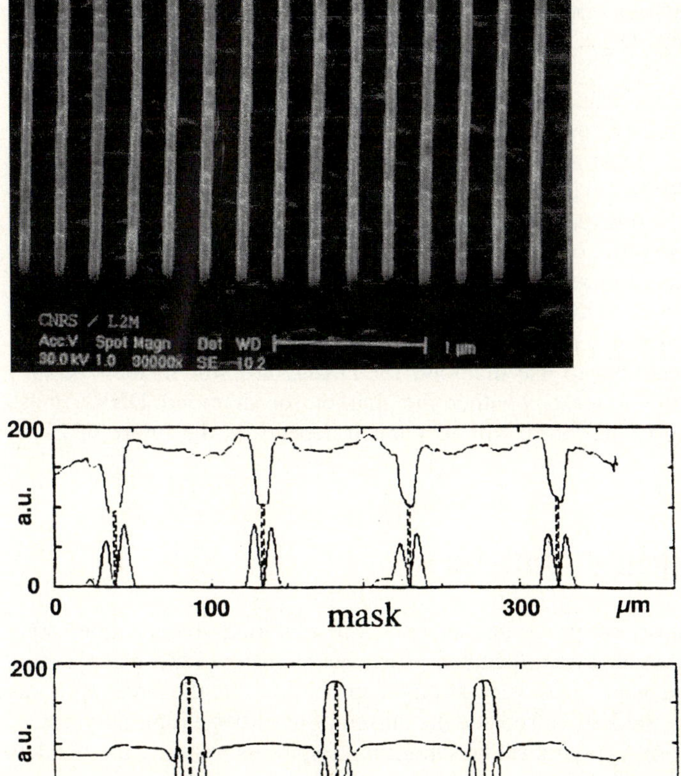

Fig. 11.11: A 40 nm line width gold grating grown on the rough growth side of an optimized CVD-diamond XRL membrane (*top*), and the optical alignment signal of a commercial X-ray stepper (*bottom*). The alignment accuracy is approximately 30 nm [11.13].

Polishing of the membrane might be important for a second reason. When investigating the radiation stability of diamond membranes for high X-ray dose levels, Tsuboi et al. [11.36] found that the Raman spectrum and optical transparency of well crystallized, fairly rough (50 nm) diamond films do not change upon irradiation with soft X-rays of up to 100 MJ/cm^3. Smoother diamond films with strong nano-crystalline Raman contributions around 1140 cm^{-1}, however, are altered by such high radiation levels. This may result in pattern distortion, and Tsuboi et al. conclude that membrane polishing is to be preferred over depositing films with higher levels of nano-crystalline compounds, in order to reduce optical scattering.

The field of XRL membrane technology is, no doubt, highly competitive. Market development is, of course, very closely connected with the questions of whether and when XRL will take over from deep-UV optical lithography. Whether diamond will proove to be cheap enough and to surpass other materials (SiC is the strongest contender) in terms of performance to win this interesting market is an open question.

For LIGA DXRL applications, thicker membranes are preferred for mechanical reasons. They have to carry much thicker (10–20 μm) metal absorber patterns that are needed for hard X-rays, and the dose levels are considerably higher than for XRL. At present, 200–300 μm thick beryllium disks or 20 μm thick silicon membranes, both not transparent in the visible, are commonly used as mask supports. Thick, phase pure, optically transparent (or at least translucent) diamond films are probably the ultimate choice for DXRL, if mask repositioning is required. A polished, 20 μm thick DXRL diamond membrane would combine mechanical stability, radiation hardness, high thermal conductivity, and optical transparency. However, to date, attempts to use diamond for DXRL are still in their infancy [11.13, 21]. Last but not least, whether the market for diamond DXRL mask supports will develop in the future will very much depend on the future of LIGA itself.

11.5 Summary and Outlook

Its unique combination of properties make CVD diamond a very interesting material for a variety of X-ray applications. However, in many of these application areas, CVD diamond also faces very strong competition. In the area of X-ray detectors, diamond is unlikely to replace the currently used beryllium, polymer, or boron nitride films. For X-ray tubes, beryllium is, despite its toxicity, the window material of choice. The development of diamond as an X-ray lithography mask support material depends very strongly on the development of X-ray lithography (XRL and DXRL) itself. European semiconductor manufacturers estimate that a few more DRAM integration steps will be covered by advanced (phase shift) optical lithography techniques. They do not expect X-ray lithography to take over for the next 5–10 years. Japanese groups expect market penetration of X-ray lithography earlier than their European counterparts and, consequently, most of the activities related to diamond XRL membranes are presently to be found in Japan. The future will tell which role X-ray lithography will be able to play, and whether or not CVD-diamond films will be of importance for this technology.

References

11.1   M. Pinneo and C. Bailey, US Patent 4939763, filed Oct. 3, 1988

11.2   L.S. Plano, K.V. Ravi, M. Peters, and M. Pinneo, European Patent Application EP 0365366 A1, priority Oct. 10, 1988

11.3   K.V. Ravi, L.S. Plano, M. Pinneo, M. Peters, and S. Yokota, in Science and Technology of NEW DIAMOND 1988, ed. S. Saito, O. Fukunaga, and M. Yoshikawa, KTK Scientific Publishers/Terra Publishing Company, Tokyo (1990), p. 29

11.4   T. Imai and N. Fujimori, European Patent Application 0476827 A1, priority Sept. 18, 1990

11.5   M.G. Peters, J.L. Knowles, M. Breen, and J. McCarthy, SPIE **1146**, 217 (1989)

11.6   K.V. Ravi, in Second International Symposium on Diamond Materials, ed. A.J. Prudes et al., The Electrochemical Society, Inc., Pennington, NJ; Proc. Electrochem. Soc. **91–8**, 301 (1991)

11.7   K.V. Ravi, in Synthetic Diamond: Emerging CVD Science and Technology, ed. K.E. Spear and J.P. Dismukes, John Wiley, New York (1994), p. 533

11.8   P.K. Bachmann, H. Lade, D. Leers, and D.U. Wiechert, presented at DIAMOND FILMS 93, Book of Abstracts, Paper 16.2 (1993)

11.9   H. Windischmann and G.F. Epps, J. Appl. Phys. **68**(11), 5665 (1990)

11.10   C.F. Mueller, US Patent Application 720 605, priority June 25, 1991

11.11   B. Löchel, H.J. Schliwinski, H.L. Huber, J. Trube, L. Schäfer, C.P. Klages, and H. Lüthje, Microelectronic Engineering **17**, 175 (1992)

11.12   M.F. Ravet and F. Rousseaux, Diamond Rel. Mater. **5**, 812 (1996)

11.13   M.F. Ravet, F. Rousseaux, Y. Chen, A.M. Haghiri-Gosnet, F. Carcenac, D. Decanini, J. Bourneix, H. Launois, P.K. Bachmann, H. Lade, D.U. Wiechert, and H. Wilson, J. Vac. Sci. Technol. **B13**(6), 3055 (1995)

11.14   G.F. Cardinale and C.J. Robinson, J. Mater. Res. **7**, 1432 (1992)

11.15   K. Suzuki, R. Kumar, H. Windischmann, H. Sano, Y. Imura, H. Miyashita, and N. Wanatabe, J. Vac. Sci. Technol. **B9**(6), 3266 (1991)

11.16   M.F. Ravet, A. Gicquel, E. Anger, Z.Z. Wang, Y. Chen, and F. Rousseaux, in 2nd International Conference on the Application of Diamond Films and Related Materials (ADC'93), ed. M. Yoshikawa et al., MYU Scientific Publishing, Tokyo (1993), p. 76

11.17   J.J. Cuomo, J.P. Doyle, K.L. Saenger, C.R. Guarnieri, and S.J. Whitehair, in 1st International Conference on the Application of Diamond Films and Related Materials (ADC'91), ed. Y. Tzeng et al., Elsevier, Amsterdam (1991), p. 169

11.18   L.M. Troilo, M.S. Owens, J.E. Butler, L. Shirey, and G.M. Wells, in 3nd International Conference on the Application of Diamond Films and Related Materials (ADC'95), ed. A. Feldman et al., NIST Special Publication **885**, 133 (1995)

11.19   L. Schäfer, A. Bluhm, C.P. Klages, B. Löchel, L.M. Buchmann, and H.L. Huber, Diamond Rel. Mater. **2**, 1191 (1993)

11.20   G. Stix, Sci. Ame. **11**, 72 (1992)

11.21   W. Ehrfeld and D. Münchmeyer, Nucl. Instrum. Methods Phys. Res. **A303**, 523 (1991)

11.22   H. Windischmann, G.F. Epps, and G.P Ceasar, in New Diamond Science and Technology, ed. R.F. Messier et al., Mat. Res. Soc. Pittsburgh, PA (1991), p. 767

11.23 H. Windischmann, BP America Inc., personal communication, Nov. 19, 1990

11.24 W. Shovin, Kevex Inc. USA, personal communication, Sept. 8, 1997

11.25 P.K. Bachmann, D. Leers, and H. Lydtin, Diamond Rel. Mater. **1**, 1 (1991)

11.26 P.K. Bachmann, in The Physics of Diamond, ed. A. Paoletti and A. Tucciarone, Societa Italiana di Fisica/IOS Press, Amsterdam (1997), p. 45

11.27 M. Joksch and P. Pongratz, TU Vienna, personal communication (1994)

11.28 P.K. Bachmann, H.J. Hagemann, H. Lade, D. Leers, D.U. Wiechert, H. Wilson, D. Fournier, and K. Plamann, Diamond Rel. Mater. **4**, 820 (1995)

11.29 H.A. Hoff, A.A. Morrish, K.A. Snail, J.E. Butler, and B.B. Rath, in New Diamond Science and Technology, ed. R.F. Messier et al., Materials Research Society, Pittsburgh, PA (1991), p. 773

11.30 M. Oda and H. Yoshihara, MRS Symposium Proc. **206**, 69 (1993)

11.31 L.E. Trimble, G.K. Celler, J. Frakoviak, and G.R. Weber, J. Vac. Sci. Technol. **B10**, 3200 (1992)

11.32 A. Moel, W. Chu, K. Early, Y.C. Ku, E.E. Moon, F. Tsai, and H.I. Smith, J. Vac. Sci. Technol. **B9**, 3287 (1991)

11.33 A.M. Hagihri-Gosnet, F. Rousseaux, B. Kebabi, F.R. Landan, C. Mayeux, A. Madouri, D. Decannini, J. Bourneix, F. Carcenac, H. Launois, B. Wiesniewski, E. Gat, and J. Durand, J. Vac. Sci. Technol. **B8**, 1565 (1990)

11.34 M.D. Levenson, Solid State Technol., February, 57 (1995)

11.35 H. Windischmann, SPIE **1263**, 241 (1990).

11.36 S. Tsuboi, H. Okuyama, K. Ashikaga, and Y. Yamashita, J. Vac. Sci. Technol. **B13**, 3099 (1995)

11.37 H. Windischmann and G.F. Epps, Diamond Rel. Mater. **1**, 656 (1992)

11.38 L. Schäfer, X. Jiang, and C.P. Klages, in 1st International Conference on the Application of Diamond Films and Related Materials (ADC'91), ed. Y. Tzeng et al., Elsevier, Amsterdam (1991), p. 121

11.39 E. Anger, A. Giquel, M.F. Ravet, Z.Z. Wang, F. Rousseaux, J. Perriere, F. Rossi, D. Fournier, and K. Plamann, Vide, Science, Technique Appl. **276**, 139 (1995)

11.40 B.S. Berry, W.C. Pritchet, J.J. Cuomo, C.R. Guanieri, and S.J. Whitehair, Appl. Phys. Lett. **57**, 302 (1990).

11.41 A. Nishiyama, H. Yamashita, H. Yoshikawa, H. Yabe, K. Kitamura, and K. Marumoto, J. Vac. Sci. Technol. (in press) (1997)

12. CVD Diamond for Cutting Tools

Benno Lux and Roland Haubner

Institute for Chemical Technology of Inorganic Materials,
University of Technology Vienna,
Getreidemarkt 9/161, A-1060 Vienna, Austria
e-mail: blux@fbch.tuwien.ac.at

Springer Series in Materials Processing
Low-Pressure Synthetic Diamond Eds.: B. Dischler and C. Wild
© Springer-Verlag Berlin Heidelberg 1998

12.1 Introduction

The excellent properties of diamond, mainly its extreme hardness, its wear resistance, thermal conductivity and transparency, predestine low-pressure diamond to become the ideal material for heat-spreaders, windows, and coatings for wear parts [12.1].

Major considerations for the development and production of appropriate high-quality low-pressure diamond for the different applications are as follows:

(a)   Finding an economic, large-volume or high growth rate synthesis method with flexible diamond growth parameters that permits a high level of reproducibility.

(b)   The determination of the optimal diamond deposition parameters appropriate for the different applications.

(c)   The application of efficient surface pretreatment to the tool substrates (diffusion barriers, surface roughness, etc.) prior to the diamond coating in order to obtain perfect adhesion, epitaxial nucleation, high nucleation rates and so on, as needed for the products.

In general, cutting tools need two important properties: wear resistance and toughness. Ceramic materials with a high wear resistance generally have low toughness. The high hardness of pure single-crystal diamond cannot fully be used, due to its easy cleavage. One way to avoid this could be through the use of thin polycrystalline "Ballas" diamond [12.2]. Another way to increase toughness is a principle also well known from hardmetals (WC–Co), whereby the WC particles are consolidated by a metal binder phase. The same type of material, called Polycrystalline-Diamond (PCD), is a diamond/Co composite. With increasing Co

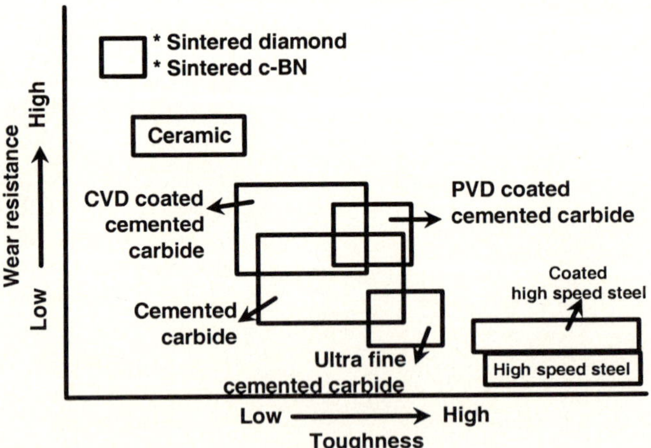

Fig. 12.1: An area map for cutting tool materials [12.5].

content, the ductility increases in the composite while the wear resistance decreases. Another successful approach to combine toughness and wear resistance for tools is to coat a hard, ductile and heat-conducting substrate with a thin surface layer, such as Al_2O_3 or TiC [12.3]. This is also possible for binder-free diamond. Its very high hardness and heat conductivity can be used for wear applications [12.4]. Fig. 12.1 shows an area map for various cutting tool materials [12.5].

12.2 Composite Tools for Wear Applications

The application of CVD-diamond coatings on to WC–Co can substantially increase their lifetime during heavy-duty cutting or milling operations on materials such as graphite, fiber-reinforced plastics, laminated wood products, timber, concrete, and various rock and stone products. [12.6–9].

Composite cutting tools, having wear-resistant superhard surface coatings on a stiff, ductile substrate as a base material, can exhibit excellent performance.

A primary requirement for the successful mass production of such *"composite tools"* is, however, an excellent interfacial adhesion of the coating on to the substrate. Essential for industrial use are also the highest reproducibility, reliability for each individual tool tip and economical mass production processes.

Two different types of tools can be produced:

- free-standing (thick) diamond layers brazed on to a carrier substrate
- thin diamond layers deposited directly on to an appropriate substrate.

12.2.1 Free-Standing Diamond Layers in Wear Applications

Industrial wear parts containing free-standing diamond layers are commercially available today. The bonding of free-standing thick CVD-diamond sheets by brazing is very applicable for tools with simple geometries. It competes with the classical PCD (Poly-Crystalline-Diamond) products for relatively simple geometries. These tools are produced by high-pressure sintering of diamond powder with a binder metal (mainly Co). PCD parts are limited in size – to about 70 mm in diameter – because of the high-pressure equipment.

The free-standing diamond layers can be produced in large sizes by various diamond-deposition methods which are carried out in three steps, as shown in Fig. 12.2 [12.10] for the manufacturing of endmills and simple cutting tools.

First by a low-pressure synthesis method, a thick, large-sized diamond layer is deposited on a dummy substrate. This layer can be removed from the substrate or manufactured as is. The diamond sheet is cut to size and trimmed into the shape needed for the tool by a laser.

Second the diamond sheets are bonded to the body of the wear part. This is done after the tool has been shaped close to its final geometry.

Third the tool and the layer are ground to their final tolerances. Low-pressure CVD-diamond layers are binder-free materials, in other words 100% diamond;

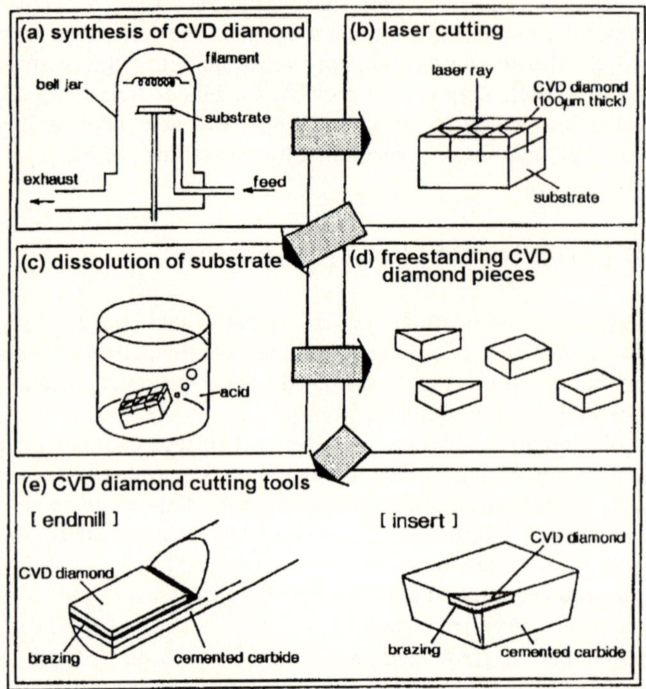

Fig. 12.2: The free-standing layer technology to produce endmills and cutting tools [12.10].

thus their wear resistance is better than that of PCD (Table 12.1). However, during machining with thick CVD-diamond layers that have sharp cutting edges, edge deterioration may occur. This is due to a cleavage of diamond crystals, as the diamond CVD structure has relatively coarse, elongated single crystals,

Table 12.1: The lifetimes of cutting tools containing free-standing diamond in selected applications (Norton Diamond Film) [12.11].

Material	Operation	Geometry	Diamond type	Parts/edge
390 T-6 Al	Finish turning OD lands	SPG-422	PCD	1500
			CVD layer	12000
390 T-6 Al	Finishing piston head face	CPG-428	PCD	4000
			CVD layer	20000

perpendicularly oriented to the substrate layers. Polycrystalline "Ballas" structures used as coatings [12.2] might provide a solution to such problems. Products using free-standing diamond layers are well known (endmills, cutting tools), but there are only few testing results available with which to compare the different products.

12.2.2 *In-situ* Diamond-Coated Tools

Hardmetal substrates are the obvious choice for most industrial applications of diamond-coated tools. A primary requirement for the industrial use of a diamond coating is its adhesion to the substrate. Therefore, the use of an appropriate hardmetal grade, and the application of an appropriate surface pretreatment – applied to the hardmetal prior to the deposition of the diamond coating – are equally important. The rigorous control of all possible interfacial interactions occurring at the diamond/hardmetal interface is necessary. The excellent bonding and coating adhesion must be reproducible. A number of parameters during diamond growth contribute to an optimal bonding of the diamond coating to the substrate. Also, the substrate can contribute largely to detrimental effects. In the case of hardmetal, the cobalt binder has detrimental effects. Negative interactions with the Co binder, such as its high mobility, must be avoided. Reproducible diamond nucleation and optimal growth conditions must be maintained, as well as a controlled interfacial roughness of the hardmetal/diamond interface, and so on.

Obviously, the diamond synthesis method applied and the correct choice of deposition parameters are also important in achieving a usable and economic product. Generally, the energy density during diamond synthesis – and thus the diamond growth rates – are lower for a tool coating (growth rate around 1 µm/h), in order to avoid thermal damage to the substrate properties, as compared to free-standing layer synthesis, in which growth rates may exceed 10 µm/h.

Diamond Deposition on Hardmetal Substrates

Hardmetals – also called cemented carbides – are the most valuable and important substrates for coated tools, due to their intrinsic properties and their wide range of mechanical properties. They consist basically of WC and Co with additions of TiC, (Ta,Nb)C, VC and so on, which change mainly their hardness and wear resistance [12.12]. The amount of Co binder is largely responsible for their ductility or brittleness. Hardmetals have been used as wear parts and cutting tools for decades, with and without coating applications.

The main coatings applied worldwide today on cutting tools to increase the tool performance are TiC, TiN and Ti(C,N) produced by CVD or PVD deposition technique. For steel cutting mainly CVD Al_2O_3 layers are used [12.13]. The thickness ranges from less than one micron to several microns. Multi-layer combinations are also frequently used.

Successful diamond coatings on WC–Co substrates having reasonable adhesion are reported in the literature on substrates that generally have no or a very low amount of cubic carbides (TiC) and also a relatively low Co content in the hardmetal. Both help to improve the adhesion of the diamond coating [12.14, 15].

Co interacts chemically with the diamond nucleation and growth during diamond synthesis. In particular, if large thermal expansion differences exist between the substrate and the coating, detrimental compressive stresses occur at the coating/substrate interface, during cooling from the deposition temperature; both Co and TiC additions increase the thermal expansion coefficient of the hardmetal [12.16, 17].

Detrimental influences of the Co binder. Vapour pressure and a high Co mobility influence diamond deposition. Near the substrate surface, Co catalyses the formation of non-diamond carbon phases in the gas phase, which can be deposited at the interface prior to the diamond formation. Thus the adhesion between the diamond and the substrate is jeopardized [12.18]. Furthermore, such carbon deposits influence diamond nucleation detrimentally; special nucleation enhancement can become necessary to achieve reasonable nucleation densities and thus rapid layer formation [12.19, 20].

By Co migration, Co droplets are formed on both the substrate surface and on top of diamond layers [12.21, 22]. How and why the Co drops reach the diamond coating surface is not yet fully understood. Surface forces might play an important role (Fig. 12.3).

The carbon balance within the WC–Co substrate. Depending on the H/O/C ratio in the gas phase during sintering, or heat treatments of hardmetals, the C and W contents in the Co binder phase change. The C content in the Co binder phase becomes most important during diamond deposition, as carbon can diffuse into the WC-Co substrate if the Co binder has a low C content.

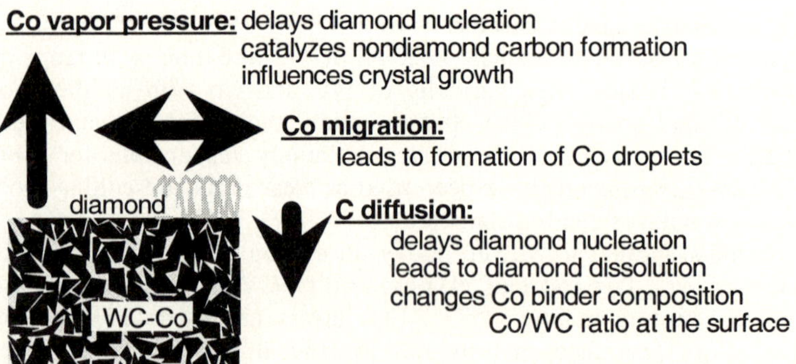

Fig. 12.3: The influence of the hardmetal Co binder phase on the diamond deposition process [12.23].

Also, adhesion of the diamond coating is influenced by carbon dissolution/diffusion processes. Industrially available hardmetal substrates can be classified in three categories. Depending on the carbon saturation of the Co binder, different reactions occur during the diamond-coating process:

- *High carbon:* Co is fully saturated with C and thus even the formation of free C precipitates (= graphite) can occur in the Co binder phase during cooling, while the W content is very low.

- *Medium carbon:* W is dissolved in the Co binder phase (up to 15 %). The Co content in the binder phase becomes lower, but the amount of binder phase increases by forming a solid solution of W in Co. The carbon content in the Co is low, but is regulated by the W content (a high amount of W results in a low C content). This type is the most common hardmetal composition used for cutting tools.

- *Low carbon:* high amounts of W are dissolved in the Co binder phase and a low C content is observed. Further C reduction in the system results in the formation of eta-carbide.

During the early diamond nucleation and growth stage, C diffusion into the substrate occurs. This continues during the whole duration of the heat treatment linked with the diamond deposition. Although the high amount of WC particles reduces the C diffusion via the Co binder, the C saturation rate of the WC–Co substrate surface occurs quite rapidly. The isolated diamond, as well as the diamond layer dissolve at the diamond/hardmetal interface, and carbon diffuse into the substrate via the Co binder as long as there is a C gradient across the WC–Co substrate [12.23, 24] (Fig. 12.4).

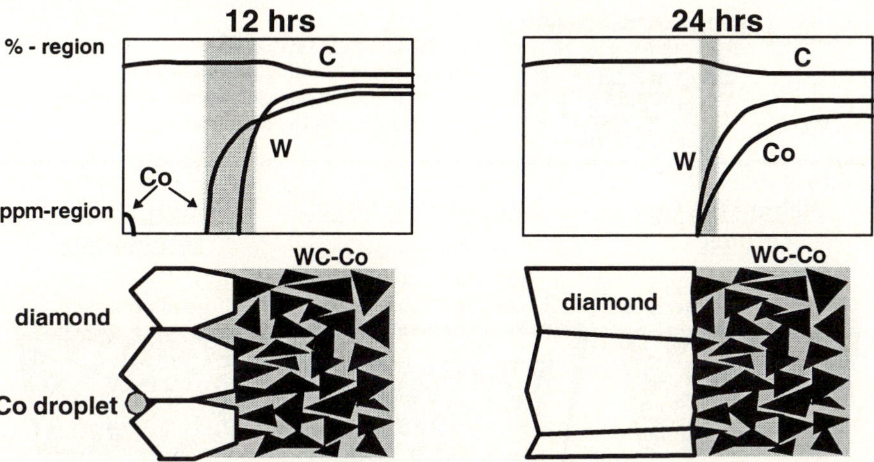

Fig. 12.4: An illustration of diamond dissolution by the Co binder phase of a hardmetal substrate, based on the depth profiles of C, W and Co concentrations, measured by SIMS [12.23]

To obtain *good adhesion of the diamond coating* on a hardmetal tool, it is necessary to *saturate the Co binder* in WC–Co hardmetal grades with C *prior to* the diamond deposition.

The diamond dissolution into the Co binder phase can also be reduced by shortening the heating time (= deposition time). This calls for diamond deposition methods that provide higher growth rates, whereby also the final diamond-layer thickness for the tool application can be reached more rapidly [12.23].

Improving adhesion of diamond coatings by surface treatments of hardmetal substrates prior to diamond deposition. In the following, several approaches to reduction of the Co vapour pressure and its migration are discussed, such as: local reduction of Co by etching surface pretreatment, [12.25–28]; intermediate layers [12.29]; reducing Co mobility by forming stable Co compounds and so on [12.30, 31] (Fig. 12.5).

Removal of the Co binder by etching. A number of pretreatment procedures aim at a depletion of the Co at the hardmetal substrate surface by selective chemical etching with acids such as HNO_3, HCl or CH_3COOH [12.25–28, 32]. This prevents Co migration. A rough surface can also increase diamond nucleation, and a rough interface can improve the coating adhesion, while formation of excessively large pores is detrimental. Thus the surface removal of Co must be carried out with care [12.32].

Another etching pretreatment consists of two steps. First, the WC is removed by a Murakami solution ($K_3[Fe(CN)_6]$ in KOH and H_2O), leaving an excess of Co on the hardmetal surface. Subsequently, the Co is treated by a H_2SO_4/H_2O_2 solution. Systematic investigations have proved that only the combination of both etching steps can give improved adhesion results [12.28, 33].

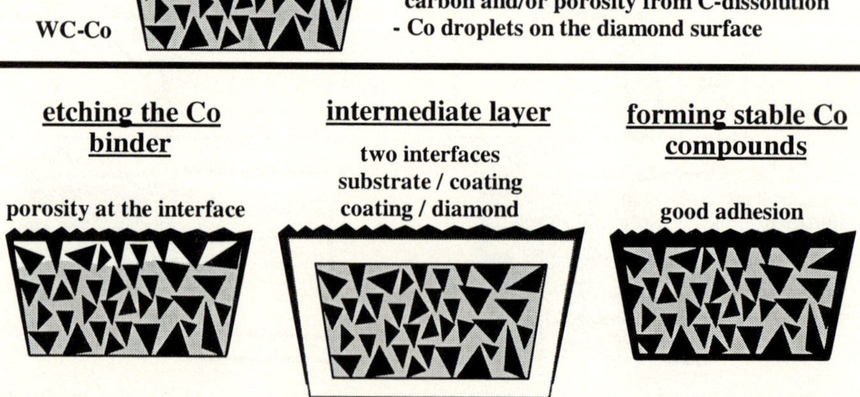

Fig. 12.5: Different types of surface treatments of hardmetal substrates leading to improved adhesion of diamond coatings.

Co leaching and replacing by Cu. Putting a WC–Co substrate into a solution of Cu salt leads to a local replacement of the Co by Cu at the substrate surface, due to a simple electrochemical reaction:

$$Co + Cu^+ + e^- \rightarrow Co^+ + Cu + e^-$$

After 1–2 minutes, the Co at the substrate surface is replaced by Cu. Adhesion of the diamond coating on such Cu-treated substrates has been reported to be much better than on substrates from which Co was removed by chemical leaching only [12.34].

Intermediate interfacial layers. The application of intermediate layers to separate the hardmetal surface and the diamond coating has frequently been described as a promising technique for the deposition of adherent diamond films. Relatively thick carbide layers have been deposited on hardmetal substrates by different CVD techniques; and thinner layers by PVD techniques [12.29, 30]. Intermediate layers fully cover the substrate and reduce thus negative Co effects. The successful adhesion of a diamond coating deposited on such intermediate layers depends strongly on the chemical nature and composition of the intermediate layers. Diamond nucleation sometimes becomes a problem, as the diamond nucleation rate can become too low, and diamond-layer formation becomes difficult [12.30]. Sometimes, acceptable adhesion of the diamond layer on the intermediate layer remains a problem. Only a few materials seem to permit a better diamond adhesion than pure WC, which could thus be used as a successful intermediate layer itself.

Forming stable Co compounds as interfacial barriers. Another successful way to improve the diamond-coating adhesion is based on the formation of Co compounds that are stable under diamond deposition conditions. Hardmetal substrates pretreated with B or Si vapour at elevated temperatures form thin Co borides or silicides on the Co binder surface [12. 31]. Also, thin sputtered layers of Ti or Cr have shown increased diamond nucleation and diamond growth rates, and can lead to excellent diamond-coating adhesions. Such surface layers on the Co binder decrease the Co vapour pressure and hinder Co migration on the substrate surface. Thus diamond layers formed on hardmetals treated in this way have well-faceted diamond crystals, and no longer show the detrimental Co influences described above [12.31].

A great number of other elements (such as B, Si, Al, Ti, Hf, and Sc) can be used to form stable Co compounds; some have prooved to be of interest and seem to be applied in industrial tools [12.35, 36].

Conclusion on adhesion of diamond coatings on hardmetal substrates
As demonstrated, many factors influence the adhesion of diamond coatings on hardmetal substrates (Fig. 12.3) [12.23]. Certainly, the diamond coating/substrate interface should be clean, and free of non-diamond carbon.

If fewer diamond nuclei are formed, the individual diamond crystals become large before they can grow together. This forms large voids between the crystals – at the interface – which can, however, sometimes be filled with Co during deposition (Fig. 12.4) [12.37].

In the case of a high level of diamond nucleation, the voids between the crystals remain much smaller and they are generally filled with Co. Simple adhesion tests have shown better adhesion for this case, even though the roughness of the interface is decreased [12.23].

As already explained above, during the heat treatment of deposited diamond, diamond dissolution into an unsaturated Co binder occurs at the diamond/substrate interface. This process leads to a reduction of the interfacial voids, but thereby the diamond interface roughness also becomes smoother and additional stresses are induced at the diamond/substrate interface. Thus the diamond dissolution can be detrimental, depending on the duration, temperature and so on applied during the diamond deposition.

Diamond dissolution by C diffusion into the substrate may reduce adhesion [12.23]; however, this dissolution does not occur with C-saturated hardmetal substrates. Thus it seems quite easy to avoid these problems at least.

Diamond Coatings on Ferrous Metals and Ceramic Materials

For cutting tools, some materials other than hardmetals are also of interest. In particular, ferrous metals such as high-speed steels (HSS) and ceramics have been investigated with a view to producing usable cutting tools [12.17, 30].

Ferrous metal substrates (HSS). The main problem for ferrous substrates is the easy diffusion of carbon into the substrates during the diamond deposition process. At the normal diamond deposition temperature, the austenitic phase is present, which has a high carbon solubility (up to 2.06 wt% C) [12.38]. Thus barrier layers have to be applied in order to inhibit carbon diffusion into the substrates [12.30], or the deposition temperature has to be decreased below the ferrite – austenite transformation temperature [12.39, 40].

Decreasing the deposition temperature below the austenitic stability raises problems with the diamond growth rate and can jeopardize proper diamond formation. Below 723°C (the ferrite–austenite transformation temperature in the pure Fe–C system), the diamond deposition rate decreases almost to zero using conventional deposition methods. Diamond deposition on HSS is possible with high-power microwave equipment and substrate cooling. Only under such conditions can carbon diffusion into the substrate – without intermediate layers – be avoided [12.39]. Although diamond coatings on HSS tools for industrial applications are of high interest, up to now no commercial tools have become available.

Diffusion barrier layers such as TiN, TiC, Ti(CN), TiAlN and CrC have been applied on various steel substrates. The production of such interlayers is expensive, as a two-step PVD process is needed to reach a fully covered surface and to prevent carbon interaction with the substrate. Such barrier coatings must cover the substrate fully, must be free of pores and cracks, and need to have excellent adhesion to the substrate and the diamond [12.30].

The combination of intermediate layers and low-temperature deposition seems to be a successful way to apply diamond coatings on to steel substrates [12.41]. Thereby, the steel substrate is coated *in situ,* prior to diamond deposition with cubic silicon carbide (3C-SiC), in order to prevent carbon diffusion and hydrogen brittleness [12.42].

In another approach, a pre-coating composite layer containing an Ni and diamond mixture is applied by <u>electro-deposition</u>. This layer serves as a nucleation layer for the diamond and provides good adhesion to the steel substrate. Diamond depositions with a duration of more than 24 hours have been made, whith the hardness of a steel substrate remaining unchanged [12.41].

Ceramic substrates. In the early days of diamond deposition, the fundamentals were already being studied using SiAlON substrates. These are stable in the diamond deposition atmosphere at high temperature, the adhesion of the diamond coatings was good and the experiments were reproducible [12.16, 17]. Diamond-deposition experiments on other ceramic materials (h-BN and ZrO_2) showed chemical reactions with the atomic hydrogen and the carbon species. However, although interactions are thermodynamically possible, no effects were observed during diamond-coating experiments with Al_2O_3, SiC and SiAlON substrates [12.17].

Various composite ceramics were tested and proposed as substrates for tool materials, such as $(Al_2O_3 + TiC)$, $(Al_2O_3 + SiC)$ and $(Al_2O_3 + ZrO_2)$, but the diamond adhesion on SiAlON substrates always was best [12.43].

Due to their relative high shock sensitivity diamond-coated ceramic tools in industrial applications are rare today. The authors only know of an Si_3N_4 round tool which is produced for graphite machining [12.44].

12.3 Reproducibility and Economic Mass Production

12.3.1 Analytical Diamond Characterization to Guarantee Reproducibility

Reproducibility and reliability are crucial points for low-pressure diamond applications. Diamond's extreme physical properties are sometimes difficult to measure. Analytical methods are used to monitor the deposition process [12.45], to characterize the diamond layer [12.46] and to measure the physical properties [12.47]. For this reason, indirect correlation between diamond quality and selected physical properties can be used to determine the interesting properties (e.g. thermal conductivity [12.48]). Most of these techniques are of interest only for research and development. For the control of diamond mass production, SEM (Scanning Electron Microscopy), and Raman and IR spectroscopy are useful

[12.49]. These methods can distinguish between different carbon phases and give information about the quality of the diamond. It would be desirable, however, to have a single or a few analytical key values (e.g. from Raman spectroscopy) which would allow determination of the overall diamond quality and thus guarantee the reproducibility and reliability of industrial diamond products.

12.3.2 Mass Production of Low-Pressure Diamond and Its Economy

Like any other new material, diamond has to fulfill several requirements prior to successful mass production. The technical criteria are acceptable product qualities resulting from the properties of the material, the process and post-processing technologies. For the applications discussed here, the product properties needed for industrial use can be achieved in many cases. On the other hand, economic requirements, such as low cost and, last but not least, the customer acceptability of these new products have to be taken into account [12.50]. Thereby, the production quantity and deposition methods for mass production determine the price, which decreases with an increasing production volume [12.51]. The appropriate production equipment for CVD-diamond mass production plays also a key role [12.52, 53].

12.3.3 Diamond Deposition Methods for Tool Coating

Today, a number of low-pressure diamond deposition methods for gas activation, permitting the production of diamond coatings and layers for tool applications, are available [12.54, 55]. Such techniques are hot-filament [12.54], microwave plasma [12.55], DC discharge [12.55] or plasma jets [12.55] (Table 12.2).

The *hot-filament* method allows easy design and up-scaling of reactors. Although the diamond deposition rates are lower (about 1 μm/hr) than with other methods [12.54], it is reliable and quite appropriate for the production of diamond coatings for wear parts, and is convenient for loading and unloading cycles (between 12 and 24 hr). Large uniform deposition areas are easy to produce with hot-filament equipment, as compared to microwave and plasma-jet methods [12.56]. The substrate surface temperature can be regulated by heating the substrates rather than by cooling, which makes it very reliable. Thus it has enormous advantages over other methods, except for the relatively low growth rate.

The high-density/low-voltage arc discharge plasma – recently introduced by Balzers and available exclusively to Sandvik – provides another very useful industrial diamond deposition method by using an "arc as a hot filament" [12.52, 57]. Thus a very high atomic H concentration allows a large number of inserts to be coated simultaneously under controlled conditions in a relatively large-volume vessel. Uniform substrate surface temperatures can be regulated by heating [12.58]. The information available so far is limited, but suggests

Table 12.2: Advantages and disadvantages of low-pressure diamond synthesis methods.

Method	Advantages	Disadvantages
Hot-filament	Low equipment costs Easy up-scaling Substrate heating	Contamination by filaments Low growth rates
Arc discharge plasma as a "hot filament"	Large deposition area No substrate cooling needed Complex 3-D coatings	Low growth rates Vertical arrangement for specimens
Microwave plasma	High diamond quality Easy temperature control	Difficult for 3-D substrates Non uniform thickness
Plasma jets	High growth rates	Forced substrate cooling
Glow discharge	High growth rates	Substrate temperature control Small deposition areas Only planar substrates
Flame	High growth rates	Inhomogeneous deposition Low reproducibility

similarities with the hot-filament method. However, due to the high arc temperature, it has the capability of producing abundant atomic hydrogen – without a need for substrate cooling. Furthermore, it seems to be rather inexpensive and reliable as far as the equipment is concerned, easy to up-scale to a practicable size, and thus appropriate for the economic and industrial mass production of coatings on small tool inserts [12.59].

Microwave plasma deposition with "ball" shaped plasmas is convenient to handle for laboratory purposes. However, it seems difficult and expensive to increase the deposition area to industrial sizes. Furthermore, plasma inhomogenities between the center and border of the discharge, as well as problems with substrates having complex irregular shapes, cause inhomogeneous diamond deposits. The main problem for industrial realization seems to be economic up-scaling for mass production [12.55, 60]. Diamond coating with high-quality diamond at deposition rates higher than with the hot-filament method are possible using high-power microwave generators [12.53].

DC- and arc-plasma jets permit large-area, high-speed growth, particularly for large-area and thick free-standing diamond, which seems easier than with micro-wave and hot-filament methods [12.61]. As the substrates are below the arc area, controlled and forced cooling is needed to regulate the substrate surface temperature, which can be a serious problem for the industrial mass production of tools.

DC-glow discharge and the flame method allow high diamond growth rates, but there can be problems with up-scaling, substrate geometry and reproducibility [12.62, 63]. It seems that neither method is easily practicable for technical mass production of coated tools.

12.4 CVD-Diamond Tools: Technological Performance

Over recent years, various diamond-coated tools that can be used for selected applications have been designed by several tool companies and are now available on the free market. Diamond-coated tools are comparable and are competing in particular with polycrystalline diamond (PCD), as well as with some ceramic tools that are well-established on the market. Improved performance – for example longer tool life, reduced total machining costs and so on, as well as low costs – are necessary to enter this market. A few technical and economic data showing the quality and potential of "new low-pressure diamond products" are based on field tests published so far. Examples for typical CVD-diamond applications concerning the machining of non-ferrous alloys and compounds are given below.

12.4.1 A Comparison of CVD-Diamond Coatings with Other Tools for Cutting Operations

The new diamond-coated tools are mainly used in turning *aluminium/silicon* alloys and non-ferrous metals (see Table 12.3 [12.64, 65] and Fig. 12.6 [12.66]). These results clearly demonstrate that, in comparison to uncoated hardmetals, diamond increases the lifetime of tools by a factor between 4 and 20.

Table 12.3: A summary of some field tests to compare CVD-diamond tools with other materials in cutting tests [12.64, 65].

Conditions	Turning test no. 1		Turning test no. 2		Turning test no. 3		Turning test no. 4	
Material	Al–7–9%Si–T6		Al–7%Si–T6		Al–11%Si		Cu	
Operation	Roughing		Roughing		Rough + finish		Finishing	
Coolant	None		Emulsion		Emulsion		None	
Component	Truck wheel		Wheel		Wheel		Armature	
Cutting data								
Cutting speed (m/min)	1850–2500		< 2150		2100–2600		300	
RPM			1800		2000			
Depth of cut (mm)	0.5–3.0		0.8–1.5		0.5–1.0		1.0	
Feed (mm/rev)	0.25–0.8		0.5		0.35–0.65		0.15	
Insert	VCGX 160412-AL		N151.4-800-60-AL		RCGX 0803Mo-AL		CCGX 120408-AL	
Grade	H10	**6 D**	H10	**6 D**	K15	**6 D**	H10	**6 D**
Results								
Pieces / edge	7	**140**	600	**2250**	200	**800**	2	**4**
Wear type	Abr.	**Fl.**	Abr.	**Fl.**	Abr.	**Fl.**	Abr.	**Fl.**

RPM = Rounds per minute   6 D = 6 µm diamond-coated hardmetal.
Abr. = Abrasion   Fl. = Flaking

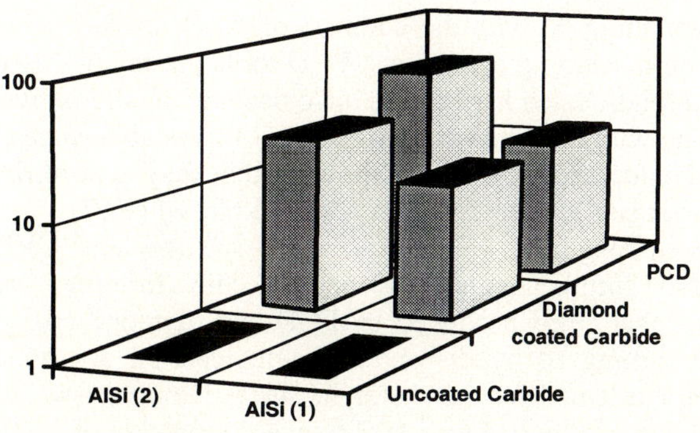

Fig. 12.6: A comparison of CVD diamond with PCD and uncoated carbide in the turning of AlSi alloys [12.66].

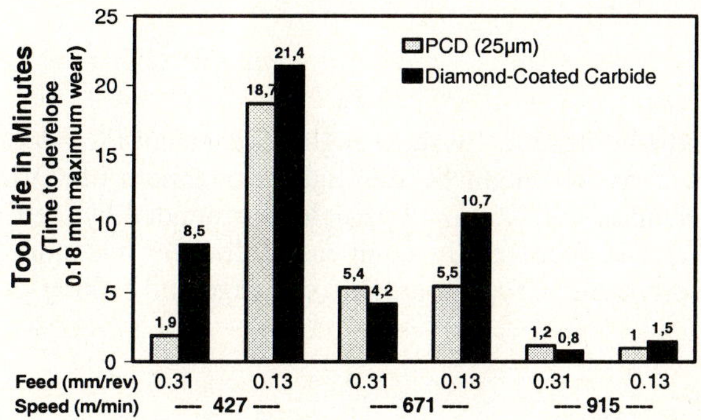

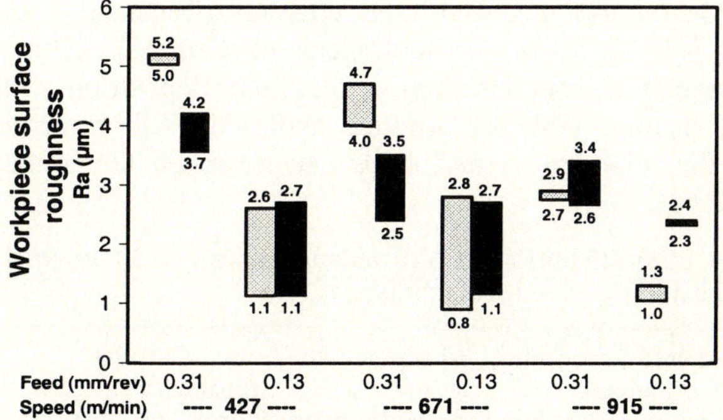

Fig. 12.7: The tool life (*top*) and workpiece roughness (*bottom*) provided by PCD (25 µm) and diamond-coated (≈ 30 µm thick) carbide tools in machining Duralcan MMC [12.67].

Metal matrix composites containing Al with the addition of Al_2O_3 or SiC show high and rapid tool wear during machining operations. PCD tools are mainly used for this application so far. Diamond-coated hardmetals have become an alternative. As the wear mechanisms change with machining conditions, tool lives also change. Fig. 12.7 compares diamond-coated and PCD tools, showing that they are similar in lifetime and also give a similar surface roughness of the workpiece [12.67].

Due to the different surface roughness of PCD (smooth surface) and CVD diamond (rough surface) also cutting operations, mainly chip forming, are different. It has been observed that chips produced by PCD and diamond-coated tools during the milling and turning of hyper- and hypo-eutectic Al are quite different [12.67]. This can be explained by the different surface roughnesses of PCD ($R_a = 0.05$ μm) and diamond coatings ($R_a = 1.0$–1.6 μm). In the case of PCD, the chip glides uniformly over a smooth polished rake surface and is undeflected unless it engages a chip-breaker. On the contrary, the chip is quickly deflected as it moves over the rough rake surface of diamond-coated tools. Therefore, the material roughness of the CVD layers acts as a microscopic chip-breaker in the cutting zone.

12.4.2 Milling

In milling operations, diamond coatings mostly show higher wear than PCD tools (Table 12.4) [12.67]. The main reason might be the higher roughness of CVD-diamond coatings and the interrupted cut. Diamond particles surrounded by the Co binder phase in PCD material can more easily compensate for the hits during milling than can pure diamond coating. Crystallographic cleavage and flaking can occur during milling.

12.4.3 Drilling

Drilling makes high demands on tool materials and coatings. Because of the complex shape of drills, it is not easy to deposit homogeneous diamond coatings. Additionally, the cutting edge gets rounded during diamond deposition. The complex shape also makes it impossible to produce drills in PCD material. Selected applications show that diamond-coated drills can be much better than uncoated ones (Fig. 12.8) [12.66].

Table 12.4: The performance of PCD (25 μm) and CVD-diamond coated (≈ 30 μm thick) carbide tools (five samples) in the milling of 383.2 aluminum [12.67].

| | Measured wear (mm), Workpiece roughness, R_a (μm) | | | |
| | 12 passes | | 36 passes | |
Tool material	Wear	Roughness	Wear	Roughness
PCD (25 μm)	0.021	0.64	0.028	0.76
CVD diamond	0.023–0.041	0.66–0.84	0.030–0.042	0.76–0.86

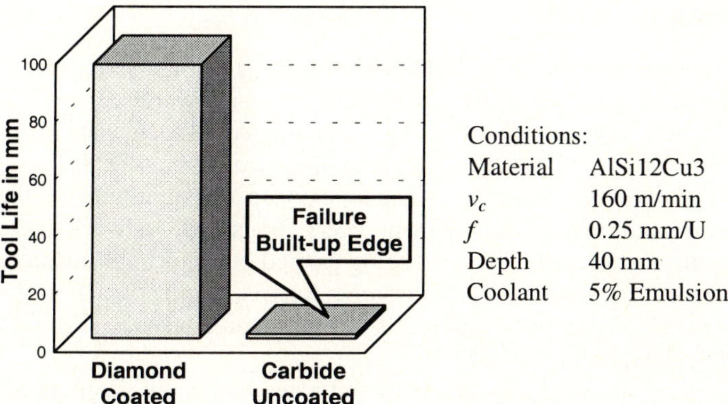

Fig. 12.8: A comparison of CVD diamond with PCD in the drilling of an Al–Si alloy [12.66].

12.5 Conclusions

More than a decade has passed since S. Matsumoto [12.68] showed, by the hot-filament method, the first technological approach to a simple low-pressure diamond synthesis process; that is, the way to form diamond layers, larger and thinner than ever seen before, on surfaces.

Although it was one of the first objectives, diamond-coated hardmetal tools, have only recently started to become a serious industrial product. The main reasons for this delay were not based on problems to do with available methods of synthesis, but on problems with adhesion and reliability of the diamond coatings on the hardmetal inserts. Last but not least, a large-scale industrial production is also problematic. Overall, the problems have turned out to be far more complicated in comparison with classical CVD and PVD tool coatings, such as TiC, TiN, Al_2O_3 and so on, which were introduced industrially two to three decades ago.

Now the industrial uses have to show the advantages of the "new diamond tools". A large domain of applications is in prospect – from the machining of non-ferrous metals, carbon products and refractory materials over fiber-reinforced plastics to hard and wear-resistant parts, and other "high-tech" structures that call for precise shaping with chip forming. However, diamond surfaces for applications such as extruder dies, heavy-duty wear and friction parts, and composite tools combining extreme hardness, wear resistance and high heat conducting properties should also profit from these developments. They also open up new areas of research and development, for better composite structures and new tools based on chipless modification of materials.

Acknowledgements

We thank the management of Sandvik, Stockholm, and in particular B. Aronsson and those who believed in these ideas, for sponsoring our early work, carried out by H. Schachner at Battelle, Geneva. We also thank Balzers AG, Liechtenstein, for sponsoring our diamond work more recently, under the management of J. Vogel and H. Kaufmann. Last but not least, we wish to express our gratitude to the Austrian Science Foundation which has, for many years sponsored our work at the Technical University in Vienna, which is now being carried out under the auspices of the trinational cooperation "D–A–CH" (Germany, Austria, Switzerland) on the "Synthesis of Superhard Materials" initiated by K. Komarek (FWF; Projects P6031, P7274, S5901, S5907 and P12055).

Finally, we thank the following persons and companies for recent discussions, and constructive dialogs, as well as for providing us with valuable and helpful information and documents related in particular to the present paper:
Steve Arnold, Kennametal, Inc., Latrobe, Pennsylvania, USA;
Kim Bigelow, Norton Saint-Gobain Company, Northboro, MA, USA;
James Herlinger, sp^3 Inc., Mountain View, CA, USA;
P.C. Lindsey, Kaley Parkinson, and Richard Koba, CMC, San Ramon, CA, USA;
Akio Nishiyama, Mitsubishi Metals Corporation, Omiya, Japan;
Toshio Nomura and Naoji Fujimori, Sumitomo Electric Industries, Itami, Japan;
Bernhard Kieffer (†), Jim Oakes and M. Ostermann, Teledyne Firth Sterling, USA, for providing – free of charge – many hardmetal substrates and performing cutting tests of tools coated in our laboratory.

References

12.1   B. Lux and R. Haubner, in Diamond Films and Coatings, ed. R.F. Davis, NOYES Publications Park Ridge, NJ (1993) pp. 184–243

12.2   B. Lux, R. Haubner, H. Holzer and R.C. DeVries, Refractory Metals & Hard Materials **15**, 263 (1997)

12.3   B. Lux and R. Haubner, in 6th Euro CVD Conference, Jerusalem, (1987) pp. 51-7

12.4   W. Kremser, J. Nedelik, and B. Lux, Diamond Films and Technology (submitted)

12.5   N. Kikuchi, H. Eto, T. Okamura, and H. Yoshimura, in Applications of Diamond Films and Related Materials, ed. Y. Tzeng et al., Elsevier, Amsterdam (1991) pp. 61–8

12.6   H. Schachner, B. Lux, K. Stjernberg, A. Thelin, and H. Tippmann, European Patent 166708 (1986)

12.7   S.J. Bull and A. Matthews, Diamond Rel. Mater. **1**, 1049 (1992)

12.8   M. Murakawa and S. Takeuchi, Surf. Coat. Technol. **49**, 359 (1991)

12.9   J. Oakes, X.X. Pan, R. Bichler, R. Haubner, and B. Lux, Surf. Coat. Technol. **47**, 600 (1991)

12.10   T. Yashiki, T. Nakamura, N. Fujimori, and T. Nakai, in 18th International Conference on Metallurgical Coatings and Thin Films (ICMCTF), San Diego, CA (1991)

12.11   NORTON Diamond Film, DIAMAPAK^{TM} (prospectus)

12.12   W. Schedler, in Hartmetall für den Praktiker, ed. Plansee TIZIT GmbH, VDI-Verlag, Düsseldorf (1988), pp. 43–6

12.13   B. Lux, R. Haubner, and C. Wohlrab, Surf. Coat. Technol. **38**, 267 (1989)

12.14   R. Haubner and B. Lux, J. Phys. **C5**, 169 (1989)

12.15   R. Haubner, A. Lindlbauer, and B. Lux, Diamond Rel. Mater. **2**, 1505 (1993)

12.16   R. Bichler, R. Haubner, and B. Lux, High Temp. – High Press. **21**, 576 (1989)

12.17   A. Lindlbauer, R. Haubner, and B. Lux, Wear **159**, 67 (1992)

12.18   H. Matsubara and J. Kihara, in Science and Technology of New Diamond, ed. S. Saito, O. Fukunaga, and M. Yoshikawa, KTK Scientific Publishers, Tokyo (1990), pp. 89–93

12.19   K. Saijo, M. Yagi, K. Shibuki, and S. Takatsu, Surf. Coat. Technol. **43/44**, 30 (1990)

12.20   J.L. Chen, T.H. Huang, F.M. Pan, C.T. Kuo, C.S. Chang, and T.S. Lin, Surf. Coat. Technol. **54/55**, 392 (1992)

12.21   A.K. Mehlmann, A. Fayer, S.F. Dirnfeld, Y. Avigal, R. Porath, and A. Kochman, Diamond Rel. Mater. **2**, 317 (1993)

12.22   D. Satrapa, Doctoral Thesis, Vienna University of Technology (1993)

12.23   S. Kubelka, R. Haubner, B. Lux, R. Steiner, G. Stingeder, and M. Grasserbauer, Diamond Films Technol. **5**, 105 (1995)

12.24   R. Haubner, W.D. Schubert, and B. Lux, in 14th International Plansee Seminar, ed. G. Kneringer et al., 12–16 May 1997, Reutte, Tirol, Austria, Vol. 3, pp. 16–30

12.25   T.H. Huang, C.T. Kuo, C.S. Chang, C.T. Kao, and H.Y. Wen, Diamond Rel. Mater. **1**, 594 (1992)

12.26   K. Shibuki, M. Yagi, K. Saijo, and S. Takatsu, Surf. Coat. Technol. **36**, 295 (1988)

12.27   A.K. Mehlmann, S.F. Dirnfeld, and Y. Avigal, Diamond Rel. Mater. **1**, 600 (1992)

12.28   M.G. Peters and R.H. Cummings, European Patent No 0519 587 A1 (1992)

12.29   T. Isozaki, Y. Saito, A. Masuda, K. Fukumoto, M. Chosa, T. Ito, E.J. Oles, A. Inspector, and C.E. Bauer, Diamond Rel. Mater. **2**, 1146 (1993)

12.30   S. Kubelka, Doctoral Thesis, Vienna University of Technology (1994)

12.31   S. Kubelka, R. Haubner, B. Lux, R. Steiner, G. Stingeder, and M. Grasserbauer, Diamond Rel. Mater. **3**, 1360 (1994)

12.32   X.X. Pan, Doctoral Thesis, Vienna University of Technology (1989)

12.33   R. Haubner, S. Kubelka, B. Lux, M. Griesser, and M. Grasserbauer, J. Phys. **IV**, C5, 753 (1995)

12.34   M. Nesladek, K. Vandierendonck, C. Quaeyhaegens, M. Kerkhofs, and L.M. Stals, Thin Solid Films **270**, 184 (1995)

12.35   W.G. Moffatt, The Handbook of Binary Phase Diagrams, Genium, New York

12.36   J. Karner, E. Bergmann, M. Pedrazzini, I. Reineck, and M. Sjöstrand, Int. Patent WO 95/12009 (4. May 1995)

12.37   H. Lux, Doctoral Thesis, Vienna University of Technology (1995)

12.38   A. Lindlbauer, R. Haubner, and B. Lux, Diamond Films Technol. **2**, 81 (1992)

12.39   G. Rosenauer, Diploma Thesis, Lehrstuhl Metalle (WTM), Universität Erlangen-Nürnberg (1997)

12.40   G. Rosenauer, T. Grögler, S.M. Rosiwal, and E.F. Singer, Diamond Rel. Mater. (in preparation)

12.41   C.-P. Klages, M. Fryda, T. Matthee, L. Schäfer, and H. Dimingen, in 14th International Plansee Seminar, ed. G. Kneringer et al., 12–16 May 1997, Reutte, Tirol, Austria, Vol. 3, pp. 1–15

12.42 L. Schäfer, A. Blum, M. Sattler, R. Six, and C.-P. Klages, in Applications of Diamond Films and Related Materials: Third Conference, ed. A. Feldman et al., NIST Special Publication **885**, Washington (1995), pp. 399–402

12.43 E. Cappelli, F. Pinzari, P. Ascarelli, and G. Righini, Diamond Rel. Mater. **5**, 292 (1996)

12.44 L.K. Bigelow, NORTON, personal communication

12.45 J.C. Cubertafon, M. Chenevier, A. Campargue, G. Verven, and T. Priem, Diamond Rel. Mater. **4**, 350 (1995)

12.46 M. Griesser, Doctoral Thesis, Vienna University of Technology (1995)

12.47 K. Plamann and D. Fournier, Diamond Rel. Mater. **4**, 809 (1995)

12.48 P.K. Bachmann, H.J. Hagemann, H. Lade, D. Leers, D.U. Wiechert, H. Wilson, D. Fournier, and K. Plamann, Diamond Rel. Mater. **4**, 820 (1995)

12.49 Product information, DeBeers Industrial Diamond Division, Charters, Sunninghill, Ascot, Berkshire SL5 9PX, England

12.50 D.S. Hoover, presentation at Diamond Films'94, 25–30 September 1994, Il Ciocco, Tuscany, Italy

12.51 N. Fujimori, presentation at Diamond Films'94, 25–30 September 1994, Il Ciocco, Tuscany, Italy

12.52 J. Karner, M. Pedrazzini, and C. Hollenstein, Diamond Rel. Mater. **5**, 217 (1996)

12.53 E. Sevillano, J. Casey, R. Gat, S. Jin, R.S. Post, and D.K. Smith, presentation at Diamond Films'94, 25–30 September 1994, Il Ciocco, Tuscany, Italy

12.54 R. Haubner and B. Lux, Diamond Rel. Mater. **2**, 1277 (1993)

12.55 P.K. Bachmann, in Properties and Growth of Diamond, ed. G. Davies, emis Datareviews Series, Short Run Press Ltd., Exeter (1994), pp. 349–67

12.56 J. Herlinger and JEH Gorham Conference, 14 March 1996

12.57 V. Böhm, V. Buck, M. Liesenfeld, T. Naubert, and J. Zeng, Diamond Rel. Mater. **4**, 33 (1994)

12.58 E. Bergmann, C. Hollenstein, J. Karner, and M. Pedrazzi, in 1st Swiss Conference on Materials Research for Engineering Systems, ed. B. Ilschner et al., 8–9 September 1994, Sion, pp. 275–9

12.59 I. Reineck, M.E. Sjöstrand, J. Karner, and M. Pedrazzini, Diamond Rel. Mater. **5**, 819 (1996)

12.60 A. Lindlbauer, R. Haubner, and B. Lux, Refract. Metals Hard Mater. **11**, 247 (1996)

12.61 Crystalline Materials Corporation CMC, personal communication

12.62 P. Hartmann, R. Haubner, and B. Lux, Diamond Rel. Mater. **5**, 850 (1996)

12.63 Y.A. Mankelevich, A.T. Rakhimov, and N.V. Suetin, Diamond Rel. Mater. **4**, 1065 (1995)

12.64 I. Reineck, M.E. Sjöstrand, J. Karner, and M. Pedrazzini, Refract. Metals Hard Mater. **14**, 187 (1996)

12.65 J. Karner, M. Pedrazzini, I. Reineck, M.E. Sjöstrand, and E. Bergmann, Mater. Sci. Eng. **A209**, 405 (1996)

12.66 D.B. Arnold, in 14th International Plansee Seminar, ed. G. Kneringer et al., 12–16 May 1997, Reutte, Tirol, Austria, Vol. 1, pp. 10–26

12.67 E.J. Oles, A. Inspektor, and C.E. Bauer, Diamond Rel. Mater. **5**, 617 (1996)

12.68 S. Matsumoto, Y. Sato, M. Tsutisimi, and N. Setaka, J. Mat. Sci. **17**, 3106 (1982)

13. CVD-Diamond Sensors for Temperature and Pressure

Matthias Werner

VDI/VDE-Technologiezentrum Informationstechnik GmbH,
Rheinstrasse 10 B, D-14513 Teltow, Germany
e-mail: mwerner@vdivde-it.de

Springer Series in Materials Processing
Low-Pressure Synthetic Diamond Eds.: B. Dischler and C. Wild
© Springer-Verlag Berlin Heidelberg 1998

13.1 Introduction

A study of the total available market for high-temperature electronics estimates that by the year 2005 roughly 4.7% of high-temperature applications will require devices operating at temperatures in excess of 300°C [13.1]. According to this study, the most important markets for high-temperature electronics are well logging, aerospace and automotive applications. Wide bandgap semiconductors such as diamond, SiC and III-nitrides are attractive materials for temperatures beyond 300°C, at which conventional devices cannot operate. The figures of merit for diamond are superior to those of all other semiconductors and diamond devices could theoretically demonstrate outstanding performance [13.2]. So far, the lack of large-area single-crystalline substrates and the lack of a significant, controllable n-type conductivity has limited the number of possible devices. Diamond Schottky diodes and active components are discussed in Chap. 17. A detailed study concerning the industrial need for diamond in general is given in Chap. 18. Apart from its ability to operate at high temperatures, diamond is chemically inert against environmental influences and is radiation hard. It may have been that the first diamond sensor application was considered over 70 years ago, with evaluations of natural diamond stones for potential use as radiation detectors [13.3, 4]. With the advent and improvement of chemical vapour deposition of large-area diamond films, new applications become possible, in combination with the outstanding physical properties of diamond. An excellent review concerning passive diamond electronic devices in general was given by Dreifus [13.4]. Currently, thermistors and pressure sensors are being tested at the engineering prototype level. It is expected that diamond sensors for high-temperature applications will penetrate into the market in 1998 [13.1]. This chapter concentrates on two kinds of promising sensing devices; namely, temperature and pressure sensors.

13.2 Diamond Sensor Technology

Many properties of diamond (e.g. its physical hardness, high Young's modulus, high tensile yield strength, chemical inertness, low coefficient of friction and high thermal conductivity) make it an excellent material for micromechanical device applications [13.5], which include temperature and pressure sensors. It is a clear requirement that processes that are part of the conventional silicon technology should be used. However, there are some important differences concerning doping, patterning and metallizations, which will be discussed in the following sections.

13.2.1 Diamond Doping

One key process for the use of diamond as a starting material for electrical components or sensors is doping. From natural diamond stones it is known that boron acts as an acceptor, with an acceptor level of 0.368 eV above the valence band edge. Boron doping can be easily achieved by *in-situ* doping during growth [13.6], by ion implantation [13.7] or by high-temperature diffusion [13.8]. Isolated substitutional nitrogen and the A centre (a pair of nearest-neighbour nitrogen atoms) have donor-like properties with energies of 1.7 eV and 4.0 eV respectively. However, even high nitrogen concentrations lead to negligible n-type conductivity at room temperature, due to the deep-lying donor level: therefore nitrogen acts mainly as a compensating donor. If the problem of a significant n-type conductivity can be solved, p–n junctions and therefore active electronic devices become possible. For most sensor applications, one type of doping is sufficient and an undoped diamond film can act as an electrical insulator. Significant diffusion of impurities is most unlikely below 1000°C due to the small diffusion coefficients of most impurities in diamond. This could be an advantage in view of high-temperature applications and device stability.

13.2.2 Patterning and Cleaning

Patterning

Important differences compared to the standard silicon technology are patterning and cleaning of diamond. Diamond is an extremely chemically inert and physically hard substance, making it difficult to etch or process it in a similar way to silicon. In principle, two basic methologies can be employed to define structures in diamond: either selective removal or selective deposition of material.

Selective-area deposition (SAD) of diamond can be achieved using an appropriate high-temperature mask, such as a oxynitride mask, on suitable substrates [13.9]. After patterning of the masking layer, conventional growth techniques can be used to achieve defined structures. A modified SAD process was reported by Taher and co-authors [13.10]. This group used a patterned diamond powder photoresist. During diamond growth, the photoresist evaporates in the reactor, and the remaining diamond particles act as seeds for the growth of a randomly oriented polycrystalline diamond film.

A completely different method is post-depositional patterning of diamond. Due to the chemical resistance of diamond to wet chemical etching, dry etching methods must be considered. Etching can be achieved through different etching procedures; for example, by reactive ion etching (RIE) in an oxygen plasma, oxygen-based electron cyclotron resonance etching, bias-assisted etching in a hydrogen-containing plasma in microwave- and DC-plasma reactors, laser processing and ion-beam-assisted etching [13.11]. In the case of RIE, suitable masking layers are oxides and nitrides. Increasing the RF power leads to an increase in the etch rate, but the etch rate is also dependent on the CVD-diamond quality, as determined by Raman spectroscopy. Typical etch rates range from 10 nm/min to 60 nm/min in a 13.56 MHz etcher at 300 W and 13.3 Pa of O_2. A comparison of the etch rate of amorphous carbon and polycrystalline diamond etched under similar conditions has revealed that the etch rate of amorphous carbon is almost one order of magnitude higher [13.12]. Each etching method has a specific drawback. For example, RIE offers the potential to pattern large areas with high resolution at low etching rates, whereas ion-beam or laser etching have much higher etch rates but are limited to small areas.

Cleaning

The characterization and control of semiconductor surfaces and interfaces plays a key role in developing any type of device. The interaction of diamond with hydrogen has received much attention recently. Hydrogen can be present in the bulk and at the surface of the as-grown material. In order to stabilize the material, it is necessary to anneal the sample in argon or dry nitrogen at temperatures around 500°C. Normally, a low-resistivity layer is present on as-grown doped and undoped diamond surfaces. This layer can be removed [13.13] by chemical cleaning with sulphochromic acid or oxygen-plasma treatment, which is necessary to obtain stable devices. However, the origin of the surface layer is not clear at present. There are three possible explanations of the surface-conducting layer:

(a) the existence of a graphite-like low-resistivity layer;

(b) hydrogen passivation of deep levels near the surface; or

(c) upwards band-bending to form an accumulation layer for holes.

Shirafuji and Sugino [13.14] have provided arguments to suggest that the band-bending model is most likely to provide an explanation of the surface conducting layer. According to their results, the surface of the as-grown hydrogen-terminated samples bends upwards for holes to give a low surface resistivity, while the oxygenated surface has a depletion layer for holes. It is therefore reasonable that the above-mentioned chemical cleaning procedures lead to oxygenated diamond surfaces and that the Fermi level is pinned at a specific energy.

13.2.3 Ohmic Contacts to Diamond

High-quality, stable, low-resistivity ohmic contacts are essential for any electrical diamond device. It has been shown that high-resistivity contacts can strongly alter measured device characteristics [13.15]. In general, the requirements for contacts can be summarized as follows:
- low contact resistivity
- good adhesion
- high thermal stability
- high corrosion resistance
- a bondable top layer
- suitable for micro-patterning.

One essential property for the contact resistivity is the metal/diamond barrier height. Following the Mead and Spitzer rule [13.16], the Schottky barrier height of covalent p-type semiconductor should be 1/3 of the bandgap. Therefore barrier heights in the range of 1.8 eV for diamond can be expected. The most conventional way to obtain low contact resistivites is heavy doping of the contact area. For boron-doped diamond, doping concentrations higher than 10^{21} cm^{-3} have been reported [13.17]. In qualitative agreement with the theory, the contact resistivity drops with increasing doping concentration [13.18]. Due to the high doping concentration, the depletion layer width is reduced so that quantum-mechanical tunnelling becomes important. At very high doping concentrations ($E_{00}/(kT) \gg 1$), field emission is the dominant current transport mechanism [13.19]. In this case, the contact resistivity can be asymptotically written in the form

$$\rho_c \approx Ce^{q\phi_b/E_{00}} \tag{13.1}$$

where C is approximately temperature independent, but is a sensitive function of the barrier height and doping concentration, and E_{00} is given as

$$E_{00} = \frac{qh}{4\pi}\sqrt{\frac{N_A}{m\varepsilon_o\varepsilon_r}} . \tag{13.2}$$

Here, m denotes the effective mass and h is Planck's constant. The other symbols have their usual meanings. It should be noted that the logarithm of the specific contact resistivity is proportional to $N_A^{-1/2}$ due to the dependence on E_{00} in the case of pure field emission; that is, for heavy doping. On the one hand, large barrier heights lead to high contact resistivities. On the other, contact resistivities as low as $\sim 10^{-7}$ Ωcm^2 have been reported for Al/Si contacts [13.20] after annealing at 450°C. The low contact resistivity has been traced back to the formation of SiC at the metal/diamond interface. Similar results were reported for carbide-forming metals such as Ti [13.21] and Mo [13.22]. Tachibana, Williams and Glass [13.21] suggested two models that may explain the drop in the contact resistance or the change in the current–voltage characteristic from rectifying to ohmic. The models are based on the assumption that the carbide acts as a defect layer that lowers the metal/diamond barrier height, enhances tunnelling, or both. Another model proposed an average decreasing amount of local disorder due to annealing [13.20]. If there is a sufficiently large density of gap states near the Fermi level, then some type of carrier transport can take place with the help of these gap states. However, none of these models has been fully proven. When the metallization of a carbide-forming metal, after carbide formation is removed small islands of carbide precipates can be identified at the diamond surface. If these carbide islands conduct better than the surroundings, where no reaction has occurred, then the current density at the metal/diamond interface must be inhomogeneous. Consequently, two different types of carrier transport mechanism operate in parallel in isolated area segments distributed uniformly across the interface. The change in the temperature dependence of the contact resistivity after carbide formation [13.23] strongly suggests that such a mechanism is responsible for the observed behaviour. Similar observations were reported by Chu and co-authors for Pt/Ti ohmic contact to p-In$_{0.53}$Ga$_{0.47}$As. [13.24]. Figure 13.1 is a plot of experimen-

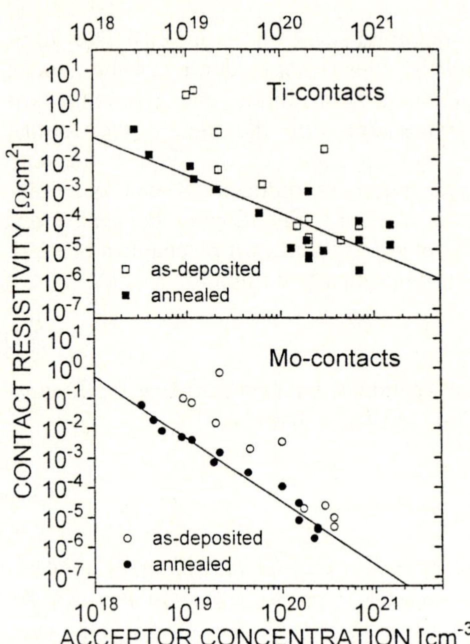

Fig. 13.1: Contact resistivities for Ti and Mo contacts on B-doped CVD diamond. The solid lines are power fits for the annealed contacts.

tally determined contact resistivities from various authors before and after annealing [13.18, 25–27]. Obviously, the data do not follow (13.1) at high doping levels. Instead, the data of the annealed samples can be fitted by a power law. Consequently, the second current contribution due to the carbide-island formation in diamond cannot be neglected.

Another important point is the use of diffusion barriers to avoid interdiffusion of the contact layer and the bondable top metallization (e.g. Au) at high temperatures. The most frequently used Ti–Au contacts display a strong interdiffusion at temperatures below 450°C [13.18]. TiWN–Au contacts are stable up to temperatures of approximately 450°C. Nitrogen is believed to saturate grain boundaries in the metallization, therefore avoiding the interdiffusion along the grain boundaries. At 600°C they show a similar degradation to Ti–Au contacts. The basic idea is thus to avoid grain boundaries. Useful concepts for high-temperature metallizations with amorphous diffusion barriers were reported by Wiley and co-authors [13.28]. However, to the author's best knowledge, such a concept has never been reported for diamond devices.

13.3 Diamond Temperature Sensors

A diamond thermistor is in its most simple form a two-terminal device with two ohmic contacts. Due to the deep-lying B-acceptor level, only a small fraction of the carriers (i.e. holes) are ionized at room temperature in lightly doped diamond.

In a typical natural semiconducting type IIb diamond, about 0.2% of the acceptors are ionized at 20°C. With increasing temperature the number of holes increases and therefore the electrical conductivity increases. Thus the electrical conductivity is a function of temperature. In general, the following demands are placed on thermistors:

- high temperature sensitivity
- good temperature stability
- a large measuring range
- stable bondable ohmic contacts
- an immunity against environmental influences
- short response times.

Optimization of these properties requires the knowledge of the current conduction mechanism to some extent.

13.3.1 Current Conduction Mechanism

The temperature dependence of the electrical conductivity σ for an extrinsic, partially compensated semiconductor can be approximated by

$$\sigma = \sigma_1 e^{-E_1/kT} + \sigma_2 e^{-E_2/kT} + \sigma_3 e^{-E_3/kT} \tag{13.3}$$

where k is Boltzmann's constant ($k = 8.62 \times 10^{-5}$ eV/K), T is the absolute temperature, σ_1, σ_2 and σ_3 are constants and E_1, E_2 and E_3 describe energies associated with three different conduction mechanisms. The first term relates to valence-band conduction and the third term to nearest-neighbour hopping, due to hopnping of holes from occupied to unoccupied acceptor sites, which is made possible by the finite overlap of the wave functions. The second term is associated with conduction in an impurity band. This conduction mechanism can only be observed in the intermediate concentration range for certain values of the compensation. Malta et al. [13.29] reported resistivity and Hall measurements on a set of nine samples with various doping concentrations and found that current transport in association with impurity band (E_2) conduction was evident in only two of the nine samples. The first term in (13.3) is simply the valence-band conductivity $q\mu p$, where q is the elementary charge, μ is the band mobility and p is the hole concentration with the activation energy $E_1 = 0.368$ eV. The carrier concentration p for an partially ionized, non-degenerated and compensated p-type semiconductor can be written as

$$p = \left(\sqrt{\frac{1}{4}\left(\frac{p_1}{2N_A} + K\right)^2 + \frac{p_1}{2N_A}(1-K)} - \frac{1}{2}\left(\frac{p_1}{2N_A} + K\right) \right) N_A \tag{13.4}$$

$$K = N_D/N_A, \qquad p_1 = N_V e^{-E_1/kT}, \qquad N_V = 4.83 \times 10^{15} \left(m/m_o\right)^{\frac{3}{2}} T^{\frac{3}{2}}$$

250   M. Werner

where K is the compensation ratio, N_A and N_D are the acceptor and donor concentration respectively, N_V is the effective density of states at the valence-band edge and m/m_o is the ratio of the effective mass to the free electron mass. The energy E_3 was shown to be dependent only on the acceptor concentration N_A, and not on the compensation ratio K [13.30]. Therefore (13.3) can be written as

$$\sigma = q\mu p + \sigma_{03}\, e^{-\left(\alpha/a_H N_A^{-1/3}\right)}\, e^{-\left(E_3/kT\right)} \tag{13.5}$$

with $\alpha/a_H = 2.16 \times 10^7\ \mathrm{cm}^{-1}$ and $\sigma_{03} = 283.3\ \Omega^{-1}\mathrm{cm}^{-1}$ [13.30].

13.3.2 Activation Energy and Temperature Sensitivity

According to (13.5), the temperature dependence of the conductivity is clearly non-Arrhenius. Therefore the activation energy of the conductivity is a function of temperature. Figure 13.2 is a plot of experimentally determined activation energies from various references [13.30]. Obviously, the activation energy drops with increasing doping concentration N_A - N_D and depends on the compensation level at low temperatures. The solid lines were calculated by using

$$E_A = -\frac{\partial \ln\sigma}{\partial\left(1/kT\right)} = -\frac{q}{\sigma}\left(\mu\frac{\partial p}{\partial\left(1/kT\right)} + \frac{\partial\mu}{\partial\left(1/kT\right)}p - \sigma_3 E_3 e^{-\left(E_3/kT\right)}\right) \tag{13.6}$$

with a temperature-dependent mobility. In engineering terms, the resistive temperature detectors (RTDs) are described by their temperature coefficient of resistance (TCR). The TCR and the activation energy are related by

$$TCR = \frac{1}{R}\frac{\partial R}{\partial T} = -\frac{E_A}{k}\frac{1}{T^2}. \tag{13.7}$$

According to (13.5), (13.4) and (13.2), the TCR depends on the carrier concentration, the mobility, the density of states at the valence band edge and on the compensation ratio. From Fig. 13.2, high TCRs can be achieved by low doping concentrations, high compensation levels and thus high resistivities which, however, limits the low-temperature performance of the thermistor.

13.3.3 Diamond NTC Thermistors

The requirements for diamond thermistors depend on the operational temperature range and desired resistance values. It has been shown that the TCR for B-doped diamond can vary from roughly $-10^{-2}\ \mathrm{K}^{-1}$ with a negative temperature coefficient (NTC) to slightly positive values with increased doping concentration [13.31].
If a diamond thermistor must operate from a few milliKelvin to temperatures around 800 K, as is necessary for cryogenic test equipment, heavily doped diamond films have to be used. For a small operating range, advantage can be

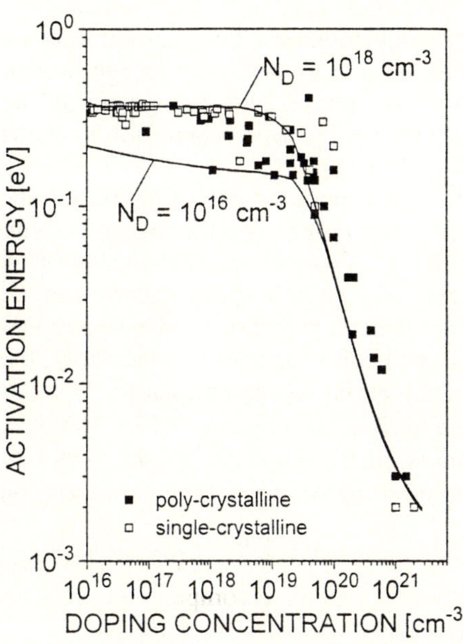

Fig. 13.2: The activation energy of B-doped poly- and single-crystalline diamond films. The solid lines were calculated for two donor concentrations at 500 K.

gained from the large TCR at low doping concentrations. A number of groups have reported different kinds of thermistors in either rectangular or serpentine arrangements [13.9, 31–41]. The impact of the geometry can be estimated from the work of Hall [13.42] and using standard textbooks, e.g. [13.43]. Figure 13.3 is an SEM photograph of a serpentine-type thermistor. The chip size is $750 \times 500\ \mu m^2$. The undoped diamond film underneath the B-doped diamond film serves as an electrical insulator to the silicon substrate. The randomly oriented undoped diamond film and the B-doped layer have thicknesses of 3.5 μm and 1.5 μm respectively. This thermistor has a TCR of $-1.5 \times 10^{-3}\ K^{-1}$ [13.31].

By lowering the doping concentration of the B-doped resistor path, the TCR can be improved. Job et al. reported a TCR of $-5 \times 10^{-2}\ K^{-1}$ [13.44] by using a planar p–i–p structure made from single-crystal insulating type IIa diamond. The

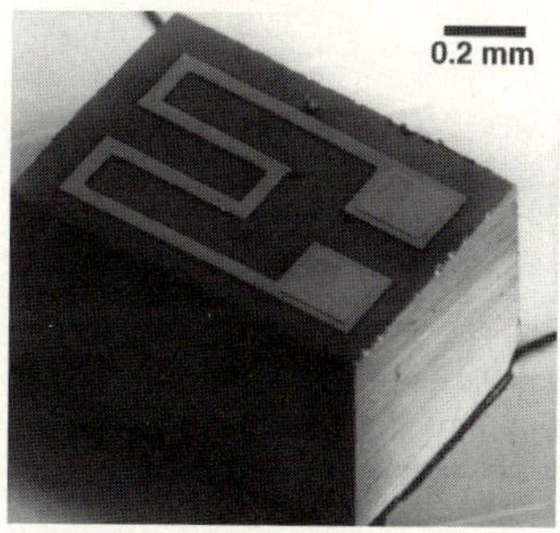

Fig. 13.3: A diamond thin-film thermistor. The B-doped resistor path is insulated from the substrate by an undoped diamond film.

p-type regions were produced by B-ion implantation. The current–voltage characteristic shows a pronounced non-linearity due to the space-charge-limited current flow in the "intrinsic" diamond region. The temperature dependence of the current–voltage characteristic is strongly influenced by traps. Therefore, it seems to be difficult to achieve reproducible results.

In general, diamond thermistors present a good opportunity for high-accuracy temperature measurements over a large measuring range. The thermal diffusivity of diamond is extremely high, and approximately 47 times faster than that of an equivalent mass of platinum [13.45]. Therefore, short response times can be expected. Measurements [13.40] on thin-film diamond thermistors have shown that the response time was greatly affected by the silicon substrate. Consequently, the authors have suggested that it is better to use thermal insulators such as Al_2O_3 or fused quartz as substrates. Furthermore, the TCR of platinum is 3.85×10^{-3} K^{-1}, whereas typical values for B-doped diamond thermistors are in the range of -5×10^{-3} K^{-1} to -1×10^{-2} K^{-1} [13.31, 40], depending on the doping concentration and compensation.

13.3.4 Related Devices

Jones [13.45] suggested that self-heated diamond thermistors can be used for hostile liquid level sensing. She pointed out that the rate of decay, or rise time in some cases, will vary according to the temperature of the environment, humidity and mass flow as well as the liquid level. Following this idea, a single-structure heater and temperature sensor was reported by Yang and Aslam [13.46] and Gluche and co-authors [13.47]. Due to its high thermal conductivity and sensitivity, the diamond resistor is superior to conventional polysilicon or platinum resistors, and can be used in harsh environments. Another approach is to use diamond as a basic material for flow sensors for corrosive or abrasive fluids [13.4, 48]. However, these devices have to be improved greatly for practical use.

A similar concept for a different application is an "intelligent" heat-spreader, with the opportunity to measure and control the temperature of a heat-spreader [13.49] for semiconductor lasers. By avoiding self-heating of a semiconductor laser, the lifetime can be markedly increased.

13.4 Piezoresistive Diamond Sensors

The piezoresistive effect in silicon technology is widely used for pressure, force and acceleration sensors. The most prominent example is piezoresistive acceleration sensors for airbags in cars. Commercially available sensors cannot operate at high temperatures because of the p–n insulation of the piezoresistors. Some industrial applications (e.g. logging applications) can require temperatures as high as 600°C and pressures of 1750 bars [13.1]. In several publications [13.10,

50–59] it has been shown that diamond exhibits a significant piezoresistive effect and can meet extreme environmental conditions. For the layout of pressure sensors, in addition to the electronic properties, knowledge of the elastic properties is required.

13.4.1 The Piezoresistive Effect in Diamond

When a force is applied to a semiconductor with a cubic lattice, its electrical resistivity changes significantly owing to the piezoresistive effect. Kanda [13.60] explained the piezoresistive effect in single-crystal silicon by the carrier-transfer mechanism and the effective mass change. The resistance change in terms of the strain ε can be expressed by the gauge factor, GF:

$$GF = \frac{\Delta R}{R}\frac{1}{\varepsilon} = 1 + 2v + \frac{1}{\varepsilon}\frac{\Delta \rho}{\rho} \tag{13.8}$$

where v is the Poisson ratio and ρ $(= 1/\sigma)$ the resistivity. In the case of anisotropic, single-crystalline semiconductors with a cubic lattice, the resistivity change depends on the longitudinal, transverse and shear piezoresistive coefficients, which are not known for diamond. However, for randomly orientated polycrystalline material, the longtudinal $<\pi_l>$ and transverse $<\pi_t>$ piezoresistance coefficients can, in principle, be estimated by averaging over all possible crystal directions. The average longitudinal and transverse piezoresistance coefficients for the polycrystalline material can be expressed in terms of the gauge factor

$$< \pi_l > = GF_l \frac{1}{<Y>}, \quad < \pi_t > = GF_t \frac{1}{<Y>} \tag{13.9}$$

where $<Y>$ denotes the Young's modulus for the polycrystalline material and the subscripts refer to the longitudinal and transverse directions respectively. The Young's modulus for a randomly orientated diamond film can be calculated by averaging over all possible crystal directions [13.61, 62]. For the tabulated stiffness and compliance constants [13.63, 64], the Young's modulus and the Poisson ratio of randomly orientated diamond are calculated to be 1143 GPa and 0.07 respectively. In the case of a fiber texture, the Young's modulus is 1166 GPa for a <111> texture, 1151 GPa for a <110> texture and 1107 GPa for a <100> texture. These values are in good agreement with the experimentally determined Young's moduli from D'Evelyn et al. [13.65], which range from 980 to 1161 GPa. These authors argued that it is most likely that not all of the grains in textured diamond have a fiber axis precisely aligned along the sample normal, which may explain the observed deviations from the theoretical values. However, the value determined for the Young's modulus depends strongly on the measurement method and the number of defects present in the sample [13.66]. Values as low as 242 GPa were determined on preferentially <110> textured 300 μm thick bulk CVD diamond, as determined by the three-point bending method; whereas on a sample from the same batch, a Young's modulus of 1198 GPa was measured by sound velocity measurements, in agreement with the calculated value [13.67]. A more detailed analysis revealed some large voids in the middle of the sample, which may have a weaker influence on the sound velocity when compared with the three-point bending measurement.

Experimentally Determined Gauge Factors

Gauge factors for B-doped poly-crystalline diamond films have been reported by several research groups [13.50–59]. Most research groups used patterned, B-doped diamond piezoresistors, usually on an insulator on a silicon substrate. Beams are normally cut from the silicon wafer containing the diamond piezoresistors. The beam is then mounted in a test fixture and deflected using a micrometer head. Another, more sophisticated, approach was reported by Gluche et al. [13.68]. They used a free-standing diamond cantilever with well-defined B-doped piezoresistors on an undoped diamond film on a back etched silicon substrate, as shown in Fig. 13.4. The beam can be deflected with a nanoindenter.

The hole at the end of the beam allows a precise positioning of the nanoindenter. This structure also allows the determination of the Young's modulus of the diamond film under test.

Dorsch et al. [13.51] reported a simultaneous determination of the longitudinal and transverse piezoresistive effects. In this case, an arrangement of the resistors at right angles was used. In the case of the longitudinal piezoresistive effect, the electrical field, the current density and the mechanical strain are in the same direction, whereas in the case of the transverse piezoresistive effect, the electrical field and current density act perpendicular to the mechanical strain. A typical measurement result, in which the relative resistance referred to the resistance at $\varepsilon = 0$ is plotted against the strain, is shown in Fig. 13.5. Obviously, the longitudinal piezoresistive effect is larger than the transverse piezoresistive effect, which also shows a pronounced non-linearity. The gauge factor of the longitudinal arrangement is 3.7. For B-doped polycrystalline CVD diamond, gauge factors

Fig. 13.4: The cantilever test structure for the determination of the gauge factor and the Young's modulus.

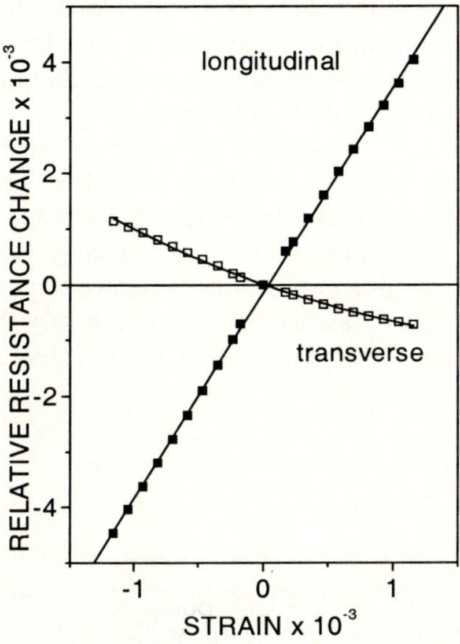

Fig. 13.5: The relative resistance change vs. strain for a heavily doped diamond film with a hole concentration of $2.5 \times 10^{20}\,\mathrm{cm}^{-3}$.

between 2 and 1000 at room temperature have been reported [13.50–59]. These values depend greatly on the doping concentration and temperature. For comparison, the gauge factors of metals and semiconductors are in the range of 2–12 and 5–175 respectively [13.10]. Silicon is most often used for piezoresistive sensors, but this material is not suitable for high-temperature and harsh-

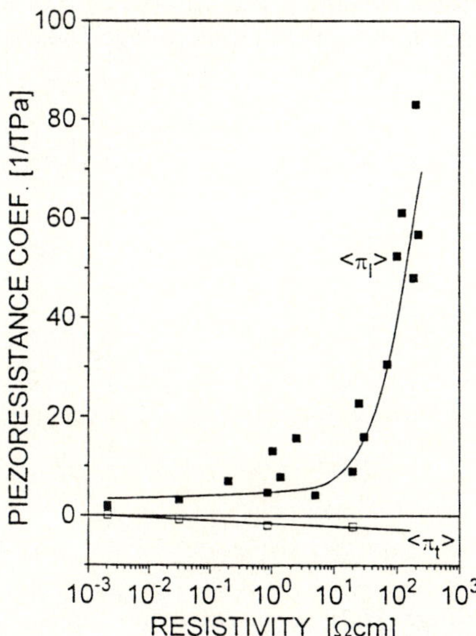

Fig. 13.6: Average longitudinal and transverse piezoresistance coefficients from various research groups.

environment applications. Also, for the technologically more developed SiC, which is suitable for high-temperature applications, maximum gauge factors around 30 have been reported [13.69]. Therefore, even polycrystalline diamond seems to be a better candidate when compared with other semiconductors.

The piezoresistive coefficients of B-doped polycrystalline diamond, calculated from the published gauge factors, versus resistivity are summarized in Fig. 13.6. The coefficients were calculated by assuming that the Young's modulus is 1143 GPa. Obviously, the piezoresistive coefficients increase strongly with increasing resistivity; that is, with decreasing doping concentration. However, care must be taken for lightly doped diamond, since it has been shown previously that piezojunction effects due to non-ohmic contacts can have a significant effect on the measurement result [13.70].

13.4.2 Diamond Pressure Sensors

The sensitivity of piezoresistive pressure sensors depends on a variety of parameters, such as the diaphragm size, the doping concentration and the location of the piezoresistors. Taking into account that the longitudinal piezoresistive effect is much larger than the transverse effect, the longitudinally piezoresistive effect should be preferred. In a piezoresistive pressure sensor, the change in resistance on exposure to pressure is determined simply by connecting four resistors in a Wheatstone bridge configuration on the diaphragm. Such a configuration with diamond piezoresistors on a silicon diaphragm is shown in Fig. 13.7 [13.71]. If hydrostatic pressure is applied to the membrane, the two outer resistors are under tensile stress and the two inner piezoresistors are under compressive stress. When the resistors are not exposed to pressure, the bridge voltage is zero. In the case of a resistance change, ΔR, of equal value but of opposite sign, the output voltage U_{OUT} is given by

$$U_{OUT} = U_S \frac{\Delta R}{R} = U_S GF\varepsilon \qquad (13.10)$$

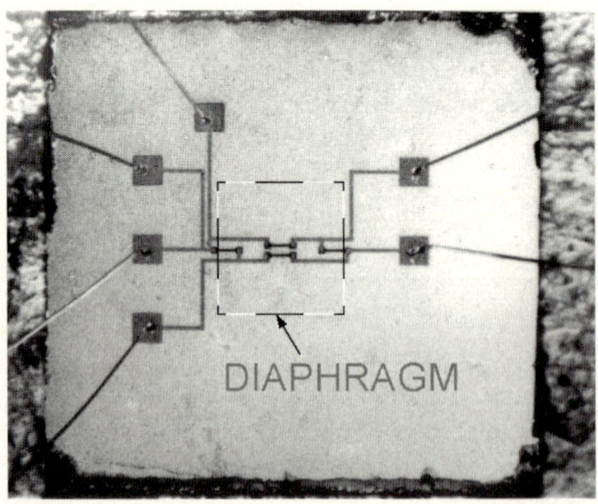

Fig. 13.7: A piezoresistive pressure sensor. The diaphragm dimensions are 1.3 mm × 1.3 mm × 30 μm.

where U_S is the source voltage. The output characteristics of such a pressure sensor with diamond piezoresistors are shown in Fig. 13.8 for various temperatures [13.71]. A similar layout with a diamond diaphragm was reported by Chalker et al. [13.9]. Wur et al. [13.55] pointed out that there are two advantages in using polycrystalline diamond as a diaphragm material. First, diamond is chemically inert and can serve as an etch-stop with a large etching time tolerance. Second, polycrystalline diamond films can be deposited with control by adjusting the deposition time and other growth parameters, resulting in selectability of the diaphragm thickness. Therefore, the full-scale pressure range can be varied with polycrystalline diamond as the diaphragm material. When compared with a circular silicon diaphragm, the silicon diaphragm must be roughly 2.7 times thicker than a diamond diaphragm with the same radius in order to achieve similar radial and tangential strains. This is mainly due to the larger Young's modulus of diamond. Furthermore, polycrystalline CVD diamond has a rupture stress that is nominally three orders of magnitude larger than that of silicon [13.57]. Hence the diamond diaphragm is more rugged than silicon; it can withstand greatly increased pressures.

Another approach is the use of an heteroepitaxial SiC diaphragm. An SiC diaphragm offers the possibility of achieving high-quality heteroepitaxial diamond growth (see Chap. 8), has a lower Young's modulus than diamond, is chemically inert and can therefore also serve as an etch-stop. Gauge factors of B-doped CVD diamond grown on SiC have been reported by Dreifus [13.4]. When lightly doped piezoresistors with high gauge factors are used, the resistivity and the output voltage are strongly temperature dependent. Therefore an on-chip recording of the temperature is required. This can be simply achieved by an integrated diamond thermistor outside the diaphragm. Currently, the highest reported operating

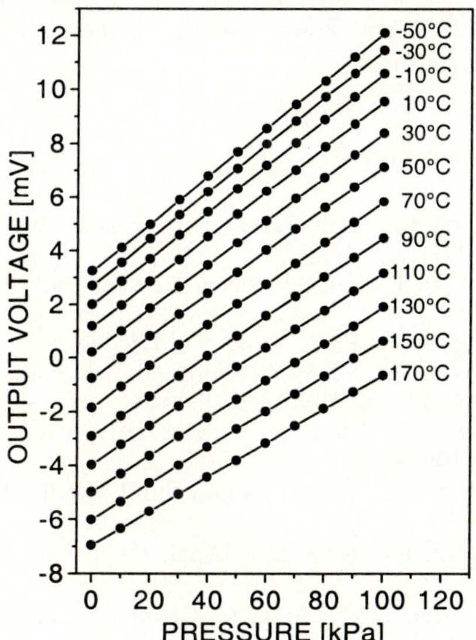

Fig. 13.8: The pressure sensor characteristic.

temperature for diamond pressure sensors is 300°C [13.57], where the maximum temperature was mainly limited by the test equipment. To achieve higher operating temperatures, the passivation of the diamond surface is necessary in order to prevent oxidation effects [13.72]. Furthermore, diamond pressure sensors with a silicon substrate are limited to operational temperatures below 600°C, where silicon undergoes plastic deformation under minimal loads. Consequently, alternative substrates have to be used [13.4].

13.5 Summary

This chapter serves as a review concerning CVD-diamond thermistors and pressure sensors. In general, CVD diamond has the potential to penetrate into a sensor niche market for high-temperature and harsh-environment applications. Currently, the basic diamond sensor technology is developed and the basic physical properties are understood. The success of CVD diamond in the market will depend on the reproducibility of the devices and on the development of competitive materials such as SiC and III-nitrides. However, for these materials, also high-temperature stable ohmic contacts and a suitable packaging technology are required.

Acknowledgements

I would like to thank Patrick Kelly (VDI/VDI-Technologiezentrum Informations-technik GmbH), Andre Vescan (University of Ulm), Otto Dorsch (TU Berlin) and Dr. Colin Johnston (AEA Technology plc) for critical reading of the manuscript and helpful comments. Furthermore, I must thank Peter Gluche (University of Ulm) for providing photographs of his test structures.

References

13.1   AEA Technology plc, The World Market for High Temperature Electronics, A HITEN Report, AEA Technology plc (1997), pp. 1–53
13.2   K. Shenai, R.S. Scott, and B.J. Baliga, IEEE Trans. Electron Devices **36**, 1811 (1989)
13.3   J.E. Field, The Properties of Diamond ,Academic Press, London (1979)
13.4   D.L. Dreifus, in Diamond: Electronic Properties and Applications, ed. L.S. Pan and D.R. Kania, Kluwer Academic Publishers, Boston (1995), pp. 371–441
13.5   J.D. Hunn, S.P. Withrow, C.W. White, R.E. Clausing, L. Heatherly and C.P. Christensen, Appl. Phys. Lett. **65**, 3072 (1994)
13.6   R. Locher, J. Wagner, F. Fuchs, C. Wild, P. Hiesinger, P. Gonon and P. Koidl, Mat. Sci. Eng. **B29**, 211 (1995)
13.7   F. Fontaine, C. Uzan-Saguy, B. Philosoph and R. Kalish, Appl. Phys. Lett. **68**, 2264 (1996)
13.8   W. Tsai, M. Delfino, D. Hodul, M. Riaziat, L.Y. Ching, G. Reynolds and C.B. Cooper, IEEE Electron Devices Lett. **12**, 157 (1991)

13.9   P.R. Chalker and C. Johnston, Phys. Status Solidi A **154**, 455 (1996)

13.10 I. Taher, M. Aslam, M.A. Tamor, T.J. Potter and R.C. Elder, Sensors and Actuators A **45**, 35 (1994)

13.11 O. Dorsch, M. Werner, and E. Obermeier, Diamond Rel. Mater. **4**, 456 (1995)

13.12 O. Dorsch, M. Werner, and E. Obermeier, Diamond Rel. Mater. **1**, 277 (1992)

13.13 Y. Mori, H. Kawarada, and A. Hiraki, Appl. Phys. Lett. **58**, 940 (1991)

13.14 J. Shirafuji and T. Sugino, Diamond Rel. Mater. **5**, 706 (1996)

13.15 V. Venkatesan, K. Das, J.A. von Windheim, and M.W. Geis, Appl. Phys. Lett. **63**, 1065 (1993)

13.16 C.A. Mead and W.G. Spitzer, Phys. Rev. **134**(3A), A713 (1964)

13.17 M. Werner, O. Dorsch, H.U. Baerwind, E. Obermeier, L. Haase, W. Seifert, A. Ringhandt, C. Johnston, S. Romani, and P.R. Chalker, Appl. Phys. Lett. **64**, 595 (1994)

13.18 M. Werner, O. Dorsch, H.-U. Baerwind, E. Obermeier, C. Johnston, P.R. Chalker, and S. Romani, IEEE Trans. Electron Devices **42**, 1344 (1995)

13.19 A.Y.C. Yu, Solid State Electron. **13**, 239 (1970)

13.20 M. Werner, C. Johnston, P.R. Chalker, S. Romani, and I.M. Buckley-Golder, J. Appl. Phys. **79**, 2535 (1996)

13.21 T. Tachibana, B.E. Williams, and J.T. Glass, Phys. Rev. **B 45**, 11975 (1992)

13.22 C.A. Hewett and J.R. Zeidler, Diamond Rel. Mater. **1**, 688 (1992)

13.23 M. Werner, R. Job, A. Denisenko, A. Zaitsev, W.R. Fahrner, C. Johnston, P.R. Chalker, and I.M. Buckley-Golder, Diamond Rel. Mater. **5**, 723 (1996)

13.24 S.N.G. Chu, A. Katz, T. Boone, P.M. Thomas, V.G. Riggs, W.C. Dautremmont-Smith, and W.D. Johnston, J. Appl. Phys. **67**, 3754 (1990)

13.25 V. Venkatesan, D.M. Malta, K. Das, and A.M. Belu, J. Appl. Phys. **74**, 1179 (1993)

13.26 H. Shiomi, Y. Nishibayashi, and N. Fujimori, Jpn. J. Appl. Phys. **30**, 1363 (1991)

13.27 A. Otsuki, J. Nakanishi, G. Kawaguchi, in Advances in New Diamond Science and Technology, ed. S. Saito, N. Fujimori, O. Fukunaga, M. Kamo, K. Kobashi, M. Yoshikawa, MYU, Tokyo (1994), pp. 713–16

13.28 J.D. Wiley, J.H. Perepezko, J.E. Nordman, and K.-J. Guo, IEEE Trans. Industrial Electronics IE-**29**, 154 (1982)

13.29 D.M. Malta, J.A. von Windheim, H.A. Wynands, and B.A. Fox, J. Appl. Phys. **77**, 1537 (1995)

13.30 M. Werner, R. Locher, W. Kohly, D.S. Holmes, S. Klose, and H.J. Fecht, Diamond Rel. Mater. **6**, 308 (1997)

13.31 M. Werner, O. Dorsch, and E. Obermeier, Electron Technology **27**, 83 (1994)

13.32 M. Werner, V. Schlichting, and E. Obermeier, Diamond Rel. Mater. **1**, 669 (1992)

13.33 K. Miyata, K. Saito, K. Nishimura, and K. Kobashi, Rev. Sci. Instrum. **65**, 3799 (1994)

13.34 J.A. von Windheim, D.L. Dreifus, B.A. Fox, J.S. Holmes, D.M. Malta, L.S. Plano, B.R. Stoner, and H.A. Wynands, HITEN News **8**, 6 (1995)

13.35 L.F. Vereshchagin, K.K. Demidov, O.G. Revin, and V.N. Slesarev, Sov. Phys. Semicond. **8**, 1581 (1975)

13.36 G.B. Rodgers and F.A. Raal, Rev. Sci. Instrum. **31**, 663 (1960)

13.37 M. Aslam, A. Masood, R.J. Fredricks, and M.A. Tamor, SPIE **1694**, 184 (1992)

13.38 N. Fujimori and H. Nakahata, New Diamond **2**, 98 (1990)

13.39 J.P. Bade, S.R. Sahaida, B.R. Stoner, J.A. Windheim, J.T. Glass, K. Miyata, K. Nishimura, and K. Kobashi, Diamond Rel. Mater. **2**, 816 (1993)

13.40 M. Aslam, G.S. Yang, and A. Masood, Sensors and Actuators A **45**, 131 (1994)

13.41 R. Job, A.V. Denisenko, A.M. Zaitsev, M. Werner, A.A. Melnikov, and W.R. Fahrner, Mater. Res. Soc. **416**, 249 (1996)

13.42 P.M. Hall, Thin Solid Films **1**, 277 (1967/1968)

13.43   S.M. Sze, Semiconductor Devices, Physics and Technology, John Wiley, New York (1985), pp. 470–71

13.44   R. Job, A.V. Denisenko, A.M. Zaitsev, A.A. Melnikov, M. Werner, and W.R. Fahrner, Thin Solid Films **290–291**, 165 (1996)

13.45   B.L. Jones, Mater. Sci. Eng. **B11**, 149 (1992)

13.46   G.S. Yang and D.M. Aslam, IEEE Electron Devices Lett. **17**, 250 (1996)

13.47   P. Gluche, R. Leuner, A. Vescan, W. Ebert, E.P. Hofer, and E. Kohn, presented at the HITEN Thematic Interest Group Meeting, Hilton Paris, France, 24 October 1996

13.48   T. Roppel, R. Ramesham, C. Ellis, and S.Y. Lee, Thin Solid Films **212**, 56 (1992)

13.49   P. Koidl, C. Wild, and E. Wörner, presented at the International Conference and Exhibition Micro Materials, Micro Mat '97, Berlin, 16–18 April 1997

13.50   M. Aslam, I. Taher, A. Masood, M.A. Tamor, and T.J. Potter, Appl. Phys. Lett. **60**, 2923 (1992)

13.51   O. Dorsch, K. Holzner, M. Werner, E. Obermeier, R.E. Harper, C. Johnston, P.R. Chalker, and I.M. Buckley-Golder, Diamond Rel. Mater. **2**, 1096 (1993)

13.52   M. Aslam, I. Taher, M.A. Tamor, T.J. Potter, and R.C. Elder, in Proceedings of the 7th International Conference on Solid-State Sensors and Actuators, Transducers '93, Yokohama, Japan (1993), pp. 718–21

13.53   D.R. Wur, J.L. Davidson, W.P. Kang, and D. Kinser, in Proceedings of the 7th International Conference on Solid-State Sensors and Actuators, Transducers '93, Yokohama, Japan (1993), pp. 722–5

13.54   J.L. Davidson, D.R. Wur, W.P. Kang, D. Kinser, J.P. Wang, and Y.C. Ling, in Advances in New Diamond Science and Technology, ed. S. Saito, N. Fujimori, O. Fukunaga, M. Kamo, K. Kobashi, and M. Yoshikawa, MYU, Tokyo (1994), pp. 693–6

13.55   D.R. Wur, J.L. Davidson, W.P. Kang, and D.L. Kinser, J. Micromech. Sys. **4**, 34 (1995)

13.56   S. Sahli and M. Aslam, in Proceedings of the 8th International Conference on Solid-State Sensors and Actuators, Eurosensors **IX**, Stockholm, Sweden (1995), pp. 592–5

13.57   J.L. Davidson and W.P. Kang, Mater. Res. Soc. **416**, 397 (1996)

13.58   M. Deguchi, M. Kitabatake, and T. Hirao, Diamond Rel. Mater. **5**, 728 (1996)

13.59   Y. Boiko, P. Gonon, S. Prawer, and D.N. Jamieson, in 7th European Conference on Diamond Diamond-like and Related Materials jointly with the ICNDST-5, Tours, France, 8–13 September 1996

13.60   Y. Kanda, Sensors and Actuators A **28**, 83 (1991)

13.61   M. Werner, S. Hein, and E. Obermeier, Diamond Rel. Mater. **2**, 939 (1993)

13.62   C.A. Klein and G.F. Cardinale, Diamond Rel. Mater. **2**, 918 (1993)

13.63   M.H. Grimsditch and A.K. Ramdas, Phys. Rev. **B 11**, 3139 (1975)

13.64   X. Jiang, J.V. Harzer, B. Hillebrands, Ch. Wild, and P. Koidl, Appl. Phys. Lett. **59**, 1055 (1991)

13.65   M.P. D'Evelyn, D.E. Slutz, and B.E. Williams, Mater. Res. Soc. Symp. Proc. **383**, 115 (1995)

13.66   M. Werner, S. Klose, F. Szücs, Ch. Moelle, H.J. Fecht, C. Johnston, P.R. Chalker, and I.M. Buckley-Golder, Diamond Rel. Mater. **6**, 344 (1997)

13.67   Ch. Moelle, TU Berlin, personal communication

13.68   P. Gluche, M. Adamschik, A. Vescan, W. Ebert, A. Flöter, R. Zachai, and E. Kohn, submitted for publication in Diamond Rel. Mater.

13.69   J.S. Shor, L. Bemis, and A.D. Kurz, IEEE Trans. Electron Devices **41**, 661 (1994)

13.70   G. Zhao, E.M. Charlson, T. Stacy, J.M. Meese, G. Popovici, and M. Prelas, J. Appl. Phys. **73** 1832 (1993)

13.71   M. Werner, O. Dorsch, and E. Obermeier, Diamond Rel. Mater. **4**, 873 (1995)

13.72   P.R. Chalker, C. Johnston, J.A.A. Crossley, J. Ambrose, C.F. Ayres, R.E. Harper, I.M. Buckley-Golder, and K. Kobashi, Diamond Rel. Mater. **2**, 1100 (1993)

14. CVD Diamond for Surface Acoustic Wave Filters

Shin-ichi Shikata

Itami Research Laboratories, Sumitomo Electric Industries Ltd,
1-1-1 Koya-kita, Itami 664, Japan
e-mail: shikata-shinichi@sei.co.jp

Springer Series in Materials Processing
Low-Pressure Synthetic Diamond   Eds.: B. Dischler and C. Wild
© Springer-Verlag Berlin Heidelberg 1998

14.1 Introduction

Amongst the varieties of applications of CVD diamond, the Surface Acoustic Wave (SAW) device, using its high SAW velocity, is supposed to be one of the practical applications, because polycrystalline film can be utilized. This is also followed by some other advantages: impurity control is not required; and thicknesses of only several microns are sufficient because the energy concentration is on the surface, which enables low-cost manufacturing.

Looking at the communications world, the demand for high frequency and high bit rates is increasing to meet the requirements for broadband, mobile, and handheld communications systems. In particular, devices with a high frequency, on the GHz range, and/or a high power capability are required, which are hard to realize via SAW devices using conventional materials such as quartz, lithium niobate ($LiNbO_3$), and lithium tantalate ($LiTaO_3$). Due to the recent progress in CVD-diamond growth technology, SAW devices fabricated by structures with diamond have been studied, and it has been found that systems involving these materials are practical for high-frequency SAW filters. Moreover, the high thermal conductivity, as well as the high elastic constant, of diamond are found to enable super-high-power-handling capabilities, which surpass the characteristics of conventional materials even at low frequency. In this chapter, the results and prospects for diamond SAW devices will be reviewed and discussed.

14.2 SAW Materials and Communications

14.2.1 SAW Materials

The Surface Acoustic Wave (SAW) is the well-known wave that occurs in earthquakes, and has been studied by seismologists for a century: the names of discoverers such as Rayleigh are familiar in the field of SAW filters. These filters utilize the transformation of RF signals into mechanical waves and vice versa

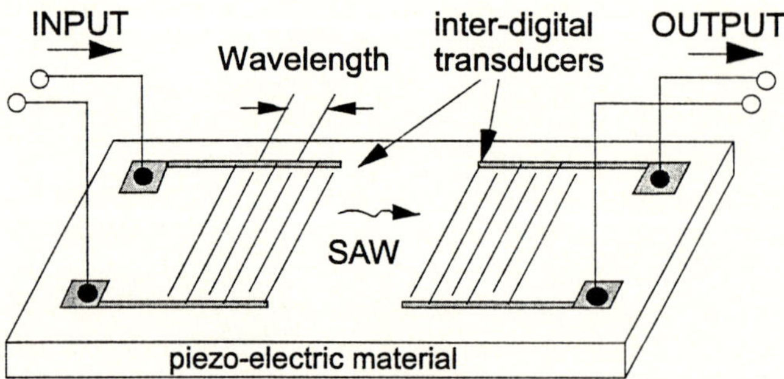

Fig. 14.1: A schematic of SAW devices.

Table 14.1: Typical elestic constant parameters of SAW-related materials (10^{11} Pa).

	Quartz	LiTaO$_3$	LiNbO$_3$	ZnO	Li$_2$B$_4$O$_7$	Sapphire	Diamond
C_{11}	0.87	2.3	2.0	2.1	1.35	4.07	10.8
C_{33}	1.07	2.8	2.43	2.1	0.54	4.98	10.8
C_{44}	0.58	0.97	0.6	0.42	0.59	1.47	5.8

using piezo-electric materials and transducers to obtain frequency filtering. The schematic of a device is shown in Fig. 14.1. In terms of a mathematical expression, the filter characteristic is given by the Fourier transform of the transducer envelope. The details can be found elsewhere. [14.1, 2]. Generally, piezo-electric bulk crystals have been used for the device, such as quartz, LiTaO$_3$ and LiNbO$_3$. However, a layered structure of a piezo-electric thin film and a substrate, such as ZnO/glass, can also be employed for this device. In this case, the wave propagates mechanically, with elliptical displacement of the layer interface. The phase velocity of this structure is influenced by both materials, and is determined by the elastic constants of the materials. Typical parameters of the elastic constants of SAW-related materials are listed in Table 14.1.

For conventional SAW devices, materials with a low phase velocity between 2500 m/s and 4500 m/s, such as quartz, LiNbO$_3$ and LiTaO$_3$, are used. For the SAW filter, the center frequency of the device is determined by the simple equation:

$$F = V/\lambda. \tag{14.1}$$

Here, F, V, and λ denote the frequency, the SAW phase velocity (hereafter, the "velocity"), and the wavelength, respectively. The wavelength is determined by the size of the lines and spaces of the interdigital transducers (IDTs). Thus, fine-pattern lithography or high-velocity material are alternatives for obtaining high-frequency devices. Recently, advanced lithography technologies have become available: 0.45 μm IDTs have been obtained [14.3] by an i-line stepper combined with a phase shift lithography process, and applied to the fabrication of a retiming filter for 2.488 GHz optical communications. However, the reduction of the IDT size suffers from problems such as reliability, power durability, and the fabrication margins in the manufacturing process. For high-velocity waves, there are three methods, as follows:

1. High-velocity waves on conventional materials.
2. High-velocity bulk crystals.
3. Piezo-electric thin films on high-velocity substrates.

For the first method, leaky waves have been studied, and crystals such as LiTaO$_3$ and LiNbO$_3$ have shown high-velocity waves from 6000 m/s to 7000 m/s [14.4]. For the second, new materials such as lithium tetraborate (Li$_2$B$_4$O$_7$), the velocity of which is 6780 m/s, have been developed. [14.5] Lastly, layered structures of piezo-electric thin films on high-velocity substrates, such as ZnO/sapphire [14.6] and

Table 14.2: The characteristics of SAW materials.

Material	Cut	Direction	Velocity (m/s)	K^2 (%)	TCF (ppm/deg)	Remarks
Quartz	ST	X	3158	0.14	0	
	36Y	Z	5088	0.11	0	SH
	LST	X	3948	0.11	0	Leaky
$LiNbO_3$	128Y	X	3992	5.50	74	
	64Y	X	4742	11.30	79	Leaky
			7400	12.30		Second leaky
$LiTaO_3$	X	112Y	3288	0.64	18	
	36Y	X	4212	4.70	45	Leaky
			6300	2.20		Second leaky
$Li_2B_4O_7$	45X	Z	3440	1.00	0	
	45X	90Z	4915	0.14	50	BG
ZnO/sapphire			5500	4.50	43	
ZnO/diamond			11600	1.20	22	
			7180	5.00	30	
SiO_2/ZnO/diamond			9000	1.20	0	
			8050	3.90	0	
$LiNbO_3$/diamond			11900	9.00	25	

AlN/sapphire [14.7] have been developed, obtaining high velocities of 5500 m/s and 6700 m/s, respectively. Diamond SAW filters belong to this third method for obtaining high velocity. By using various types of piezo-electric thin films and insulators, a variety of SAW characteristics can be expected on diamond substrates. [14.8–19]. Various kind of materials systems with high-velocity waves are listed in Table 14.2.

14.2.2 Communications and SAW Devices

Here, high frequency communications systems related to SAW devices are briefly introduced to provide ideas about applications for diamond SAW devices.

Wireless Communications Systems and SAW Devices

The increasing demands for large-volume data transmissions and mobile communications have spurred numerous plans for communications systems, with worldwide standardization. Recently, the radio spectrum has been recognized as a limited and valuable natural resource, and reallocation of the frequency bands is under way. It was an historical moment when the radio frequency bands were sold at auction by the United States Government in 1994. Figure 14.2 shows the frequencies of the major communications systems in the world. For mobile communications systems, high frequencies of 900 MHz to 1.9 GHz are being used for second-generation digital systems. In third-generation systems, well known as

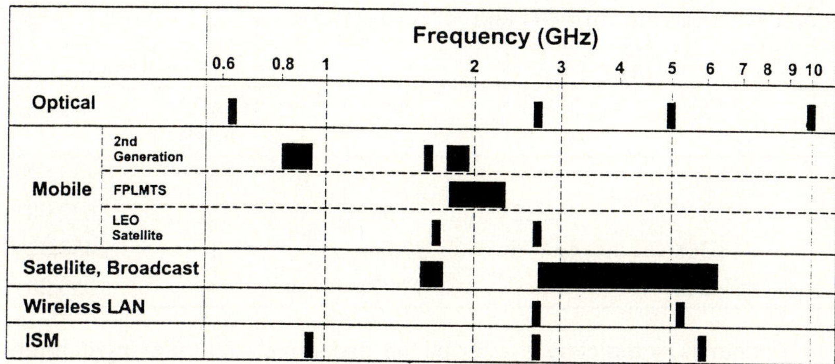

Fig. 14.2: The allocated frequencies for major communications systems.

IMT-2000, high frequencies in the range of 1.8–2.2 GHz have been allocated in the world for international roaming. For satellite communications, the global positioning system (GPS) created by the US Department of Defense, using 1.575 GHz, is now provided for civilian users. Recently, navigation services have become available not only for marine and airborne applications, but also for terrestrial systems. Additionally, new satellite communications services utilizing low Earth orbit (LEO) are under way. As shown in Fig. 14.2, frequencies ranging from 1.6 GHz to 2.5 GHz will be used for these applications. For mobile communications, high-frequency SAW filters are used as bandpass filters for radio frequency (RF). Local wireless communications systems, such as wireless LANs, and vehicular information communications systems, such as electric toll collection systems (ETC), will also be carried out in the ISM (Industrial, Scientific, and Medical) band at 2.5 GHz and 5.8 GHz. High power durability is the largest problem to be solved in order to use SAW filters as transmission RF. For this purpose, diamond, which has high power-handling capabilities (which will be discussed in Sect. 14.5.2), is expected to be a hopeful candidate of material.

Optical Communications Systems and SAW Devices

Remarkable technological progress has been made in optical fiber communications systems during the past decade, and telecommunication using this technology is now the most important infrastructure of the information society. The networks already installed are based on the Plesiochronous Digital Hierarchy (PDH). This system has different communication rates in North America, Europe and Japan. Associated with the recent demand for large-volume data transmission, the system is now being upgraded to the Synchronous Digital Hierarchy (SDH). This is well known as the Synchronous Optical Network (SONET) in the United States, and has been adopted as the high bit rate communications network standard across the

Table 14.3: The hierarchies of PDH and SDH (in Mbps)

PDH				SDH	
Hierarchy level	EC	North America	Japan	Hierarchy level	World
1	2.048	1.544	1.544	1	155.52
2	8.448	6.312	6.312	2	622.08
3	34.368	44.736	32.064	3	2488.32
4	139.264		97.728	4	9953.28

world. The system is completely synchronized and direct multiplexing/demultiplexing of signals is possible. Thus, in this system, a high bit rate of 156 Mbps, and its higher hierarchical levels of 622 Mbps and 2.5 Gbps, are possible. The hierarchies of PDH and SDH are shown in Table 14.3. 10 Gbps is the standard higher hierarchy for communications infrastructure, and 5 Gbps is used for submarine lightwave systems. Finally, in the near future, the Broadband Integrated Services Digital Network (B-ISDN) will cover all of the world, by SDH networks associated with Asynchronous Transfer Mode (ATM) switching technology, to combine all of the information systems looking for Fiber To The Home (FTTH) age. This network will cover the Wide Area Network (WAN), including various types of high bit rate LANs, such as ATM, fiber channel, gigabit ethernet and so on, and wireless LANs such as HIPERLAN.

In these optical communications systems, SAW filters are used in retiming clock generation, as shown in Fig. 14.3. The pulses transmitted through the optical fiber are amplified after O/E conversion, and the signal is regenerated by a retiming circuit with a narrowband high-Q SAW filter. For an higher hierarchy such as 2.5 Gbps to 10 Gbps, high-frequency retiming filters are required, and diamond SAW filters have an advanatage over conventional quartz filters, both in frequency and performance characteristics. An example of the 2.5 GHz filter is shown in Sect. 14.5.1.

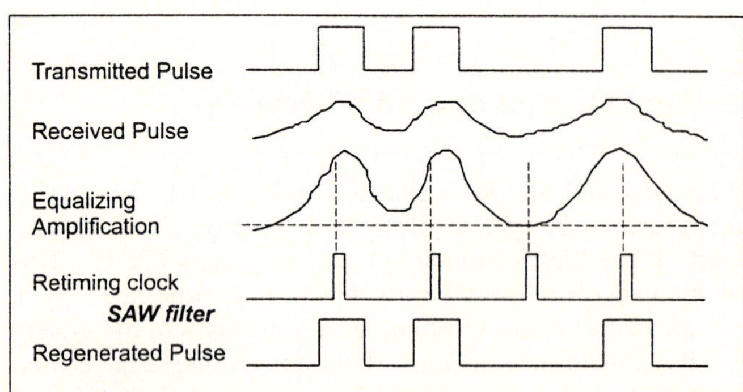

Fig. 14.3: The concept of retiming using a SAW filter.

Other Applications

Another application of the diamond SAW filter is Intermediate Frequency (IF) filters for higher-frequency systems. Hopeful candidates for these applications are microwave and millimeter-wave systems, such as traffic information systems, vehicle anti-collision systems and high bit rate wireless LANs. In these systems, a 76 GHz signal is converted down to the baseband frequency by two IFs, and the 1–2 GHz band is used for the first IF signal.

The high-frequency resonators with low loss and high Q are also potential application for the diamond SAW filter, utilizing the zero temperature coefficient and high coupling coefficient of the SiO_2/ZnO/diamond structure. High-frequency SAW resonators are required for clock generation in high speed computers, wireless key entry systems, and other systems, such as aviation, satellite, and military.

Sophisticated devices utilizing SAW based technology, such as identification tags (ID Tag) and sensor devices, can also be considered as future applications.

14.3 The Diamond Wafer Technology and SAW Filter Fabrication Process

14.3.1 Diamond Wafer Technology

The most significant requirement of diamond for SAW device applications is a "wafer" type substrate, which enables fabrication processing using semiconductor device processing equipment. An example of the requirements for SAW applications is shown in Table 14.4.

Because impurity control is not required in SAW applications, various types of deposition systems can be employed for the preparation of the diamond wafer: 2 and 3 inch wafers on silicon have already been prepared by hot-filament CVD, and

Table 14.4: The requirements of a diamond wafer for SAW application.

Item	Requirement
Wafer size	2, 3 and 4 inch
Wafer thickness	300 μm–1000 μm (< 600 μm for SMT device)
Wafer bow	< 30μm (Total Thickness Value)
Diamond thickness	Approx. > 15 μm (depends on frequency)
Defects density	< 100 cm^{-1}
Surface Roughness	< 5 nm
Resistivity	> 10^6 Ω cm
Velocity	As expected from theory (with piezo-electric film)
Propagation loss	Depends on application

(a)

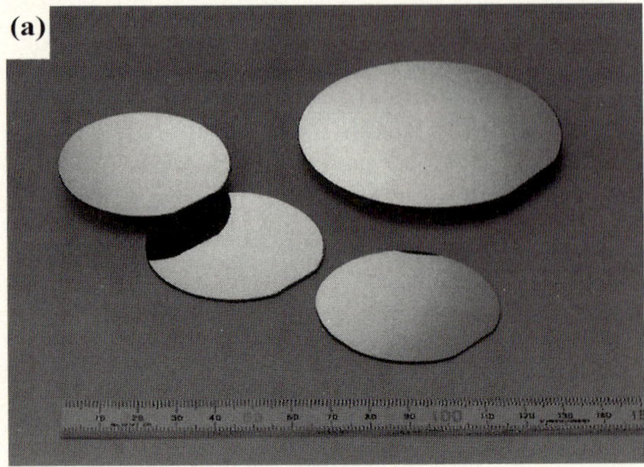

(b)

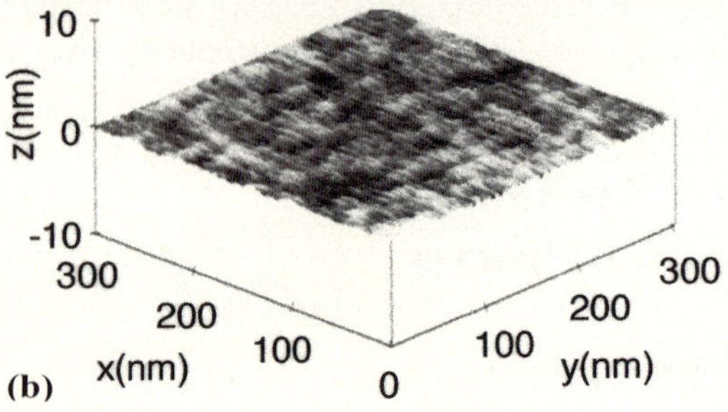

Fig. 14.4: Two and three inch diamond-on-Si wafers: **(a)** wafers; **(b)** surface morphology by AFM

devices have been fabricated. In this case, randomly oriented polycrystalline diamond films approximately 20–30 µm thick, were deposited using 2150°C tungsten filaments with deposition conditions of $CH_4/H_2 = 1–2\%$ and 80 Torr pressure. The most significant requirement for SAW applications is the smoothness of the surface, and surface polishing is another key technology for SAW wafers. As the IDT size is very small (sub-microns) and also very thin (several hundred angstoms), the surface roughness must be minimized. Figure 14.4 shows 2 and 3 inch wafers and a typical surface, observed by atomic force microscopy (AFM). The roughness of the wafer surface is under several nanometers. Another important requirement for wafer is the defect. For these wafers, low defect density below 100/cm have been realized. A characteristic of diamond peculiar to SAW applications is the propagation loss from the piezo-electric thin film and diamond, as well as IDT aluminum. Although the origin of this loss has not been analyzed in detail, it depends highly on the surface morphology of the wafer. With a rough surface, a part of the SAW will be converted to a bulk acoustic wave, resulting in a large loss. With the smooth surface wafer described above, the propagation loss is below 0.03 dB/λ at 1.8 GHz, which is comparable with that of conventional materials [14.20].

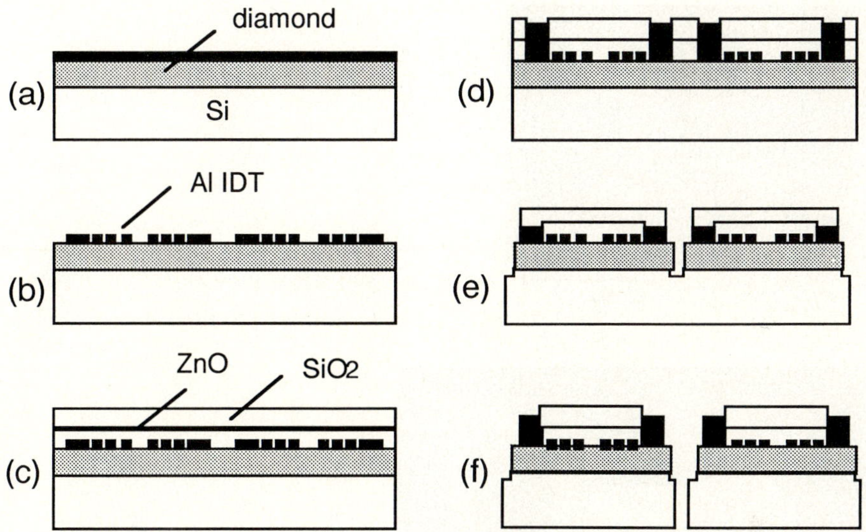

Fig. 14.5: The fabrication process flow for diamond SAW filters.

14.3.2 The Fabrication Process of the SAW Filter

The fabrication process flow of SAW filters is shown in Fig. 14.5 for the SiO_2/ZnO/IDT/diamond/Si structure as an example. First, a 40–100 nm thick layers of Al, or of an Al-based alloy such as AlCu, was deposited by sputtering, and fine line IDTs were fabricated by conventional photo-lithography and etching processes. Here, since the diamond is inert to all of the wet treatments in device fabrication processes, a lift-off process for the IDT can be applied, as well as a dry etching process using reactive ion etching (RIE). This is followed by the deposition of ZnO and SiO_2 thin films, which serve as a piezo-electric material to generate SAW and a film for cancelling the temperature coefficient, respectively. The deposition of ZnO and SiO_2 have been carried out by conventional radio frequency (RF) magnetron sputtering. The ZnO thin film is widely known to be c-axis oriented, regardless of the substrate. The standard deviation of the (001) peak in the X-ray diffraction rocking curve was reported to be less than 1°, which indicates that a ZnO thin film on diamond is highly c-axis oriented, similar to other substrates, such as glass. As ZnO has an hexagonal crystal structure, the film deposited on the (111) plane of diamond was reported to have fine orientation, and an X-ray rocking curve deviation of up to 0.27° has been obtained, even with high lattice mis-match of 28.8% by epitaxial relationship of $[11\bar{2}0]$ ZnO // $[\bar{1}01]$ diamond. [14.15] After deposition, ZnO and SiO_2 deposited on the pad area are removed by lithography and etching.

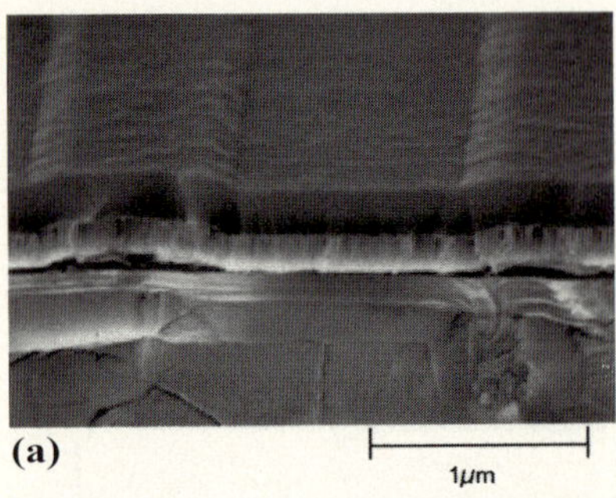

Fig. 14.6: Fabricated filter of an SiO$_2$/ZnO/diamond structure: **(a)** an SEM of the cross-section; **(b)** an overview of the wafer.

For chip processing, the wafer is cut into filter chips with a YAG laser and following dicing of the Si substrate. A frequency tuning process is required for some applications, such as narrowband filters and resonators. This proecss is carried out by slightly etching the top film and, here, conventional dry etching of SiO$_2$ or ZnO can be applied on the SiO$_2$/ZnO/diamond or ZnO/diamond structure, respectively. In Fig. 14.6, a cross-sectional SEM photograph of an SiO$_2$/ZnO/IDT/diamond SAW filter and an overview of a fabricated 2 inch wafer are shown.

14.4 Theoretical Results for Various Structures

14.4.1 ZnO/Diamond

The SAW velocities of several modes of the Rayleigh wave in the ZnO/diamond structure have been calculated by solving Maxwell's equation, the piezo-electric equation and the equation of motion associated with the strain field formation

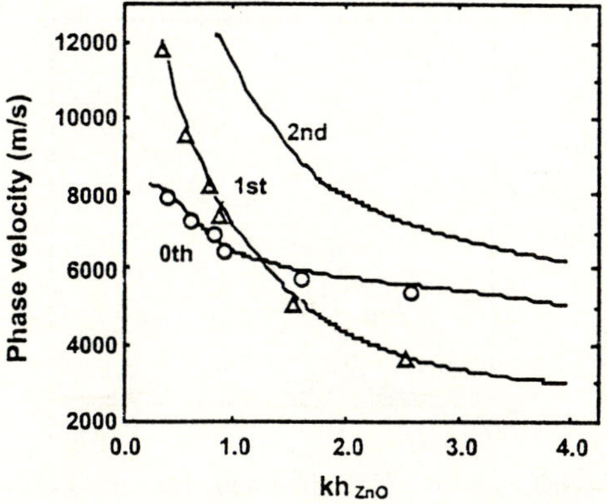

Fig. 14.7: The thickness dependence of the velocity. (k: $2\pi/\lambda$; h_{ZnO}: thickness of ZnO).

[14.16]. The computer simulation calculations were carried out for a typical crystal plane of ZnO (001) / diamond (001) / Si (001) with various thicknesses of ZnO and diamond. This calculation is presumed to be adequate, because of the isotropic characteristics of diamond, and this is confirmed experimentally using poly-crystalline diamond film. Figure 14.7 shows the ZnO thickness dependence of the velocity. The thicknesses of ZnO shown here are expressed by a normalized value in conjunction with the wavelength, which is determined by the interdigital transducer pitches. Here, k and h denote $2\pi/\lambda$ and the ZnO thickness, respectively. The diamond thickness was taken to be large enough to eliminate the effect of the silicon substrate. As the SAW energy concentration is at the surface, the thickness of diamond is enough for 10 µm. As can be seen from Fig. 14.7, the velocity decreases with increasing ZnO film thickness: however, a velocity higher than 9000 m/s can be obtained with diamond, which is significant compared with other materials.

On the other hand, one of the most important factors in SAW devices is the electromechanical coupling coefficient, generally denoted as K^2, the characteristic that expresses the efficiency of generating the SAW. K^2 is generally calculated using the following equation [14.2, 21]:

$$K^2 = 2\,(V_0 - V_m)/V_0 \tag{14.2}$$

where V_0 and V_m are the velocities at the layer boundaries, with the IDT open or grounded respectively. According to the spatial configurations of the IDTs and ground planes in the layered structures, 12 types of SAW layer structures are available. The simulation results on ZnO/IDT/diamond and IDT/ZnO/diamond structures are shown in Fig. 14.8 as a function of the ZnO thickness. The plotted K^2 was calculated by the following equation, using measured data for the S parameter S_{11}:

$$K^2 = G/(8f_0CN^2). \tag{14.3}$$

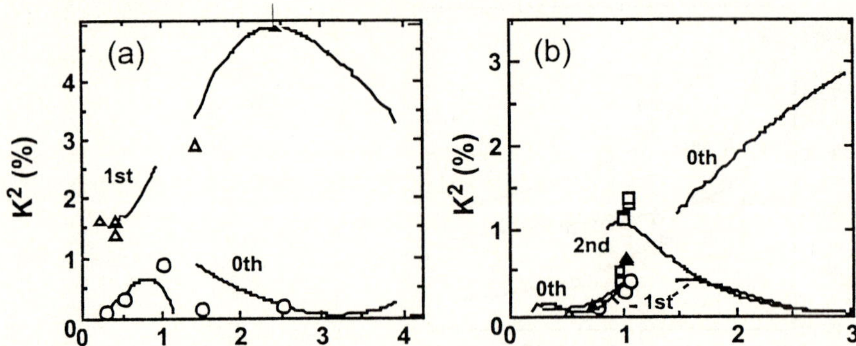

Fig. 14.8: The electromechanical coupling coefficient of a ZnO/diamond structure: **(a)** the ZnO/IDT/diamond configuration; **(b)** the IDT/ZnO/diamond configuration.

Here, G, f_0, C, and N denote the maximum conductance, the center frequency, and the capacitance between a pair of IDTs, and the number of IDT pairs, respectively. As can be seen from this figure, ZnO/diamond SAW shows a practical K^2 value from 1.4% to 4.8%, which is close to that of $LiTaO_3$ with high velocity. A high K^2 is also expected for this structure, and a pseudo-SAW was studied by theoretical calculations. Extremely high K^2 values of 23% for the first mode and 15% for the second mode were obtained by this calculation [14.22]. In Figure 14.7 and 14.8, experimental data measured on polycrystalline diamond are plotted, and it can be seen that they coincide well with the theoretical results for single-crystal diamond. This indicates that the polycrystalline film is adequate for SAW use, because of the small deviation of the velocity in the different crystal directions.

The temperature characteristics of the frequency (TCF) of this structure is in the range 25–35 ppm/°C, depending on ZnO film thickness. This value is also close to that for $LiTaO_3$, and is practical for most of the filters used for wireless applications.

14.4.2 SiO$_2$/ZnO/Diamond SAW

The temperature coefficient of frequency (TCF) is one of the important characteristics of SAW devices, especially for narrow bandwidth devices. In order to meet requirements on a wide range of applications, such as filters for optical communication and resonators, a lower TCF is needed. Amorphous SiO_2 is well known as a positive temperature coefficient material, and by combining it with negative materials, zero TCF can be expected. It has already been reported that this is possible for $LiTaO_3$. [14.23]. This technique had been applied to ZnO/diamond structures by theoretical calculations using Campbell's method [14.1] and the finite element method. As can be seen in Fig. 14.9, the first order of temperature coefficient can be set to zero for each kh_{ZnO}, at the cost of a slight

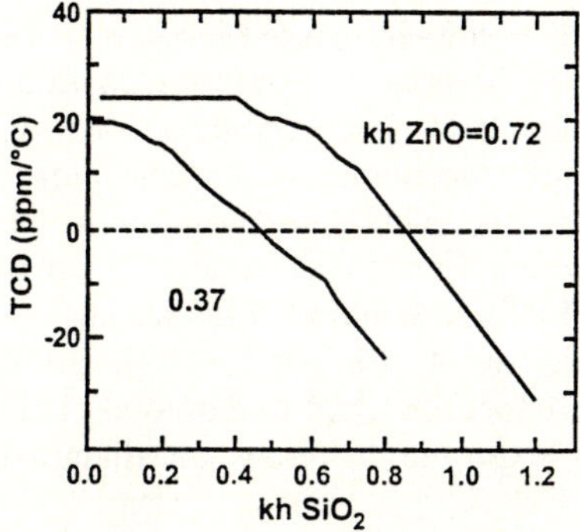

Fig. 14.9: The temperature cancellation of ZnO/diamond by an SiO_2 overlay.

reduction of the values for K^2 and velocity. The SAW velocities and electro-mechanical coupling coefficients also vary with $hSiO_2$, but practical values are expected to be obtained. The SAW parameters have already been listed in Table 14.2.

14.4.3 LiNbO₃/Diamond and LiTaO₃/Diamond

For the diamond SAW structures mentioned above, only ZnO has been considered as the piezo-electric thin film, because of its high orientation to the c-axis regardless of the substrate. Clearly, various kinds of piezo-electric thin films deposited on diamond can also have large advantage of a high SAW velocity, and wide range of SAW material systems based on diamond can be considered. The typical piezo-electric materials such as $LiTaO_3$ and $LiNbO_3$ have already been considered by computer simulation for layered structures with diamond to obtain fine characteristics such as a high coupling coefficient [14.16].

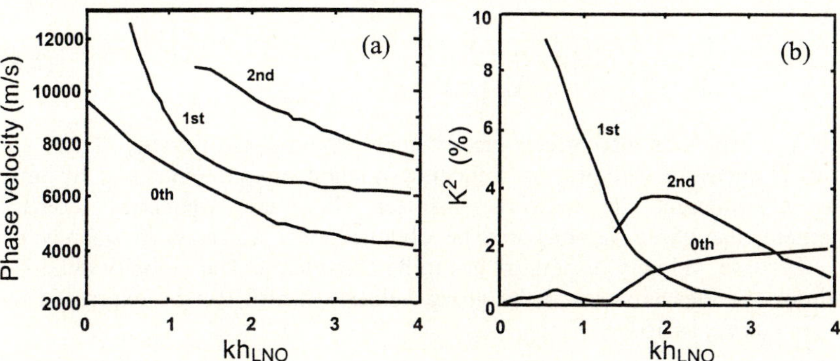

Fig. 14.10: The phase velocity **(a)** and coupling coefficient **(b)** of c-axis oriented $LiNbO_3$/diamond.

Both crystals are classified into the 3m group in the Hermann–Mauguin space group. However, considering thin film growth on diamond, c-axis oriented films are expected because of the difficulties encountered in single-crystal growth on hetero-substrates. With this assumption, the elastic constant C_{14} and the piezo-electric constant e_{22} can be set to zero, which makes calculations possible in the 6 mm crystal group. The results for the LiNbO$_3$/diamond structure are shown in Fig. 14.10. As can be seen from the figure, the LiNbO$_3$/diamond structure has the potential for a high electro-mechanical coefficient of 9% with a high SAW velocity of 11 890 m/s. LiNbO$_3$ (001) is pseudo-lattice matched to diamond (111) with a value of 102%, and thus LiNbO$_3$ is expected to grow on diamond epitaxially.

14.4.4 Theoretical Remarks on the Layered Structure

The major difference of the layered structure SAW to bulk crystal SAW is velocity dispersion, which originates from the phase velocity difference between layered materials. Thus, for example, the frequency responce function $H(f)$ of the Fourier transform in the delta function model of the layered structure SAW can be written as

$$H(f) = \sum_{n=-(N-1)/2}^{(N-1)/2} (-1)^n A^n \exp\left(-j\frac{2\pi f}{V(f)} x^n\right). \tag{14.4}$$

Here, A^n and $(-1)^n$ denote the amplitude and polarity of the IDT, respectively. For a diamond SAW filter, $V(f)$ should be calculated for each structure, whereas V can be used for conventional SAW devices on bulk crystals. $V(f)$ can be written as (14.5) and thus $1/V(f)$ is rewritten as (14.6). This causes the bandwidth narrowing. This is very important in the design of SAW filter.

$$V(f) \approx V_0 + \Delta V(f) \tag{14.5}$$

$$\frac{1}{V(f)} \approx \frac{1}{V_0 + \Delta V(f)} \approx \frac{1}{V_0}\left(1 - \frac{\Delta V(f)}{V_0 + \Delta V(f)}\right). \tag{14.6}$$

The velocity dispersion also affects measurement of the phase velocity. There are two ways to measure velocity; by frequency domain measurement and by time domain measurement in network analyzer. For the frequency domain measurement, the phase velocity can be obtained only at center frequency by (14.1); otherwise, velocity dispersion has to be considered. The velocity obtained by time domain measurement is the group velocity, which can be expressed by using $V_p = V_p(kh)$ [14.15]:

$$V_g = d\omega/dk = dkV_p/dk = V_p + dV_p/dk = V_p + kh\, dV_p/dhk. \tag{14.7}$$

Here, V_g and h denote group velocity and piezo-electric film thickness. This has to be considered for the analysis of SAW filter characteristics by time domain measurement.

14.5 Features of Diamond SAW and Its Applications

14.5.1 The Zero Temperature Coefficient

The zero temperature coefficient characteristic of the SiO_2/ZnO/diamond structure is one of the most promising applications of the diamond SAW filter, especially for narrowband fiters and resonators. The diamond SAW of the SiO_2/ZnO/diamond structure has the advantages of zero TCF with a high phase velocity, which can provide a high-frequency device, and also a high electro-mechanical coupling coefficient, which enables low-loss filters. This structure has been successfully applied to a 2.5 GHz retiming filter for optical fiber communications systems. [14.19] The design of the fabricated retiming filter was a two-port resonator type with 70 pairs of fingers for the IDTs using a single type of electrode. The first mode SAW was adopted for the filter. The phase velocity of the wave was 9000 m/s and the IDT size was 0.9 μm. The frequency was tuned by etching SiO_2 slightly by a dry process. A 2.488 GHz narrowband SAW filter for the retiming circuit of the STM-16 system in SDH (OC-48 for SONET) was successfully obtained. The typical frequency response of the retiming filter S_{21} is given in Fig. 14.11. The filter shows fine characteristics, such as a 14 dB insertion loss and a Q value of 650. The phase characteristics were linear for the passband and its slope was as small as –0.038°/kHz. By optimizing the filter IDT design, the rejection from the main peak to zeroth mode and second mode waves was as high as 20 dB. The temperature characteristics of the retiming filter are shown in Fig. 14.12. As can be seen from the figure, the temperature coefficient is smaller than that of 36 Y cut and ST cut quartz, and a parabolic characteristic is observed.

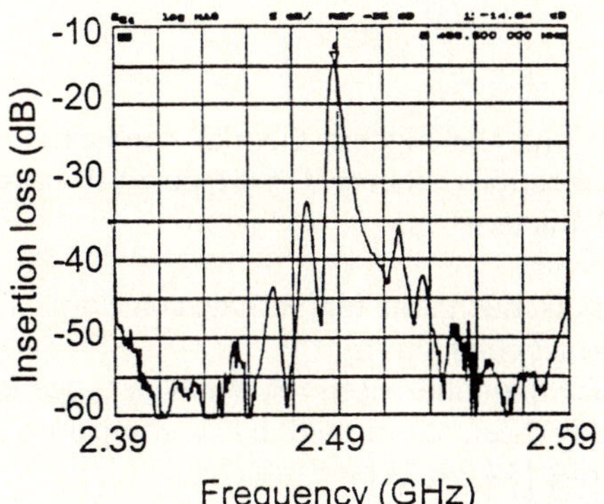

Fig. 14.11: The typical frequency response of S_{21} of the 2.5 GHz retiming filter.

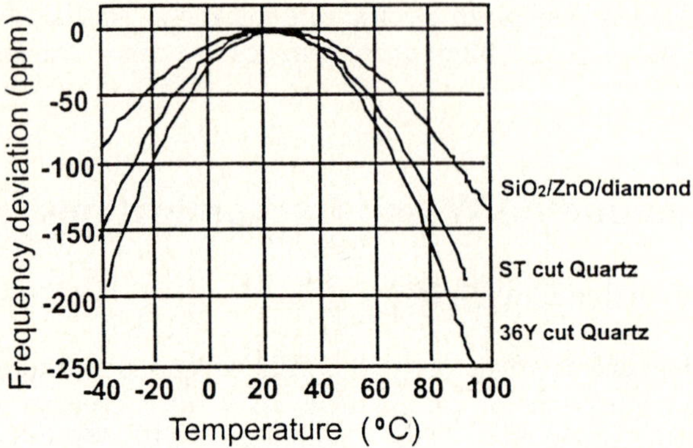

Fig. 14.12: Temperature characteristics of the SiO$_2$/ZnO/diamond SAW.

Using this structure, a low-loss filter was also obtained. The loss of the filter is about 5–10 dB smaller than that of a quartz filter. Reliability testing has also been carried out for these devices and passed all the specifications required in SDH systems. This indicates that the diamond SAW filter has high stability to stress migration, even with the multi-layered structure. As described above, the effectiveness of this diamond SAW structure compared to quartz has been confirmed for high-frequency optical devices.

This SAW filter is expected to be applied in 2.5 Gbps optical systems and optical links. Also, this promising data indicates the possibility of diamond SAW devices for further applications in optical fields, such as 10 Gbps for STM-64 in an SDH system, 5 Gbps for transocean transmission systems, and other high bit rate systems for LANs. The potential applications of this structure are not only in optical applications, but also for filters in wireless systems of narrowband and high-frequency resonators, due to the high reflectivity of SAW and high power durability, which will be discussed later in subsection 14.5.2.

14.5.2 High Power Durability

Diamond has the highest elastic constant and the highest thermal conductivity amongst all materials. Thus, for SAW applications, diamond is expected to have the following features due to its material characteristics:
1. a large IDT, due to the high elastic constant;
2. a small displacement amplitude of materials, due to the high elastic constant;
3. high heat diffusivity, due to the high thermal conductivity.
Thus, a high power durability is expected for diamond-based SAW devices, due to these advantages. This feature has already been confirmed by a comparative experiment with a LiTaO$_3$ (X-112Y) substrate [14.24, 25].

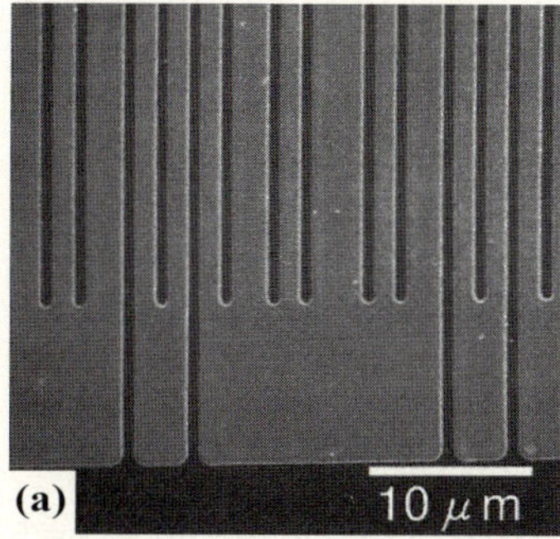 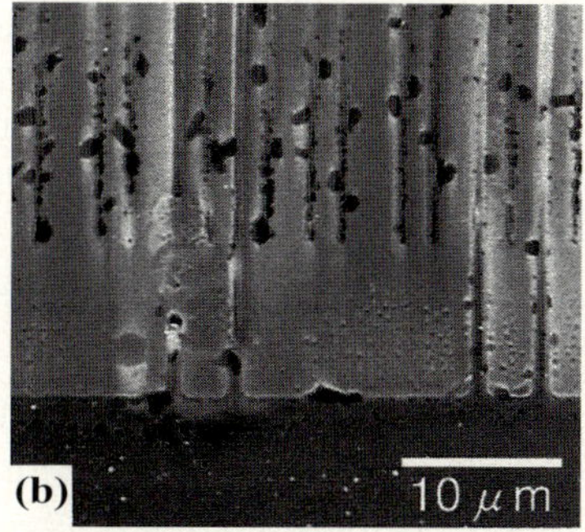

Fig. 14.13: SEM photographs of diamond and LiTaO$_3$ SAW devices after high-power application. **(a)** ZnO/diamond (36 dBm applied); **(b)** LiTaO$_3$ (27.7 dBm applied).

The experiment was carried out for an IDT/ZnO/diamond structure of transversal filters with open strip reflectors, using an electrode size of 1 μm (wavelength = 4 μm). As the phase velocity of the ZnO/diamond SAW filter is about three times larger than that of an LiTaO$_3$ SAW filter, a filter response appeared at 2.90 GHz for the ZnO/diamond SAW filter and at 822 MHz for the LiTaO$_3$ SAW filter. The power durabilities of these SAW filters were measured under the severe conditions of openair at 120°C. It was found that there was no significant change in the frequency response of the ZnO/diamond SAW filter with an input power of up to 36 dBm, which was the upper limit of the measurement system. However, in the case of the LiTaO$_3$ SAW filter, the insertion loss increased with increasing input power, and significant damage in the frequency response was observed at an input power of 27.7 dBm. Figure 14.13 shows SEM photographs of these filters after the high-power application. There was no change on the IDT of the ZnO/diamond SAW filter. However, for the LiTaO$_3$ SAW filter, melting of the electrodes and cracks on the surface of the LiTaO$_3$ substrate were observed, mainly at the edge of the IDT.

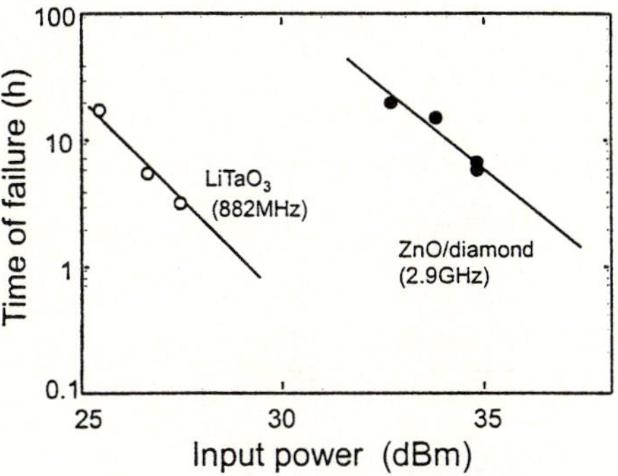

Fig. 14.14: The input power dependence of the Time of Failure of these filters.

The input power dependence of the Time of Failure (TF) (defined as 0.5 dB down) was also investigated with these SAW filters, which were mounted in ceramic packages. Figure 14.14 shows the input power dependence of the TF of these filters measured at 120°C. It was found that the ZnO/diamond SAW filter was durable for an input power 8 dB higher than that of the LiTaO$_3$ SAW filter, even at 3.5 times higher frequency. To investigate the frequency dependence of the power durability of both materials, the power of TF at 10 hours was plotted and is shown in Fig. 14.15. As can be seen from this figure, the power durability of diamond is extremely high compared with that of conventional materials, when devices operating at the same frequency are made.

As is common knowledge, the degradation of SAW devices due to high-power application is induced by stress migration of Al-based metal IDTs, which is accelerated by the temperature rise and the large displacement of the substrate surface that occurs with SAW propagation. The temperature rise of the device is related to the RF power, which is converted to thermal energy in the IDT area by IDT resistance and propagation loss. Diamond film has very high thermal diffusivity, such as approximately 5 cm^2/s for a thickness of around 20 μm, and the converted thermal energy is presumed to be spread out from the IDT area to the region of the whole device, which results in only a small temperature rise in the IDT area. For comparison, the thermal diffusivity of LiTaO$_3$ was measured and observed to be 0.013 cm^2/s. Thus, for LiTaO$_3$, the thermal diffusivity is about 1/400th that of diamond, and the thermal energy is assumed to be confined in the IDT area, which induces a large temperature rise and accelerates the stress migration of the IDT. For the spatial displacement of the substrate surface associated with SAW propagation, diamond has a higher elastic constant, and the displacement amplitude of ZnO/diamond is smaller compared with that of LiTaO$_3$, which results in a small mechanical stress in the IDTs.

For the high power durability of this material system, diamond-based SAW devices will find various fields of application, not only in high-frequency systems, but also for high-power applications such as RF filters for transmission.

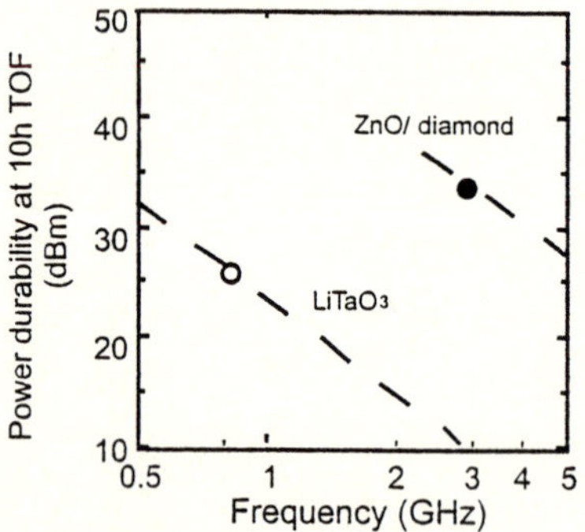

Fig. 14.15: The frequency dependence of the power durability.

14.6 Summary

In this chapter, the recent progress in the field of SAW filter of thin-film piezo-electric material on high-velocity diamond has been described. As the characteristic of a high elastic constant is utilized for SAW applications, the requirements for diamond wafers have mostly to do with the mechanical specifications, such as the surface smoothness and defects as well as the propagation loss, whereas the impurity controll and high crystallinity are less important. Therefore, the SAW application of CVD poly-crystalline diamond film is easier campared to other applications, such as semiconductor device and ICs.

SAW filters using CVD-diamond wafers have already been studied at a practical level. Two and three inch wafers of poly-crystalline diamond on silicon have been made with small wafer bow and smooth diamond surface such as several nanometers of surface roughness. The small propagation loss, which is sufficient for low-loss filters, is obtained. For SAW filters, several structures and configulations have been investigated with piezo-electric thin film of ZnO, insulator SiO_2 and IDTs. In addition to the high velocity of SAW, the durability for extremely high-power handling capability was confirmed for ZnO/diamond SAW devices. This is supposed to be due to the small displacement amplitude of the materials due to the high elastic constant and the rapid heat diffusivity by the high thermal conductivity. For practical filters, the 2.5 GHz retiming filter for optical communications systems was constructed using a zero temperature coefficient structure of SiO_2/ZnO/diamond, and it has ben confirmed that excellent filter properties can be obtained by diamond SAW compared to the quartz filter. In particular, the temperature deviation is smaller than that of quartz, and this characteristic is promising for further applications such as narrowband filters and resonators. By the computer simulation of various materials systems for SAW filters, several structures, such as $LiNbO_3$/diamond have been proposed, to give high coupling coefficient characteristics. Based on ZnO/diamond technologies, a variety of materials systems on diamond is expected to be realized and to form a high-frequency diamond SAW filter family.

In forthcoming communications systems of the high frequencies and high bit rates, the demand for new type of devices is increasing to meet the requirements for broad applications both in wireless and optical communications. Not only in the the high-frequency field, but also in high power-handling capability, diamond will become the unsurpassed material for SAW applications in the near future.

References

14.1   C. Campbell, Surface Acoustic Wave Devices and Their Signal Processing Applications, Academic Press (1989)
14.2   D.P. Morgan, Surface Wave Devices for Signal Processing, Elsevier, Amsterdam (1985)
14.3   K. Asai, A. Isobe, T. Tada, and M. Hikita, Proc. IEICE Spring Meeting Japan, A-**378** (1995)

14.4   S. Tonami, A. Nishikata, and Y. Shimizu, Jpn. J. Appl. Phys. **34**, 2664 (1995)
14.5   T. Sato and H. Abe, IEEE Ultrasonics Conference Proceedings (1994)
14.6   J. Koike, K. Shimoe, and H. Ieki, Jpn. J. Appl. Phys. **32**, 2337 (1993)
14.7   T. Kobayashi, Proceedings of the 15th Symposium on Ultrasonic and Electronics, Japan (1994), p. 187
14.8   K. Yamanouchi et al., IEEE Ultrasonics Symposium Proceedings (1989), p. 351
14.9   H. Nakahata, A. Hachigo, S. Shikata, and N. Fujimori, IEEE Ultrasonics Conference Proceedings (1992), p. 377
14.10   S. Shikata, H. Nakahata, A. Hachigo, and N. Fujimori, Diamond Rel. Mater. **2**, 1197 (1993)
14.11   H. Nakahata, A. Hachigo, S. Shikata, and N. Fujimori, in 2nd International Conference on Application of Diamond Films and Related Materials, Omiya (1993), p. 361
14.12   S. Shikata, H. Nakahata, K. Higaki, A. Hachigo, N. Fujimori, Y. Yamamoto, N. Sakairi, and Y. Takahashi, IEEE Ultrasonics Conference Proceedings (1993), p. 277
14.13   H. Nakahata, A. Hachigo, S. Shikata, N. Fujimori, Y. Takahashi, T. Kajiwara, and Y. Yamamoto, Jpn. J. Appl. Phys. **33**, 324 (1994)
14.14   S. Shikata, H. Nakahata, K. Higaki, S. Fujii, A. Hachigo, and N. Fujimori, in Proceedings of the 4th International Conference on New Diamond Science and Technology, Kobe (1994), p. 697
14.15   A. Hachigo, H. Nakahata, K. Higaki, S. Fujii, and S. Shikata, Appl. Phys. Lett. **65**, 2556 (1994)
14.16   H. Nakahata, K. Higaki, S. Fujii, A. Hachigo, S. Shikata, and N. Fujimori, IEEE Trans. Ultrason., Ferroelect., and Freq. Contr. **42**, 362 (1995)
14.17   S. Shikata, H. Nakahata, K. Higaki, S. Fujii, A. Hachigo, H. Kitabayashi, Y. Seki, K. Tananbe, and N. Fujimori, in 3rd International Conference on Applications of Diamond Films and Related Materials, Gaithersberg, MD (1995), p. 21
14.18   H. Nakahata, S. Shikata, K. Higaki, S. Fujii, A. Hachigo, H. Kitabayashi, Y. Seki, and K. Tananbe, in IEEE Ultrasonics Symposium, Seattle (1995), p. 361
14.19   H. Nakahata, H. Kitabayashi, S. Fujii, K. Higaki, K. Tanabe, Y. Seki, and S. Shikata, in IEEE Ultrasonics Symposium, San Antonio (1996), p. 285
14.20   J. Koike and Y. Ieki, Jpn. J. Appl. Phys. **34**, 2678 (1995)
14.21   M. Feldman and J. Henaff, Surface Acoustic Wave for Signal Processing, Artech House, MA (1989)
14.22   E.L. Adler and L. Solie, in Proceedings of IEEE Ultrasonics Symposium (1995), p. 341
14.23   R. Inaba and K. Wasa, Jpn. J. Appl. Phys. **20**, Suppl. 20-3, 153 (1981)
14.24   K. Higaki, H. Nakahata, H. Kitabayashi, S. Fujii, K. Tanabe, Y. Seki and S. Shikata IEEE MTT-S International Microwave Sympsium (1997) p829
14.25   K. Higaki, H. Nakahata,H. Kitabayashi, S. Fujii, K. Tanabe, Y. Seki, and S. Shikata, in IEEE Trans. Ultrason., Ferroelect., and Freq. Contr. **44**, 1395 (1997)

15. Electron Emission from CVD-Diamond Cold Cathodes

Peter K. Baumann[1] and Robert J. Nemanich[2]

[1] Materials Science Division, 212; Argonne National Laboratory,
9700 S. Cass Avenue, Argonne, IL 60439-4838, USA
e-mail: peter_baumann@qmgate.anl.gov

[2] Department of Physics, North Carolina State University,
Raleigh, NC 27695-8202, USA

Springer Series in Materials Processing
Low-Pressure Synthetic Diamond Eds.: B. Dischler and C. Wild
© Springer-Verlag Berlin Heidelberg 1998

15.1 Introduction

Electron beams are fundamental to many electronic applications, ranging from cathode ray tube (CRT) displays to microwave or power amplifiers. While hot cathodes are suitable for many applications, the development of cold cathodes could lead to improved performance in many existing applications and, more importantly, to new technologies including vacuum micro-electronics, flat-panel display technologies, and new types of microwave amplifiers. A new approach being considered for these applications is to employ semiconducting materials in which electrons in the conduction band can be emitted directly into vacuum without overcoming an energy barrier. This property of the semiconductor has been termed as negative electron affinity (NEA).

The first evidence of this possibility was reported for diamond. Himpsel et al. [15.1] and Pate [15.2] reported a high quantum efficiency for photoelectron emission from (111) surfaces of natural diamond samples. It was concluded that these surfaces did indeed exhibit a true negative electron affinity, and hydrogen termination was found to induce this effect on the (111) surface. These studies first highlighted the potential of diamond as a cold cathode source material.

As diamond-film deposition techniques have been developed since the first photoemission studies, there has been substantial interest in studying the potential of diamond as the emitting material in electron emission structures and devices.

Cold electron emission from metals by means of high electric fields (i.e., field emission) has been studied for many years [15.3], and it has been known for some time that low work function metals emit electrons more readily than metals with a higher work function. The emission process for a semiconductor is more complicated than for a metal. Considering a semiconductor electron emission structure, electrons must be supplied to the semiconductor and then extracted by an electric field at the surface. The field emission process then involves injection of electrons from an electrical contact into the semiconductor, transport of the electrons through the bulk to the emitting surface, and finally the emission from the surface into vacuum. The initial studies of Himpsel et al. [15.1] and Pate [15.2] demonstrated that, unlike metal surfaces, the emitting surface will not limit the emission for hydrogen-terminated (111) natural diamond surfaces.

In this chapter, recent studies pertaining to electron emission based on diamond are reviewed. The relation of the electron affinity to the surface properties is presented, and the determination of the electron affinity by means of photoemission and secondary electron emission is described. Results for diamond surfaces are reported. Since field emission will be required for most device applications, representative measurements are presented. The complexity of the measurements is discussed.

15.2 Electron Affinity and Negative Electron Affinity

15.2.1 Definition

The electron affinity of a semiconductor is defined as the energy difference between the vacuum level and the conduction-band minimum, both extrapolated to the surface. This corresponds to the energy necessary to excite an electron from the conduction-band minimum to the vacuum. We also note that the vacuum level is the energy of an electron at rest in vacuum. The free-electron model essentially describes the band structure of the vacuum. It is worth noting that the electron affinity of the semiconductor is essentially the heterojunction band offset between the semiconductor and vacuum.

In general, the electron affinity is independent of the position of the Fermi level. We make this point since the work function of semiconductors is sometimes quoted; but, in general, the work function may be different for p- and n-type doping. For most semiconducting materials, the vacuum level lies above the conduction-band minimum, and by convention this has been defined as the electron affinity. For wide bandgap semiconductors such as diamond, the conduction-band minimum is near to the vacuum level; and, in fact, in some instances the vacuum level is below the conduction-band minimum. This case has been termed a negative electron affinity, or simply NEA. In this case, electrons present in the conduction band have sufficient energy to overcome the work function of the surface and can be emitted into vacuum.

There are many ways in which to view the energetics of a semiconductor surface, but the following is helpful in understanding some effects to be described here. The electron affinity of a semiconductor may be determined by (1) the

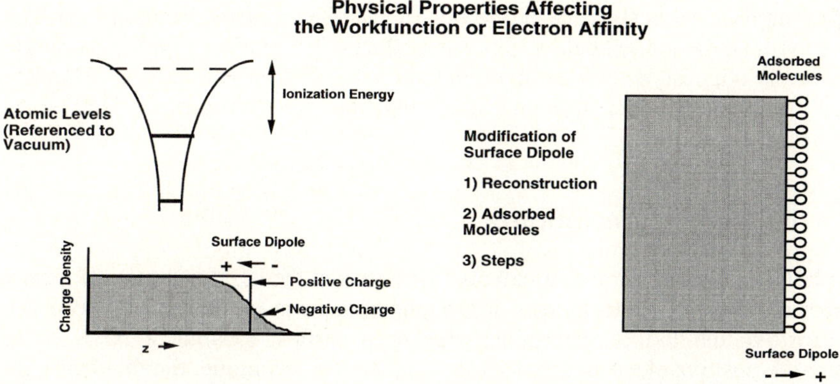

Fig. 15.1: A representation of the effects that contribute to the work function (or electron affinity) of any material. While the atomic levels are an intrinsic property of the material, changes in the surface bonding can substantially affect the work function or electron affinity [15.5].

properties of the material itself, as well as (2) the surface termination, including adsorbates, reconstructions, and steps [15.4]. A schematic illustration of these effects is shown in Fig. 15.1 [15.5]. The atomic levels essentially reference the ionization energy of the atom to the levels that broaden into the valence band. Since the atomic levels are more or less intrinsic to a material, they cannot be changed (but alloys may provide a degree of variation). At the surface of any material several effects can lead to the formation of a surface dipole (we note that it is difficult to define the surface dipole exactly, but in much of what follows we will examine how various processes increase or decrease the value of the surface dipole). For a simple free-electron metal, the surface dipole would arise from the quantum-mechanical extension of the electron wave functions into the vacuum beyond the surface. This also results in a positive layer due to the loss of this electron density. The combination of the electron density away from the surface with the positive charge layer results in a surface dipole that effectively holds electrons in the material.

However, the surface dipole can be influenced substantially by surface adsorbates, surface reconstructions, and steps on the surface. These effects may either increase or decrease the electron affinity of the semiconductor. Ignoring the specifics of bonding and charge distribution, a molecular adsorbate that pulls electrons from the surface toward the adsorbate will increase the electron affinity, while an adsorbate that contributes electrons to the material will result in a lower electron affinity.

To illustrate the magnitude of this effect, consider a hydrogen-passivated surface. Let us assume that the average nuclear and electronic charges are point charges separated by 0.5 Å. Then, for a surface density of $1 \times 10^{15}\,cm^{-2}$, a surface dipole induced energy shift of about 9 eV can be calculated. (Certainly complete charge transfer is never a reasonable possibility, but this simple calculation demonstrates the significance of the surface dipole.) Full quantum-mechanical calculations have addressed the properties of the H-terminated diamond surface, and they will be mentioned below. Since the effect of the surface dipole is so large, it is basically impossible to determine if a material is "intrinsically NEA." Thus the surface termination is critical in describing the electron affinity (or NEA) properties of a material.

15.2.2 Measuring Techniques

While UV-photoemission measurements first detected the high quantum efficiency of electron emission, the technique of UV-photoemission spectroscopy (UPS) is a very sensitive method to determine whether a surface exhibits a NEA or to measure the positive electron affinity [15.1, 2]. In this technique, the incident light excites electrons from the valence band into states in the conduction band. Some of these electrons quasi-thermalize to the conduction-band minimum. For NEA surfaces, these secondary electrons may be emitted into vacuum and are detected as a sharp feature at the low-energy end of photoemission spectra. A careful

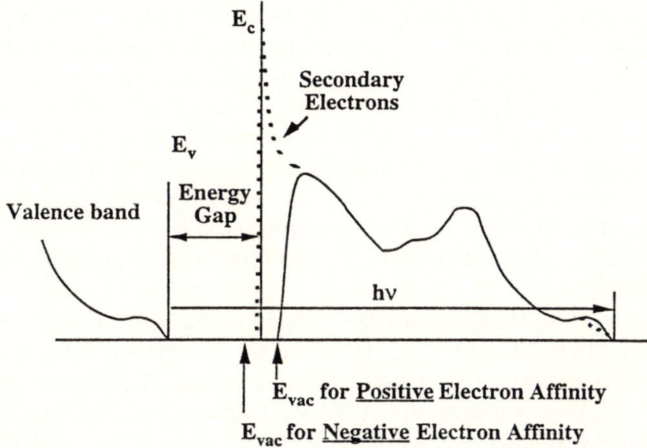

Fig. 15.2: A schematic of how NEA affects the photoemission spectra. For a NEA surface the spectrum is broadened to lower kinetic energy and a peak due to quasi-thermalized electrons is also detected at the lowest kinetic energy [15.6].

measurement of the width of the photoemission spectrum can be used to determine if the low-energy emission occurs from the conduction-band minimum. The width represents the energy difference from the photoemission onset to the low-energy cutoff. For a positive electron affinity, the low-energy cutoff will be determined by the vacuum level, and emission from the conduction-band minimum will not be detected. A schematic of photoemission spectra of a semiconductor with a NEA or a positive electron affinity is shown in Fig. 15.2 [15.6]. The electron affinity (χ) or the presence of a negative electron affinity can be deduced from the width of the spectrum (W) as follows:

$$\chi = h\nu - E_g - W \quad \text{for a positive electron affinity}$$

$$0 = h\nu - E_g - W \quad \text{for a negative electron affinity} \tag{15.1}$$

where $h\nu$ is the photon energy and E_g is the bandgap. It is evident that, for a positive electron affinity, the value of the electron affinity can be deduced from the measured width of the spectrum. However, the absolute value of the electron affinity for a NEA surface cannot be measured by means of photoemission spectroscopy.

By carefully measuring the spectral width, one can determine whether the low-energy emission originates from the conduction-band minimum. In fact, recent measurements have indicated emission that extends several tenths of an eV below the conduction-band minimum. Bandis and Pate [15.7] have ascribed this emission to excitons for the C(111) surface exhibiting an NEA. It was found that the observation depended on the band bending near the surface. For flat band and upward band bending, exciton emission was observed; while for downward band bending the emission was ascribed to electrons in the band [15.7]. The band bending may be due to states in the bandgap that cause Fermi-level pinning. Another possibility is H passivation of the boron acceptors near the surface, which will lead to different band bending for the different regions on the surface [15.8].

Photoemission spectroscopy can also be used to determine the position of the surface Fermi level. For a grounded sample, the Fermi level of the sample will be the same as that of the metal sample-holder – and the Fermi level of the metal can easily be determined. This measurement can be employed to monitor the semiconductor Fermi level or to detect band bending. We note that care must be taken to avoid photovoltage shifts that may be observed in highly resistive samples.

Secondary electron emission (SEE) is another technique that can be used to characterize the surface [15.9, 10, 11]. To facilitate the secondary emission experiments, the sample is exposed to a monochromatic source of high-energy electrons. Typical accelerating voltages are 1–5 kV. Electron–hole pairs are generated in the conduction and valence band of the semiconductor by the incident electrons. The electrons then move toward the surface and may be emitted as described in the photoemission process. In general, SEE is less surface sensitive than UPS, since the electrons generated in SEE are distributed deeper in the sample than those from UPS. A typical application of this technique is to measure the electrons emitted per incident electron. We note that the gain is obtained since a single high-energy electron can excite numerous electron–hole pairs that can be emitted and detected. A negative electron affinity surface will enhance the emission of electrons.

An alternative measurement is to obtain the energy spectrum of the emitted electrons. Similar to photoemission spectra, a negative electron affinity would be indicated by the presence of a sharp low-energy peak in the SEE spectra (corresponding to the one displayed in Fig. 15.2). The energy of the incident electron beam depends on the work function of the electron gun relative to that of the target material. This effect makes an analysis based on a measurement of the width of the spectrum more difficult than for UPS.

It also needs to be emphasized that, since the electron–hole pairs are excited deep in the sample in comparison to UPS, any band bending at the semiconductor surface is expected to influence the SEE gain. In particular, an upward band bending will inhibit electron transport to the surface, while a downward band bending will sweep electrons towards the surface. Therefore, UPS is more suitable and easier to use than SEE to determine the electron affinity of a semiconductor surface.

15.2.3 Surface Termination Effects

Different surface terminations can shift the positions of the bands with respect to the vacuum level and, therefore, induce a NEA or remove it. Such changes have been found for molecular surface adsorbates. Different molecular surface adsorbates result in changes of the surface dipole layer, and for a wide bandgap material the surface dipole layer can lead to a positive or negative electron affinity.

For example, hydrogen has been reported to induce a NEA on the diamond (111) surface [15.1, 2, 6, 12]. More recently, an NEA effect has been shown for the hydrogenated C(100) and C(110) surfaces [15.8, 13–15]. By comparison, oxygen leads to a dipole such that a positive electron affinity is observed on these surfaces.

Various surface treatments designed to remove nondiamond carbon result in an oxygen-terminated surface. These include acid etching and an electrochemical etch process. It has been suggested that different treatments can lead to different bonding configurations (bridge versus double bonding) of oxygen on the C(100) surface [15.8, 16, 17]. Vacuum annealing of cleaned C(100) to 500°C does not remove a significant portion of the oxygen from the surface as detected by means of Auger electron spectroscopy (AES), indicating that most of the oxygen was chemisorbed. The UPS spectra of as-loaded samples, as well as those heated to 500°C, showed a positive electron affinity. A value for the electron affinity of 1.5 eV was detected following the 500°C anneal. Following an anneal to about 1000°C, the amount of oxygen on the surface dropped below the detection limit of the AES instrument. A reconstructed (2 × 1) LEED pattern appeared. In addition, the width of the UPS spectra increased to ~15.7 eV, and a sharp low-energy feature appeared, indicating a NEA following the last annealing step. It was presumed that the surface was hydrogen-terminated after this step.

Annealing the (100) surface at temperatures above 1150°C results in the desorption of the remaining hydrogen, and a clean surface is obtained that exhibits a 2 × 1 reconstruction. The UPS measurements of this surface display a positiveelectron affinity. Figure 15.3 shows UV-photoemission spectra for oxygen-terminated, clean, and hydrogen-terminated diamond (100) surfaces. The NEA character of the H-terminated surface is evident in both the width of the spectrum and the presence of the sharp peak at the lowest binding energy.

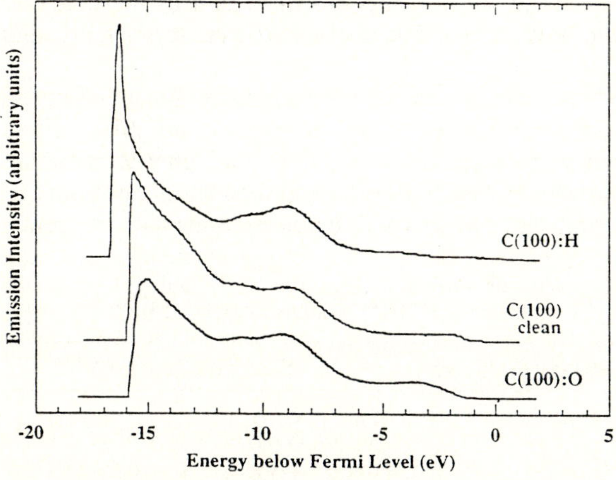

Fig. 15.3: Photoemission spectra of oxygen-terminated, hydrogen-terminated, and clean diamond (100) surfaces. The broadening of the spectral width and the sharp feature at high (negative) binding energy (i.e. low kinetic energy) are indicative of a NEA.

Furthermore, surfaces exposed to hydrogen plasma exhibited an NEA and 2×1 reconstruction indicating a monohydride termination. These studies were actually preceded by *ab initio* calculations suggesting an NEA for the monohydride-terminated 2×1 reconstructed (100) surface [15.13]. The same theoretical studies reported a positive electron affinity for the clean 2×1 surface, in agreement with experimental observations. In fact, the difference in the electron affinity between the two surfaces was ~3 eV, indicating the magnitude of the change in the surface dipole.

Diamond (110) surfaces also exhibited oxygen termination after surface cleaning, and they were found to have a positive electron affinity as evidenced by means of UPS [15.8]. After annealing the samples at 700°C, the oxygen concentration on the surface dropped to below the detection limit of the AES instrument. The low-energy cutoff of the UPS spectrum shifted to lower energies, indicating a reduction of the electron affinity. In addition, a sharp low-energy peak attributed to an NEA appeared. An 800°C anneal removed the sharp NEA feature, and the width of the spectrum was reduced by 0.7 eV. Exposing the surfaces to an H plasma resulted in the re-appearance of the NEA characteristics. By employing an 800°C anneal, the NEA could be removed again. The results for the (100) and (110) hydrogen-terminated, oxygen-terminated, and clean surface are summarized in Table 15.1.

In Fig. 15.4, a schematic of the band alignments for a clean and hydrogen-terminated diamond surface is shown, with band bending consistent with p-type doping. Hydrogenation changes the surface dipole layer. This causes a shift of the bands with respect to the vacuum level. Oxygen is bonded the strongest to diamond (100), and the weakest to diamond (111) surfaces. The bond strength of O on the C(110) surface falls between these two values [15.8]. Following this approach, according to the annealing temperatures necessary to remove an NEA from a hydrogen-terminated diamond surface, the hydrogenation of the diamond (100) surface appears to be the most stable, that of the (110) the least stable, and C(111) falls inbetween [15.8].

Recent work has explored the effects and stability of deuterium bonding on diamond. Deuterium termination has been found to induce an NEA effect comparable to that of hydrogen, but the surface was stable to a higher temperature than a hydrogen-terminated surface. Future studies should focus on the stability of the hydrogen- or deuterium-terminated surfaces and the interactions with oxygen.

Table 15.1: The UPS spectral width for different diamond (100) and (110) surface terminations. The electron affinity or presence of a NEA is deduced using (15.1).

Surface	UPS spectral width (eV)	Electron affinity (eV)
C(100):H	15.7	NEA
C(100) (clean)	15.05	0.65
C(100):O	14.2	1.50
C(110):H	15.7	NEA
C(110) (clean)	15.1	0.60
C(110):O	14.3	1.40

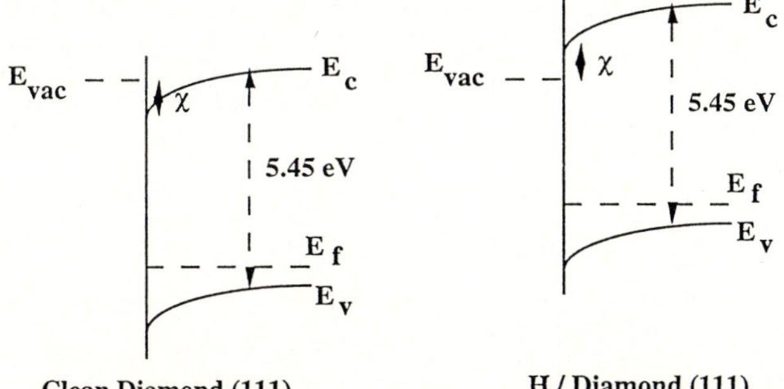

Fig. 15.4: The band alignments of clean diamond and H-terminated diamond surfaces. Note that the figures have been aligned at the vacuum level.

15.2.4 Metallic Layers on Diamond

Deposition of thin metal films is another possibility of inducing an effective NEA on diamond. It has been known for many years that low work function metals such as Cs can induce NEA type characteristics on III-V semiconductors such as GaAs. High-efficiency photocathodes are based on such structures. Cs deposition on diamond has also been demonstrated to induce an NEA effect [15.18, 19]. Since diamond has a large bandgap other, higher work function metals may be suitable to establish a NEA. A few years ago, Ti [15.20] or Ni [15.21] were found to induce an NEA on diamond (111). More recently, NEA characteristics have been reported for Co, Cu, or Zr on diamond (100), (111), and (110) surfaces [15.16, 22–24]. Figure 15.5 shows photoemission spectra of diamond surfaces coated with a few monolayers of these metals. A sharp peak indicative of an NEA is detected for all of these spectra.

Deposition of a thin metal layer was found to change the effective electron affinity of the diamond surface. This effect is illustrated in Fig. 15.6 for Ti on the clean diamond surface. This structure was found to exhibit an NEA. The effective electron affinity for a thin metal layer on the diamond can be modeled in terms of two interfaces: the vacuum/metal and the metal/diamond. Equation (15.2) gives an expression for the effective electron affinity:

$$\chi_{eff} = (\Phi_M + \Phi_B) - E_G. \tag{15.2}$$

In this model, a lower Schottky barrier height would result in a lower effective electron affinity, and this is consistent with the experimental results. For each metal, the surface termination prior to metal deposition appears to be have a significant effect on the Schottky barrier height. For metals deposited on clean surfaces, lower values for the Schottky barrier height and a greater likelihood of inducing an NEA are expected than for metals on nonadsorbate free surfaces.

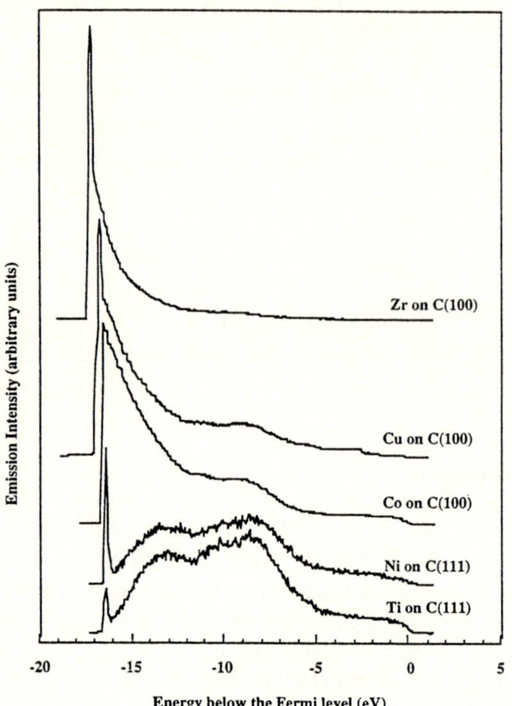

Fig. 15.5: UV-photoemission spectra of diamond surfaces with thin metal overlayers. The metal thicknesses correspond to several monolayers.

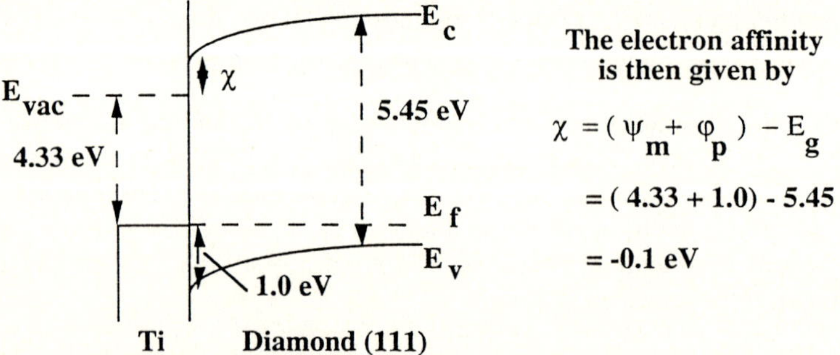

Fig. 15.6: The band structure at the surface of diamond with a thin metal coverage. The electron affinity can be deduced from the Schottky barrier, the metal work function and the diamond bandgap. The numbers for Ti on diamond (111) are illustrated.

Figure 15.7 shows the band diagrams for Ti on a clean or a H-terminated diamond surface. Metal/diamond interfaces exhibiting an NEA have a lower Schottky barrier height than those exhibiting a positive electron affinity.

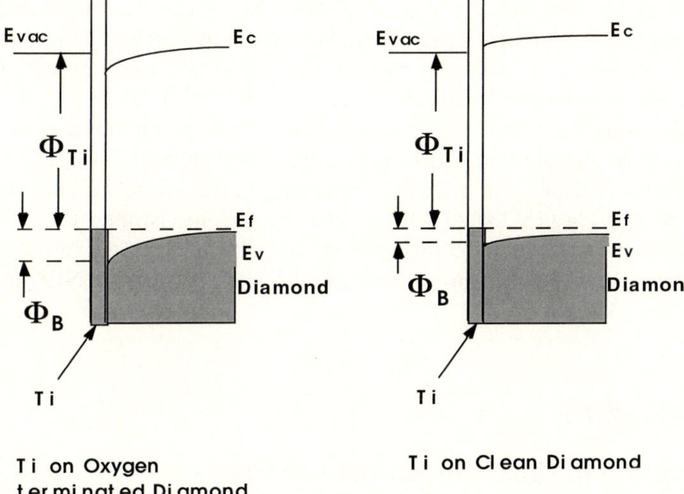

Fig. 15.7: A schematic band diagram of Ti on clean and H-terminated diamond surfaces. Note that the figures have been aligned at the vacuum level.

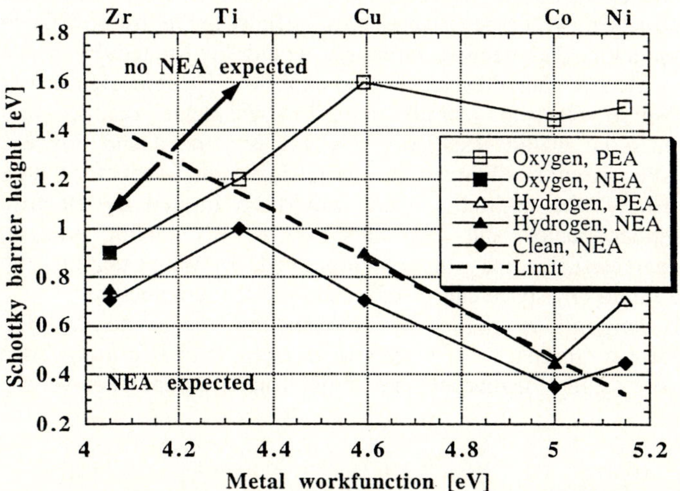

Fig. 15.8: The Schottky barrier height versus the metal work function for Ti, Zr, Cu, Co, and Ni. The dashed line represents the limit for which an NEA is expected for metal/diamond interfaces according to (15.2). Thus an NEA is expected for data points below this dashed line and a positive electron affinity for those above. The filled markers correspond to an experimentally observed NEA and the empty markers indicate an experimentally observed positive electron affinity (PEA) [15.24]

In Fig. 15.8, the experimentally (by means of UV photoemission) measured Schottky barrier heights are plotted versus the metal work function [15.24]. It is indicated whether an NEA or a positive electron affinity was observed by means of UV photoemission. The experimental data are also compared to (15.2).

Among the results obtained to date the most significant may be the observation of an NEA of Co and Zr on diamond (100), (111), and (110) [15.23, 24]. These films have been shown to be uniform, with little tendency to the islanding that has been observed for Ni and Cu [15.23, 25]. It has also been demonstrated that the NEA was stable following air exposure for Cu, Co, and Zr on diamond surfaces [15.23, 24]. In addition, it has been demonstrated that Ti oxide induces an NEA on diamond (111) surfaces [15.26].

15.3 Field Emission

15.3.1 A Description of the Field Emission Process

Most practical applications of electron emitters will require field-induced emission. The actual field-emission process from a semiconductor combines four effects: (1) electron supply to the semiconductor; (2) transport through the semiconductor; (3) emission into the vacuum; and (4) transport in vacuum to the anode. For an ideal structure, with a negative electron affinity and a low-resistance contact and semiconductor, the electron emission would be limited by space charge in the vacuum, the I–V characteristics would exhibit a $V^{3/2}$ dependence. For the case of a positive electron affinity and low-resistance contact and semiconductor, the material would essentially respond as a metal and Fowler–Nordheim characteristics would be obtained. For an intrinsic semiconductor with a negative electron affinity, the emitted current would be limited by electron injection into the semiconductor, and since this may be a tunneling process, Fowler–Nordheim characteristics may also be obtained. In this situation, the current could also be limited by space charge effects in the semiconductor.

Field (electron) emission is not only a promising means to develop intense/controlled electron currents for a variety of devices. It is also a powerful and rapid research tool to study the mechanisms of electron emission, whatever the means of stimulation.

15.3.2 Measuring Techniques

The field-emission characteristics of diamond surfaces are most commonly determined by using a movable probe as an anode that can be stepped toward the surface of the specimen [15.23, 27, 28], a large flat anode kept at a certain distance (typically in the mm range) away from the sample by a spacer [15.29–31], or a probe that is used to scan across the surface [15.31, 32]. In the case of a movable anode, the current voltage characteristics are measured at different distances. The distances are of the order of a few µm to a few 10 µm. To avoid emission from edges of the anode, it often has a spherical shape with a diameter of a few mm. This method has the advantage that the emission can be determined as a function of distance, which may be important for the characterization of rough or highly resistive samples. The technique employing the large anode has the advantage of exploring the entire surface area. Also, if a fluorescing anode is used, the location of various emission sites can be observed. However, in this method field-emitted electrons with keV energies are striking the anode and can desorb material from there. In turn, the cathode is exposed to keV energy ions from the anode. As a result, the surface of the cathode becomes contaminated and discharges can occur. These effects can lead to changes in the emission behavior. Also, when employing a phosphor screen as an anode, problems can occur since small amounts of phosphor powder may get pulled to the cathode due to the high electric fields. The use of a concentrated binder with the phosphor may alleviate this problem. To avoid the problems with phosphor, an indium tin oxide (ITO) layer can be used instead [15.30]. However, a higher voltage is necessary for ITO, increasing the likelihood of changing the emission characteristics due to contamination of the cathode surface or the striking of a discharge. Also, the spatial resolution of an ITO-coated anode is worse due to the higher energy of the electrons hitting the anode. Overall, to avoid discharging effects, a better vacuum needs to be maintained for the technique using a large anode kept at a large distance, than for a method employing a probe in close proximity of the emitting surface. A variation of this technique is the arrangement of an anode grid close to the surface facilitating the field emission. A collector screen can then be placed at a larger distance away from the sample surface [15.31]. The third technique employs a tip that is placed a few µm from the emitting surface and can be scanned across this surface. This method can be useful for correlating the distribution of emission sites with the surface morphology.

15.3.3 Field-Emission Results from Diamond

There have now been several studies that have indicated electron emission from flat diamond surfaces at relatively low fields. Kordesh and co-workers used field-emission microscopy and showed a uniform emission from p-type diamond films at low fields (~3 V/µm) [15.33]. Using a scanning probe with µm resolution, Talin et al. reported turn-on fields of 3 V/µm for about one half of the surface area of nanocrystalline diamond [15.32]. Raman spectra of the emitting areas exhibited lower-quality sp^3 bonding. Latham and co-workers [15.34] measured field emission from flat diamond films at fields of ~20 V/µm. They showed that the emission was from very specific point sites, which may be correlated with defect structures rather than sharp features of the films. Zhu and co-workers found a correlation with specific sp^2 bonding structures in the films [15.28, 35]. As shown in Fig. 15.9, higher electric fields were necessary for emission from high-quality p-type diamond than for defective or particulate diamond [15.35]. To date, most of these measurements have been on p-type boron doped diamond. Miskovsky and

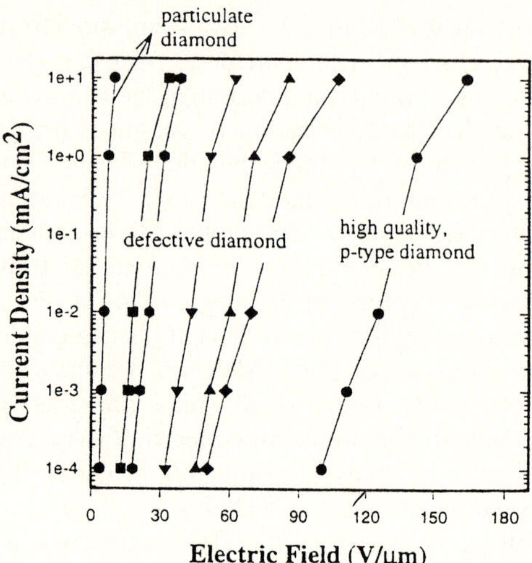

Fig. 15.9: A plot of the emission current densities versus the field required for emission from p-type CVD diamond, defective undoped diamond, and nanometer-size diamond powder [15.35].

Cutler have analyzed the emission possibilities from these films, and have suggested that states in the bandgap must be present and must participate in the emission process to account for the low field emission from the p-type material [15.35, 36].

It has been found that thin metal films of Cu, Co, or Zr on natural crystal p-type diamond result in a decrease in the field emission threshold compared to the oxygen-terminated diamond surface. The results of the field emission threshold and electron affinity showed a similar trend, in which the field emission threshold decreased as the electron affinity decreased [15.36, 37].

Prins has developed a method of ion implanting diamond to produce nominally n-type material [15.38, 39]. Geis and co-workers have used this technique to fabricate a diamond cold cathode emitter structure based on an all-diamond p–n

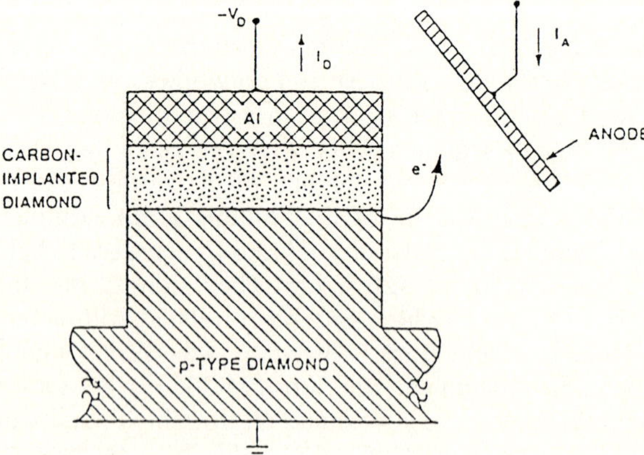

Fig. 15.10: Schematic of the cold cathode based on an all-diamond p–n junction [15.40].

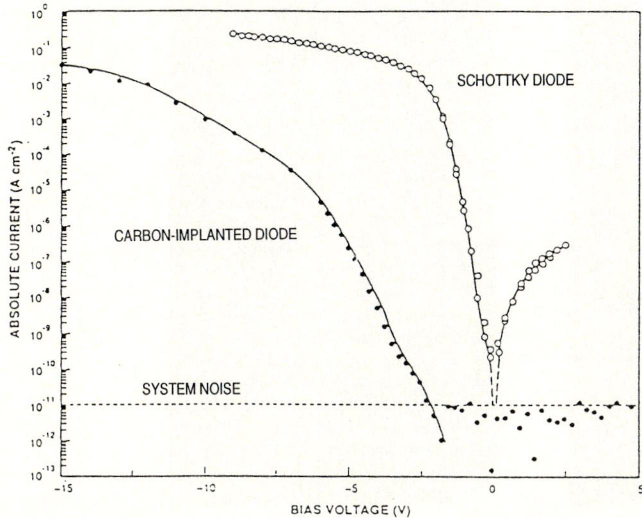

Fig. 15.11: Diode current versus applied voltage (*I–V* characteristics) for the carbon ion implanted diode as compared to that of an aluminum Schottky contact on p-type diamond [15.40].

junction [15.40]. Carbon ion implantation was employed to induce the n-type like characteristics in diamond. Figure 15.10 shows a schematic of the cold cathode device [15.40]. The diode current versus the applied voltage for the carbon ion implanted diode was compared to that of an aluminum Schottky contact on p-type diamond. Figure 15.11 illustrates the measured *I–V* characteristics for both structures [15.40]. Geis et al. have also obtained field emission from nitrogen-doped diamond [15.27]. There are several possible configurations for nitrogen impurities in diamond, but for nitrogen occupying single substitutional sites, the impurities exhibit n-type characteristics with a donor level at ~1.7 eV below the conduction band. These materials also showed low field emission, but the spatial variation has not been reported.

There have been several recent studies of the field emission from diamond-coated field emitters (i.e., pointed structures) made of silicon or metals [15.33, 41–46]. In these measurements, the electron emission was found at significantly lower fields than from uncoated surfaces. However, many studies have shown nonuniform growth of CVD diamond on the sharp tips. In particular, distinct diamond particles have been observed on the tips, as shown in Fig. 15.12 [15.47]. Possible explanations for the electron emission based on negative electron affinity diamond surfaces or based on different radii of silicon tips underneath the diamond have been proposed [15.46, 47]. The growth of nanocrystalline CVD diamond has been reported [15.32, 48–50]. This material may be suitable for coating field emitter tips (or flat emission surfaces) with a smooth, uniform diamond layer.

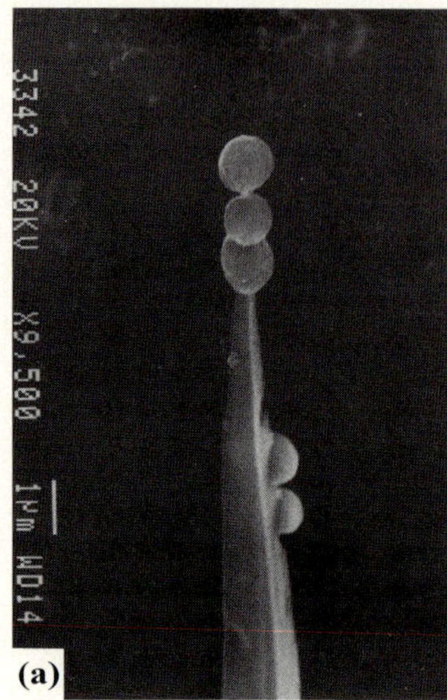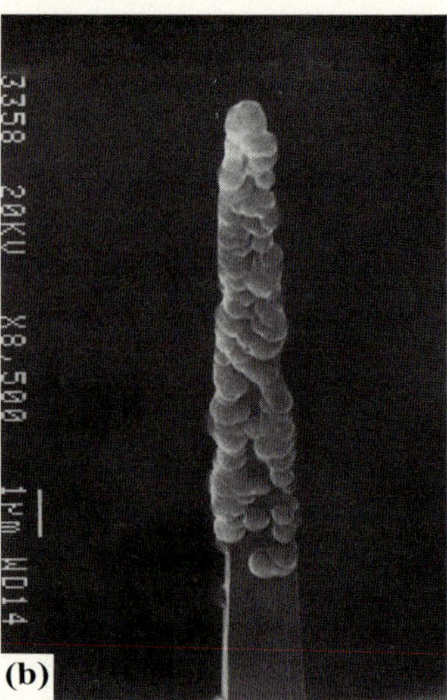

Fig. 15.12: Distinct diamond particles on silicon tips. **(a)** Single-particle-type diamond coverage. **(b)** A conglomerate of diamond particles [15.47].

Preliminary results of emitter tips coated with nanocrystalline diamond indicate a significant (three-fold) reduction in field emission threshold compared to the uncoated tips [15.51].

It is evident that the complicated processes involved in field emission have impeded the advancement of our understanding of these measurements. The measurements themselves require care and, in fact, some early reports may have been dominated by artifacts attributable to poor vacuum, or by other effects due to the high fields present in the measurements. The residual gas in the measuring chamber results in a background ion current. Moreover, the strong fields can sometimes cause a plasma to ignite and severely damage the surface. Following arc discharges, crater formations and molten areas with debris were observed on the surfaces of CVD-diamond and amorphous-carbon films [15.52]. A corresponding improvement in the field emission has been reported. It has been suggested that this improvement may be due to the formation of protrusions or sharp edges that could act as emission sites.

Even under ultrahigh vacuum conditions, sputtering or desorption of material from the anode can be caused due to the high-energy electrons emitted form the sample. This may result in deposition of material on the sample or in a destructive discharge. Positive ions will be accelerated toward the surface of the sample and may damage the surface, even if no plasma discharge occurs. One way to avoid these damaging effects is the use of lower voltages. This implies that smaller distances need to be employed to facilitate the field emission. At smaller distances, effects due to the surface roughness of the sample or the anode become more significant.

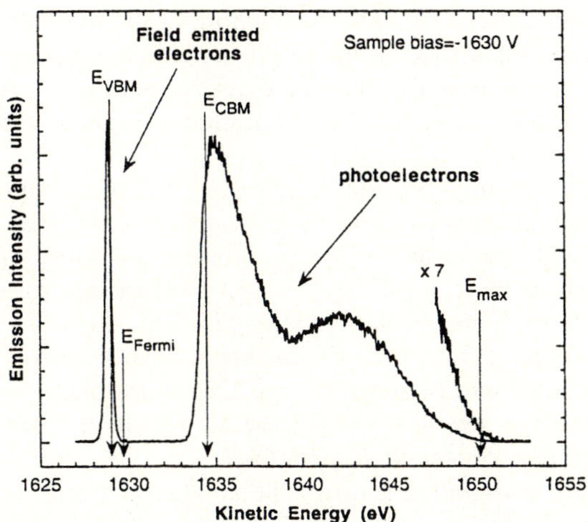

Fig. 15.13: The electron energy distribution for the simultaneous field emission and photoemission measurements on C(111) 1×1:H. The energy position of the field emitted electrons corresponds to the valence band maximum [15.53].

Simultaneous field emission and photoemission measurements from a (111) 1×1:H p-type natural diamond surface were reported by Bandis and Pate [15.53]. The electron energy distributions are shown in Fig. 15.13. From these experiments, it was found that the field-emitted electrons originated from the valence band maximum. In contrast, the photoemission process involved emission from the conduction band. For boron-doped polycrystalline diamond films, Glesener and Morrish found no dependence of the field emission on temperature, and they also suggest that the electrons originate from the valence band [15.54]. To date, there have been no studies that confirm field electron emission originating from the conduction band of diamond. Most studies on emission from diamond may be explained as electron emission from the valence band or from defect states. Only UV-photoemission spectroscopy measurements, where UV radiation is employed to excite electrons from the valence band into the conduction band, have studied the electron emission from the conduction band.

The supply of electrons to the conduction band remains a significant problem. Ideally, a shallow n-type dopant could solve this issue. Then, a field-emission structure could consist of a highly n-type doped region at the electron injecting contact and lower doping in the bulk of the material. Nitrogen is a substitutional dopant and has been reported in concentrations of 10^{19} cm^{-3} in diamond [15.55]. High concentrations of incorporated nitrogen were found to enhance electron emission [15.27, 29]. According to theoretical [15.56] and experimental [15.57, 58] studies the "shallowest" level of nitrogen is located at about 1.5–2.1 eV below the conduction-band minimum. Extremely high nitrogen concentrations would be necessary to obtain a n-type doping effect and to facilitate the electron injection at the back contact. There has been great difficulty in incorporating such large amounts of nitrogen into diamond. Alternatively, roughening of the interface at the contact may help to circumvent this difficulty and lower the effective barrier for electron injection due to field enhancement at the rough interface [15.27]. A novel approach for CVD-diamond deposition has been reported to enable the incorporation of large amounts of nitrogen into diamond.

Phosphorus is another substitutional impurity that may be a potential n-type dopant in diamond. For substitutional phosphorus, an activation energy of 0.20 eV has been calculated [15.56]. Although the value for the equilibrium solubility of phosphorus in diamond is expected to be low [15.56], a doping effect due to a shallow level of phosphorus has been found experimentally [15.59]. An n-type conductivity and a corresponding value of 0.20–0.21 eV for the activation energy was measured for phosphorus implanted into high-purity type IIa natural diamond [15.60]. Deeper levels with activation energies of 0.84–1.16 eV have been calculated [15.61, 62] and measured experimentally [15.63, 64].

Lithium is an interstitial impurity in diamond and may be another potential n-type dopant. A donor level of 0.1 eV below the conduction-band minimum has been reported [15.56]. However, diffusion of lithium would result in an undesirable deterioration of the doping characteristics over time [15.56]. Also, the solubility of lithium in diamond is predicted to be low.

The commonly observed nonuniformity in the emission from diamond surfaces [15.27] may also be a significant obstacle for the potential use of diamond in emission devices. This implies that the emission site density becomes a crucial characteristic to determine whether an emitting surface is suitable for practical applications. It has been estimated that an emission site density of the order of at least 10^6–10^7 sites/cm^2 is necessary for applications as field emission displays [15.65]. Figures 15.14a, b show emission images of a carbon layer for 6.5 V/μm

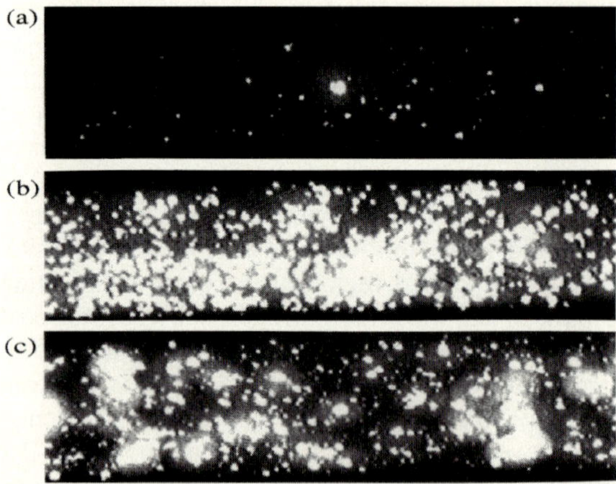

Fig. 15.14: The field electron emission from **(a)** cathode #1, 6.5 V/μm, **(b)** cathode #1, 10 V/μm, and **(c)** cathode #2, 10 V/μm [15.65].

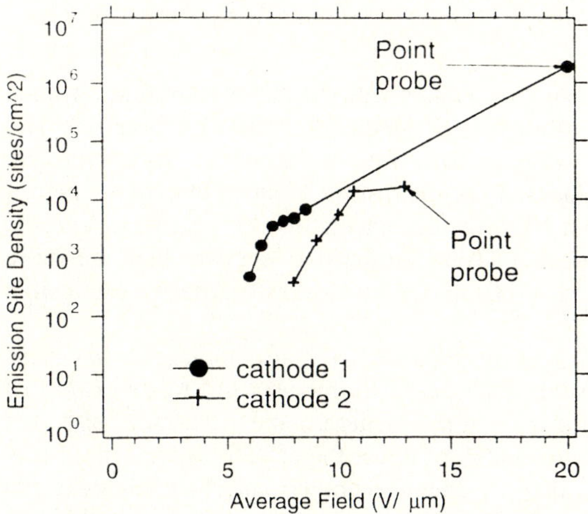

Fig. 15.15: A plot of the emission site density versus the average field for cathodes #1 and #2 [15.65].

and 10 V/µm [15.65]. An image of a differently prepared carbon film for 10 V/µm is shown in Fig. 15.14c [15.65]. A comparison of the emission site density for these two carbon films can be seen in Fig. 15.15 [15.65]. A reported increase in emission site density of CVD-diamond films has been attributed to a pre-growth treatment [15.66]. Figure 15.16 shows the emission site density of a treated and an untreated area [15.66].

However, uniform emission from diamond films has been reported in one study using a novel surface electron microscope that was operated in the field emission mode [15.67, 68]. The instrumental limit for the field was 2 V/µm. Since typical threshold fields in other reports are about an order of magnitude higher, it may be conceivable that there is a uniform emission at low fields, which becomes nonuniform at higher fields. This issue will need to be resolved for most applications.

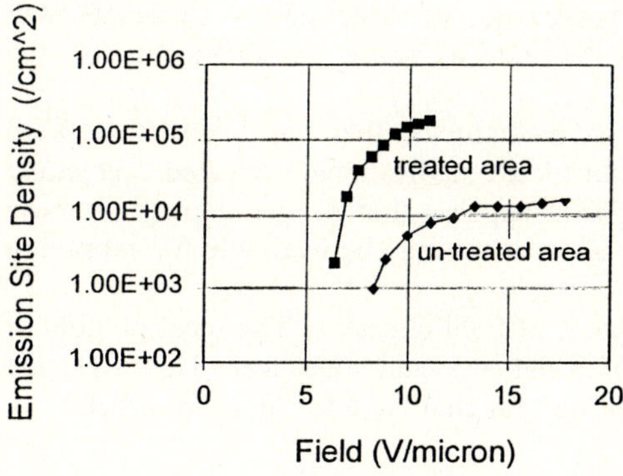

Fig. 15.16: A plot of the emission site density versus the field for treated and untreated areas of the substrate [15.66].

15.4 Conclusions

It is now evident that it is possible to obtain a true NEA for diamond. Results indicate that a positive electron affinity is obtained for both adsorbate-free and oxygen-terminated surfaces. However, an NEA is obtained for hydrogen termination of all low-index surfaces. It is interesting to note that more than a decade transpired between the first NEA measurements on (111) surfaces and the discovery of a NEA for H-terminated (100) surfaces. It appears that the more tenacious bonding of oxygen to the (100) surface is at least partially responsible for the delayed observation.

The properties of thin metal layers on diamond have also indicated that these may be suitable for obtaining an NEA surface. The model used to describe this effect is based on two interfaces – the vacuum/metal interface and the metal/diamond interface. Within this model, to lower the effective electron affinity of this structure it is necessary that the metal-diamond interface changes the surface dipole of the diamond. A particularly encouraging result was found for Zr deposited on clean diamond surfaces and surfaces with oxygen or H adsorbates. Here it appears that the Zr displaces the oxygen or hydrogen termination.

The field emission is the most complicated and potentially least understood process. The combined photo- and field emission measurements indicate that field emission from p-type diamond originates from electrons in the valence band. Similarly, many studies have indicated lower field emission thresholds for diamond with a significant defect density. This again suggests that the emission does not involve electrons in the conduction band. While conduction-band emission has been suggested for nitrogen-doped diamond, the role of defects needs to be further explored.

The two most pressing problems related to field emission are (i) the verification of whether field emission from the conduction band has been obtained and (ii) an understanding of the nonuniform emission. Certainly, the development of a process to obtain shallow n-type doping would go a long way toward solving each of these problems.

Even without the development of an n-type dopant, it seems likely that diamond with defects may substantially improve the emission characteristics of various pointed and flat surfaces. The preparation of actual device structures may be necessary to determine whether the current status is sufficient for the applications.

Diamond has been the wide gap semiconductor that has been most often considered for emission applications, but recent studies have indicated a negative electron affinity for both BN and AlN. It appears that n-type doping of these materials is also problematic, but other approaches may be available for supplying electrons to the conduction band.

This review has neglected some aspects of field emission. The most notable of these may be the beneficial effects of roughness and interfaces. However, the complexities of the processes may provide a real challenge for future research.

References

15.1   F.J. Himpsel, J.A. Knapp, J.A. van Vechten, and D.E. Eastman, Phys. Rev. **B20**, 624 (1979)

15.2   B.B. Pate, Surf. Sci. **165**, 83 (1986)

15.3   R. Gomer, Field Emission and Field Ionization, Harvard University Press, Cambridge, MA (1961)

15.4   A. Zangwill, Physics at Surface, Cambridge, MA (1988)

15.5   R.J. Nemanich, P.K. Baumann, M.C. Benjamin, S.W. King, J. van der Weide, and R.F. Davis, Diamond Rel. Mater. **5**, 790 (1996)

15.6   J. van der Weide, and R. J. Nemanich, Appl. Phys. Lett. **62**, 1878 (1993)

15.7   C. Bandis and B.B. Pate, Phys. Rev. Lett. **74**, 777 (1995)

15.8   P.K. Baumann and R.J. Nemanich, submitted to J. Appl. Phys.

15.9   G.T. Mearini, I.L. Krainsky, J.A. Dayton, Jr., Y. Wang, C.A. Zorman, J.C. Angus, and D.F. Anderson, Appl. Phys. Lett. **66**, 242 (1995)

15.10   D.P. Malta, J.B. Posthill, T.P. Humphreys, R.E. Thomas, G.G. Fountain, R.A. Rudder, G.C. Hudson, M.J. Mantini, and R.J. Markunas, Appl. Phys. Lett. **64**, 1929 (1994)

15.11   R.E. Thomas, T.P. Humphreys, C. Pettenkofer, D.P. Malta, J.B. Posthill, M.J. Mantini, R.A. Rudder, G.C. Hudson, and R.J. Markunas, Mater. Res. Soc. Symp. Proc. **416**, 263 (1996)

15.12   R.J. Nemanich, L. Bergman, K.F. Turner, J. van der Weide, and T.P. Humphreys, Properties of Interfaces of Diamond, Trieste Semiconductor Symposium on Wide-Band-Gap Semiconductors, Physica B **185**, 528 (1993)

15.13   J. van der Weide, Z. Zhang, P.K. Baumann, M.G. Wensell, J. Bernholc, and R.J. Nemanich, Phys. Rev. B **50**, 5803 (1994)

15.14   J. van der Weide and R. J. Nemanich, J. Vac. Sci. Technol. **B 12**, 2475 (1994)

15.15   P.K. Baumann and R. J. Nemanich, Diamond Rel. Mater. **4**, 802 (1995)

15.16   P.K. Baumann, T.P. Humphreys, and R.J. Nemanich, Mat. Res. Soc Symp. Proc. **339**, 69 (1994)

15.17   M.J. Rutter and J. Robertson, submitted to Phys. Rev. B

15.18   M.W. Geis, J.C. Twichell, J. Macaulay, and K. Okano, Appl. Phys. Lett. **67**, 1328 (1995)

15.19   O.M. Küttel, O. Gröning, E. Schaller, L. Diederich, P. Gröning, and L Schlapbach, Diamond Rel. Mater. **5**, 807 (1996)

15.20   J. van der Weide and R.J. Nemanich, J. Vac. Sci. Technol. **B 10**, 1940 (1992)

15.21   J. van der Weide and R.J. Nemanich, Phys. Rev. **B49**, 13629 (1994)

15.22   P.K. Baumann and R.J. Nemanich, Appl. Surf. Sci. **104/105**, 267 (1996); Mater. Res. Soc. Symp. Proc. **416**, 157 (1996); J. Vac. Sci. Tech. **B 15**(4) (1997) in press

15.23   P.K. Baumann and R.J. Nemanich, submitted to Phys. Rev. B

15.24   P.K. Baumann and R.J. Nemanich, submitted to J. Appl. Phys.

15.25   P.K. Baumann, T.P. Humphreys, R.J. Nemanich, K. Ishibashi, N.R. Parikh, L.M. Porter, and R.F. Davis, Diamond Rel. Mater. **3**, 883 (1994)

15.26   C. Bandis, D. Haggery, and B.B. Pate, Mater. Res. Soc. Symp. Proc. **339**, 75 (1994)

15.27   M.W. Geis, J.C. Twichell, N.N. Efremow, K. Krohn, and T.M. Lyszczarz, Appl. Phys. Lett. **68**, 2294 (1996)

15.28   W. Zhu, G.P. Kochanski, S. Jin, L. Seibles, D.C. Jacobson, M. McCormack, and A.E. White, Appl. Phys. Lett. **67**, 1157 (1995)

15.29   K. Okano, S. Koizumi, S.R.P. Silva, and G.A.J. Amaratunga, Nature **381**, 140 (1996)

15.30   P.V. Latham, High Voltage Vacuum Insulation, ed. R.V. Latham, Academic Press, San Diego (1995)

15.31   Z. Feng, I.G. Brown and J.W. Ager III, J. Mater. Res. **10**, 1585 (1995)

15.32   A.A. Talin, L.S. Pan, K.F. McCarthy, T.E. Felter, H.J. Doerr, and R.F. Bushah, Appl. Phys. Lett. **69**, 3842 (1996)

15.33   C. Wang, A. Garcia, D.C. Ingram, M. Lake, and M.E. Kordesch, Electron. Lett. **27**, 1459 (1991)

15.34   N.S. Xu, R.V. Latham, and Y. Tzeng, Electron. Lett. **29**, 1596 (1993)

15.35   W. Zhu, G.P. Kochanski, and S. Jin, Mater. Res. Soc. Symp. Proc. **416**, 443 (1996)

15.36   Z.-H. Huang, P.H. Cutler, N.M. Miskovsky, and T.E. Sullivan, Appl. Phys. Lett. **65**, 2562 (1994)

15.37   N.M. Miskovsky, P.H. Cutler, and Z.-H. Huang, J. Vac. Sci. Technol. **B 14**, 2037 (1996)

15.38   J. Prins, Thin Solid Films **212**, 11 (1992)

15.39   J. Prins, Mater. Sci. Rep. **7**, 271 (1992)

15.40   M. Geis, N. Efremow, J. Woodhouse, M. Mcalese, M. Marchywka, D. Socker, and J. Hochedez, IEEE Electron Devices Lett. **12**, 456 (1991)

15.41   N.S. Xu, Y. Tzeng, and R.V. Latham, J. Phys. D: Appl. Phys. **26**, 1776 (1993)

15.42   J. Liu, V.V. Zhirnov, A.F. Myers, G.J. Wojak, W.B. Choi, J.J. Hren, S.D. Wolter, M.T. McClure, B.R. Stoner, and J.T. Glass, J. Vac. Sci Technol. **B 13**, 422 (1995)

15.43   W.B. Choi, J.J. Cuomo, V.V. Zhirnov, A.F. Myers, and J.J. Hren, Appl. Phys. Lett. **68**, 720 (1996)

15.44   W.B. Choi, J. Liu, M.T. McClure, A.F. Myers, V.V. Zhirnov, J.J. Cuomo, and J.J. Hren, J. Vac. Sci. Technol. **B 14**, 2050 (1996)

15.45   V. Raiko, R. Spitzl, B. Aschermann, D. Theirich, J. Engemann, N. Puteter, T. Habermann, and G. Müller, Thin Solid Films **290–291**, 190 (1996)

15.46   E.I. Givargizov, V.V. Zhirnov, A.N. Stepanova, E.V. Rakova, A.N. Kieselev, and P.S. Plekhanov, Appl. Surf. Sci. **87/88**, 24 (1995)

15.47   V.V. Zhirnov, J. Phys. IV **6**, C5-107 (1996)

15.48   D.M. Gruen, X. Pan, A.R. Krauss, S. Liu, J. Luo, and C.M. Foster, J. Vac. Sci. Technol. **A 12**, 1491 (1994)

15.49   D.M. Gruen, C.D. Zuiker, and A.R. Krauss, in SPIE **2530**, 2 (1995)

15.50   C.D. Zuiker, A.R. Krauss, D.M. Gruen, X. Pan, J.C. Li, R. Csencsits, A. Erdemir, C. Bindal, and G. Fenske, Thin Solid Films **270**, 154 (1995)

15.51   A.R. Krauss, personal communication

15.52   O. Gröning, O.M. Küttel, E. Schaller, P. Gröning, and L Schlapbach, Appl. Phys. Lett. **69**, 476 (1996)

15.53   C. Bandis and B.B. Pate, Appl. Phys. Lett. **69**, 366 (1996)

15.54   J.W. Glesener and A.A. Morrish, Appl. Phys. Lett. **69**, 785 (1996)

15.55   T. Evans, The properties of Natural and Synthetic Diamond, ed. J.E. Field, Academic Press, London (1992), p. 259

15.56   S.A. Kajihara, A. Antonelli, J. Bernholc, and R. Carr, Phys. Rev. Lett. **66**, 2010 (1991)

15.57   W.J.P. Enkevort and E.H. van Versteegen, J. Phys.: Condens. Matter **4**, 2361 (1992)

15.58   R.G. Farrer, Solid State Commun. **7**, 685 (1969)

15.59   K. Okano, H. Kiyota, T. Iwasaki, T. Kurosu, M. Iida, and T. Nakamura, in New Diamond Science and Technology, MRS Int. Conf. Proc. (1991), p. 917

15.60   J.F. Prins, Diamond Rel. Mater. **4**, 580 (1995)

15.61   K. Jackson, M.R. Pederson, and J.G. Harrison, Phys. Rev. **B 41**, 12641 (1990)

15.62   V.V. Tokiy, N.D. Samsonenko, D.L. Savina, and S.V. Gorban, in Proceedings of the 2nd International Conference on the Applications of Diamond Films and Related Materials, 25–27 August 1993, Japan, MYO, Tokyo, p. 757

15.63   M.I. Landstrass, M.A. Plano, D. Moyer, S.P. Smith, and R.G. Wilson, Diamond Materials, Electrochemical Society (1991), p. 574

15.64   M. Kamo, H. Yarimoto, T. Ando, and Y. Sato, in New Diamond Science and Technology, MRS International Conference Procedings (1991), p. 637

15.65   A.A. Talin, B.F. Coll, E.P. Menu, J. Markhalm, and J.E. Jaskie, in Proceedings of the 1st Specialist Meeting on Amorphous Carbon, Cambridge, UK, 31 July 1997
15.66   Z.L. Tolt, R.L. Fink, and Z. Yaniv, in Proceedings of the IVMC'97, J. Vac. Sci. Technol. A
15.67   J.D. Shovlin and M.E. Kordesch, Appl. Phys. Lett. **65**, 863 (1994)
15.68   J.D. Shovlin, M.E. Kordesch, D. Dunham, B.P. Tonner, and W. Engel, J. Vac. Sci. Technol. **A13**, 1111 (1995)

16. CVD Diamond for Ultraviolet and Particle Detectors

Richard B. Jackman

Electronic and Electrical Engineering, University College London,
Torrington Place, London, WC1E 7JE, UK
e-mail: r.jackman@eleceng.ucl.ac.uk

Springer Series in Materials Processing
Low-Pressure Synthetic Diamond Eds.: B. Dischler and C. Wild
© Springer-Verlag Berlin Heidelberg 1998

16.1 Introduction

Conventional solid-state photodetectors, such as those made using silicon, are typically diode structures operated with a reverse bias placed across them. Low dark currents result; carriers photo-generated in the depletion region form a drift current which is the basis for light detection. Light with an energy greater than the band gap of the material can be seen. Silicon has a 1.1 eV band gap and devices fabricated from this material therefore react to both ultra-violet and visible wavelengths. The high resistivity of diamond suggests that diode structures may not be needed to achieve low dark currents if this material were used for the fabrication of photodetectors. Metal/diamond/metal devices can be considered which simply rely upon photoconductivity for their operation; such a device could possess high gain, since many carriers may be able to flow around the detector circuit during the lifetime of a photo-generated electron–hole pair. The wide band gap (5.5 eV, 225 nm) of diamond implies that this form of photodetector will be capable of detecting deep UV light while being essentially "blind" to visible wavelengths. This property is highly desirable; filtering conventional devices to make them visible blind significantly reduces their sensitivity to UV light. The physical and chemical robustness of diamond also suggests that such devices may be suitable for operation in hostile environments. Diamond can exhibit high carrier mobilities, saturated carrier velocities and electric field breakdown strength; these properties suggest that fast detectors may be realisable. Many industrial, military and environmental applications can thus be envisaged for diamond UV sensors. The emergence of commercially accessible thin-film diamond grown by chemical vapour deposition (CVD) has enabled reliable devices to be developed. In this chapter, the intrinsic and extrinsic photoconductivity of thin-film diamond is reviewed; the design, fabrication and performance of diamond UV photodetectors is then discussed, along with some of the uses for these devices. For some applications, it may be desirable to fabricate diamond photodiodes; the realisation of this type of device from p-type material is also addressed.

Solid-state devices can also be used for the detection of high-energy particles and ionising radiation. The detector structure again comprises metal electrodes placed upon a substrate, which are used to create an electric field within the material. Irradiation of the structure creates free carriers which can either be detected as a current or as an electrical charge within the device if the rate of creation of the carriers is low. The substrate should not be materially modified during operation. Silicon is not a good choice for the fabrication of these devices as it is readily damaged; the wide band gap, high ionisation energy and high atomic displacement energy of diamond suggest that long-life robust devices could be made from this material. The extreme electrical properties of diamond also suggest that it should be possible to achieve high-performance levels from such detector structures. In addition to being "radiation hard", diamond has a low atomic number which is similar to that of human tissue, of interest in medical dosimetry applications. In this chapter, the radiation hardness of diamond is reviewed; the use of diamond for alpha- and beta-particle detection is then considered, as well as the use of diamond structures for X-ray, gamma- and neutron-radiation detection.

16.2 UV Detectors

16.2.1 Intrinsic and Extrinsic Photoconductivity in Diamond

In a real material, defect states will enable light with an energy lower than the band gap energy to be absorbed; some of these processes may lead to enhanced electrical conductivity and it is therefore important to understand the extrinsic, as well as the intrinsic, photoconductivity that is possible in diamond. The photoconductivity of natural diamond was investigated by Nahum and Halperin some years ago [16.1]. A strong photoconductive response with a threshold of around 225 nm originating from band-to-band transitions was recorded for a stone with properties between types I and II. The presence of peaks in photoconductivity at longer wavelengths was found to be strongly affected by temperature; features present in a 300 K spectrum were often absent when investigated at 80 K. Indirect transitions to the conduction band by phonon interaction and thermal ionisation of carriers from states in the band gap (as opposed to luminescence) were suggested to be the cause of this difference. Denham, Lightowlers and Dean investigated the influence of the nitrogen content on photoconductivity [16.2]. In Fig. 16.1, the low-temperature (90 K) photocurrent measured by these researchers as a function of incident photon energy has been plotted against the photon wavelength. For a stone with little nitrogen (IIb), strong intrinsic absorption with a threshold of ~225 nm is apparent, but nitrogen-containing stones (Ia and "intermediate") have a threshold around 300 nm; this was attributed to a nitrogen-based donor level.

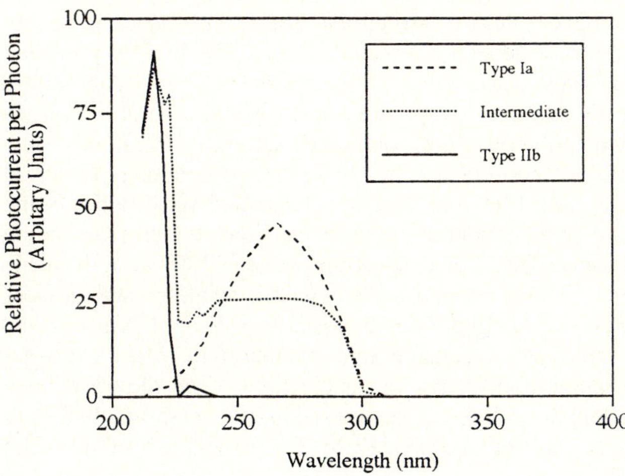

Fig. 16.1: The measured photocurrent as a function of illumination wavelength for differing types of single-crystal diamond (after [16.1]).

Photoconductivity associated with the population of states by longer-wavelength light (3.5 µm–350 nm) has also been characterised [16.3–6]. Sharp peaks within the spectral range 440–364 nm are thought to be caused by the presence of vacancies, and these features have been widely observed following high-energy electron irradiation [16.7–9].

Thin-film diamond, grown by CVD techniques on non-diamond substrates [16.10], is a polycrystalline material that can include differing quantities of non-diamond carbon depending upon the growth conditions used. The photoconductive properties of this form of diamond have also been investigated by a number of groups [16.11–15]. Allers and Collins [16.14] have found that the dominant photoconductive response in the near infra-red (IR) and visible spectral regions was a broad band with a threshold of ~825 nm (1.5 eV). While it was not clear whether this was a donor or acceptor state, the authors concluded that this defect was exclusive to diamond grown by CVD methods. Gonon and co-workers [16.13, 15] identified three distinct peaks in this spectral region when they studied thin (10 µm) films supported on silicon, at 1127 (1.1 eV), 886 (1.4 eV) and 653 nm (1.9 eV). The latter were believed to be acceptor states associated with bulk diamond. It was suggested that the 1127 nm feature was associated with ionisation from acceptor states responsible for dark conductivity within the film, thought to be associated with disordered regions at the grain boundaries within the polycrystalline film.

16.2.2 Photoconductive UV Detectors

An ideal UV photodetector will be highly responsive to deep UV light, blind to visible wavelengths, support only a low dark current and rapidly reset after operation. It is apparent from the discussion above (Sect. 16.2.1) that significant extrinsic photoconductivity can be measured in most forms of diamond; the control, or eradication, of this property will therefore be necessary for the realisation of high-performance device structures from this material. Early attempts to fabricate UV photoconductive detectors from both single-crystal and thin-film diamond were only moderately successful [16.16]. Aluminium electrodes, separated by 300 µm, were evaporated on to type IIa natural, type Ib synthetic and CVD-grown polycrystalline diamond samples. Relatively low sensitivity to 200 nm light was reported for the single-crystal structures (external quantum efficiency < 0.39); thin-film diamond structures had sensitivities some ~10^3 times lower. Extrinsic photoconductivity prevented the device from being blind to visible light, with only 100 : 1 discimination between 200 nm and visible wavelengths being achieved. Moreover, even the type IIa single-crystal devices took >10^4 seconds to reset once they had been exposed to UV light. Vaitkus and co-workers [16.17] also measured a considerable photocurrent when exposing thin-film polycrystalline diamond structures to visible light, once they had been exposed to deep-UV wavelengths. More recent reports have also shown that dark current device levels can be strongly influenced by UV irradiation of thin-film diamond structures [16.18].

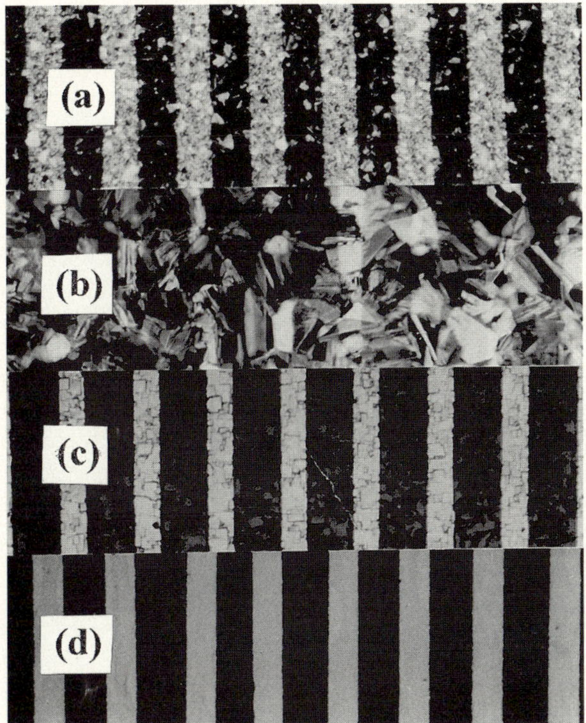

Fig. 16.2: Optical micrographs of differing forms of CVD diamond films. **(a)** and **(b)** display randomly aligned crystallites of differing sizes, while **(c)** reveals (100) textured material with a high degree of alignment between the grains and **(d)** is a homoepitaxial film. The stripes are 20 µm wide metal lines.

Work in the author's laboratory has led to the realisation of device designs that are more appropriate for use with polycrystalline material, and we have succeeded in producing high-performance diamond UV detectors [16.19–24]. CVD-diamond films can comprise relatively large grains on the upper surface, while the substrate side of the film is fine-grained material with significant graphitic (sp^2) inclusions. This region can therefore be expected to exhibit high dark conductivity through states that are likely to cause photoconduction from visible light. Furthermore, carriers may be trapped, leading to recombination and reduction in carrier lifetimes. To reduce the influence of this region on the photoconductive response, a structure in which both electrodes are placed on the large grains of the top surface can be used. Carrier mean free paths in CVD diamond have been found to be in the range 1–10 µm [16.25]; optimal electrode spacing will therefore be of similar dimensions. An interdigitated electrode structure can be used to provide a large collection area for the device. The crystal structure within CVD-diamond films can be controlled by variation of the growth conditions; Fig. 16.2 shows optical micrographs of free-standing material displaying randomly oriented crystallites with differing grain sizes (a, b) compared with a (100) textured film (c) in which the grains are aligned; a micrograph of the featureless surface of a homoepitaxial film (d) is included for comparison.

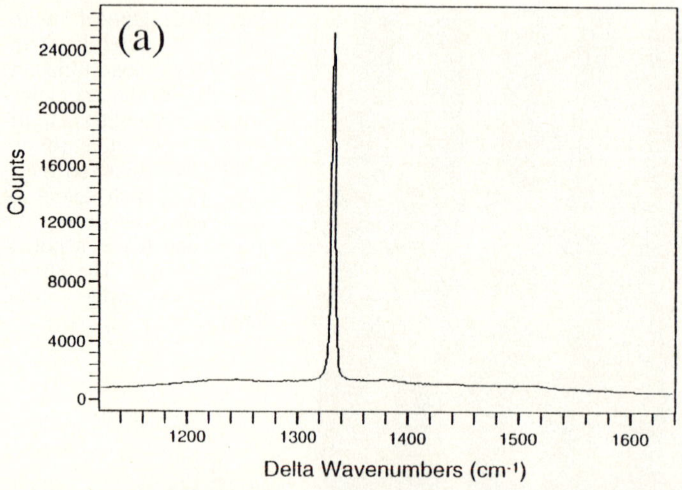

Fig. 16.3: (a) A Raman spectrum (He–Ne laser, 632 nm) recorded from a free-standing CVD-diamond film used for photodetector fabrication. **(b)** An optical micrograph of the same material.

An idealised arrangement would thus comprise planar electrodes the spacing of which is well matched to the grain size within the film, which is of the order of 10–20 µm; the regions between the electrodes are then effectively single crystallites and carrier generation and recombination characteristics should be dominated by the CVD diamond as opposed to the grain-boundary regions; Fig. 16.3 shows a micrograph typical of the free-standing CVD diamond used, along with a Raman spectrum taken from this film (using a Renishaw system 2000). The prominent

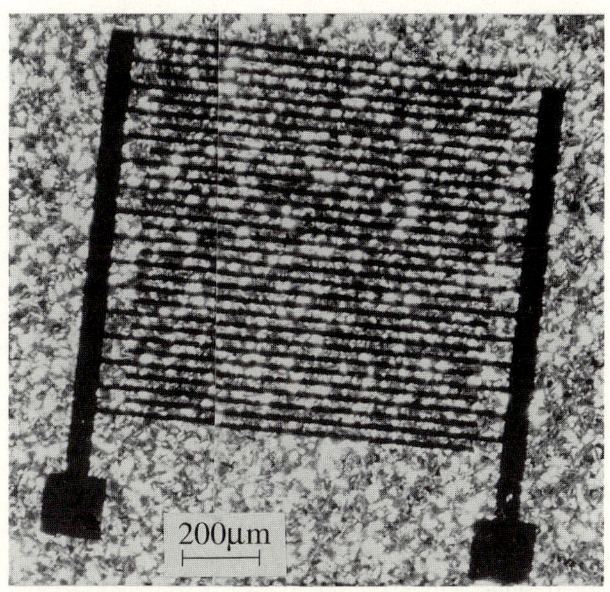

Fig. 16.4: An optical micrograph of a UV photoconductive device fabricated in the author's laboratories. The interdigitated electrodes have a 40 µm pitch and 20 µm spacing.

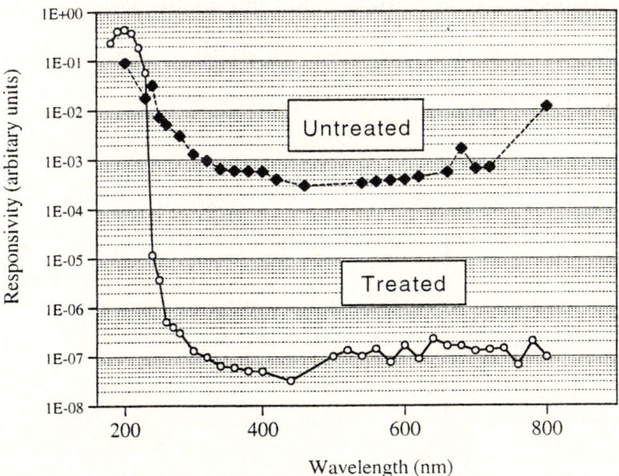

Fig. 16.5: The responsivity of the device shown in Fig. 16.4, plotted as a function of illumination wavelength, compared to the responsivity of a similar device that has not been gas treated.

peak at 1332 cm^{-1} and the absence of other structure is indicative of a high-quality film with low sp^2 (non-diamond) content. Figure 16.4 shows an optical micrograph of an interdigitated device structure formed on this material. In practice, grains vary in size and orientation, and the influence of grain boundaries cannot be completely ignored. In addition to careful device design, it was necessary to treat the diamond in well-chosen gaseous environments to achieve a low photo-conductive response to visible light and a high sensitivity to UV [16.19]. Figure 16.5 shows the spectral responsivity reported for these photoconductive devices before and after application of the methane–air gas treatment developed in our laboratories. The response to visible light is dramatically reduced by the treatment. A sharp cut-off in responsivity at the band-gap edge (225 nm) is apparent. In addition to becoming visible blind (10^6 : 1 discrimination between 200 nm and visible wavelengths), the device also has a low dark current (< 1 nA), improved turn-off characteristics, low operational bias level (1–10 V) and is highly sensitive [16.19–22]. An external quantum efficiency of 6 was demonstrated (i.e. 600%, indicating gain), suggesting that the material properties had been considerably improved, even when compared to single-crystal structures [16.16]. The changes are believed to be associated with passivation of photoconductive defects during the gas treatments used [16.23].

Further work [16.24] has led to the demonstration of devices with gains as high as 10^6 (Fig. 16.6). Photoconductive devices can show gain since the number of electrons that will travel round the external circuit following the photogeneration of an electron–hole pair will depend upon the lifetime of the carriers. High gain requires high mobility, long carrier lifetime and closely spaced electrodes. The mobility–lifetime product, denoted $\mu\tau$, is a useful indicator of the potential for

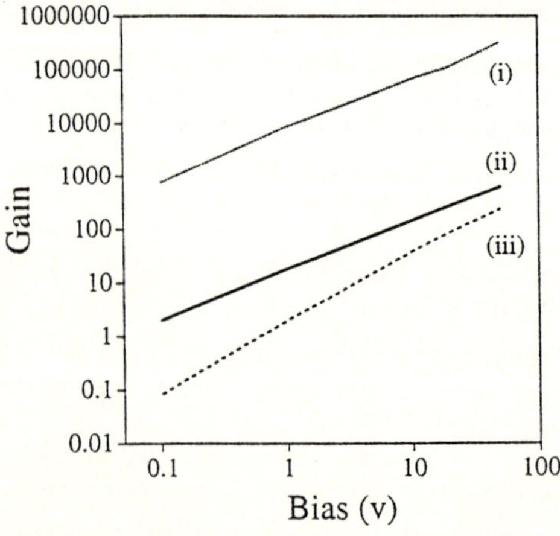

Fig. 16.6: The gain (external quantum efficiency) plotted as a function of the device bias for photo-conductive UV photodetectors fabricated from free-standing CVD-diamond films with differing grain sizes: (*i*) 40–60 µm, (*ii*) 20–40 µm and (*iii*) 10–30 µm (electrode pitch and spacing constant at 40 µm and 20 µm respectively).

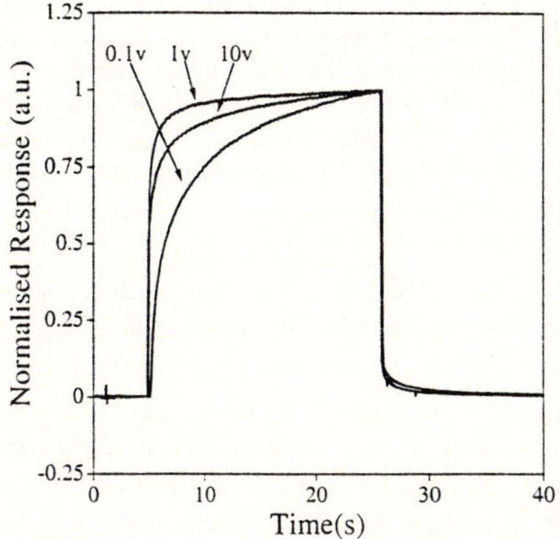

Fig. 16.7: The transient response of a photo-conductive device at 200 nm, plotted as a function of the device bias level (1–10 V).

high-performance operation of a diamond structure. The values measured [16.24] following gas treatments for diamond films with differing grain sizes (at 10^{-2}–10^{-5} $cm^2\,V^{-1}$) are very high compared to those previously reported. For example, Salvatori et al. [16.26] measured a value of 1×10^{-6} $cm^2\,V^{-1}$ for silicon-supported polycrystalline diamond films; this low value was attributed to the presence of a high concentration of defects and impurities within the diamond film. These researchers fabricated planar photoconductive devices on silicon-supported thin-film diamond and recorded gains up to 400, although a sharp cut-off in responsivity at the band edge was not measured. There are applications, such as in the fabrication of fast high-voltage optoelectronic switches, where a short carrier lifetime is actually desirable. Indeed, Yoneda and co-workers [16.27] have cited the short carrier lifetimes measured in most polycrystalline CVD diamond as a reason why this material may be preferable to natural diamond for the fabrication of such devices.

The devices described above reset comparatively quickly after exposure, enabling these devices to be used within a number of applications. An example of the transient response of a typical device, plotted as a function of the bias level, is shown in Fig. 16.7. Improved gain comes at the expense of speed, since the gain–bandwidth product is essentially constant. However, we have been able to show [16.24] that further gas treatments enable the fabrication of devices that are fast enough for imaging applications (where a standard CRT output is used) to be produced with high gain (~50). Figure 16.8 b shows the response of such a device to a (nominal) 10 ms 200 nm light pulse, compared to the circuit response of the test system (Fig. 16.8 a) and an (unimproved) device with a gain of 10^6 (Fig. 16.8 c).

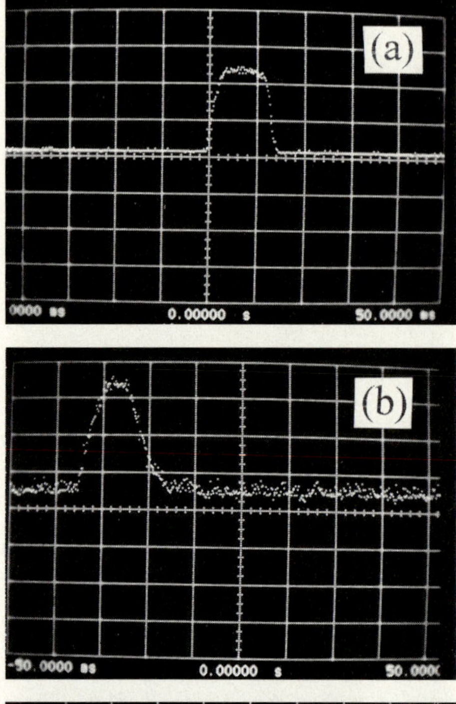

Fig. 16.8: The transient response curve following exposure to a (nominal) 10 ms 200 nm pulse: **(a)** the response of the test circuit; **(b)** the response of an improved CVD-diamond detector; and **(c)** the response of an unimproved very high gain device.

16.2.3 Photodiodes as UV Detectors

In normal operation, photodiodes do not display gain, although avalanche breakdown can be used to promote current amplification. However, since the photo-generated carriers need only traverse the depletion region to form a photocurrent, photodiodes can exhibit high-speed characteristics. The absence of a reliable n-type dopant for diamond (see, for example, Chap. 17) means that Schottky barrier diodes must be used based on p-type material. Marchywacka and co-workers [16.28] have studied the UV response of an Al Schottky diode on p-type (boron-containing) single-crystal material; at reverse bias levels greater than 5 V the leakage current exceeded the photocurrent detected; at lower voltages 100 : 1 discrimination between UV and visible light was reported.

The first purposefully designed thin-film diamond photodiode structures utilised a two-level Au Schottky contact on boron-doped ($\sim 10^{17}$ cm^{-3}) silicon-supported thin-film polycrystalline CVD diamond (~ 6 µm thick) [16.29]. A thin Au region enabled penetration of light through the contact into the depletion layer within the device while a thicker Au was used as the contact pad. Ti–Ag–Au ohmic contacts were formed in the same plane (Fig. 16.9). We measured low leakage currents (< 2 pA) for reverse bias levels of up to 100 V. With a reverse bias level of 50 V, the spectral response of the device showed a sharp cut-off at the band edge (225 nm) and a discrimination of greater than 10^5 between deep-UV and visible wavelengths (Fig. 16.10). A photoconductive device fabricated on similar material (without boron doping) showed a softer cut-off at the band edge and considerably higher dark currents, indicating the advantages offered by diode structures when using this type of fine-grained thin-film diamond. Salvatori and co-workers [16.26] have also fabricated Schottky barrier photodiodes, along with metal–semiconductor–metal (MSM) structures (where both contacts form Schottky barriers). A field-dependent photocurrent was measured at 185 nm, but the spectral response of the devices was not reported.

It has been demonstrated that near-surface hydrogen can promote the formation of shallow acceptor states in diamond, leading to p-type conductivity [16.30–35]. We have recently shown [16.36] that thin Al Schottky contacts to as-grown (hydrogen-containing) CVD-diamond films can also be used for the formation of

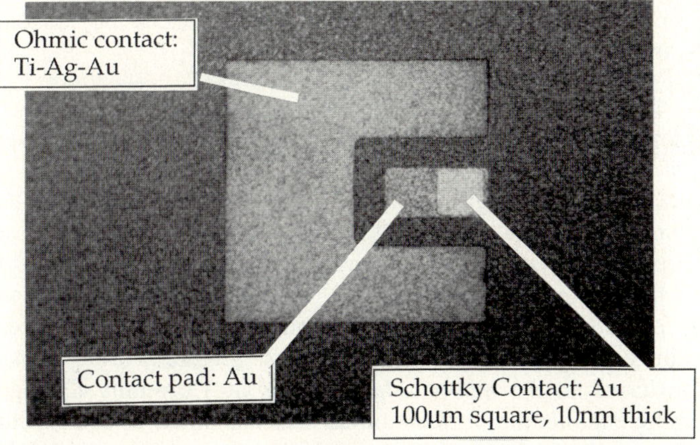

Fig. 16.9: An optical micrograph of a planar photodiode structure fabricated on silicon-supported thin-film CVD diamond. The central Schottky barrier metallisation can be seen to consist of two regions of differing thickness, while the outer ohmic contact is a Ti–Ag–Au alloy.

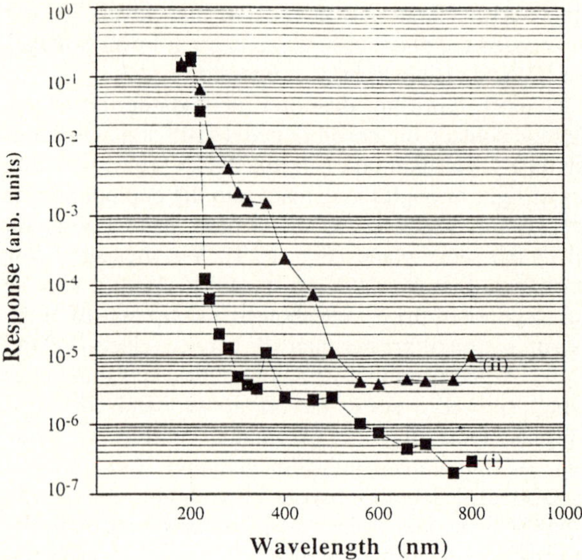

Wavelength (nm)

Fig. 16.10: The spectral response for (*i*) an silicon-supported thin-film diamond photodiode, reverse biased at 50 V; and (*ii*) an interdigitated photoconductive device with 20 μm electrode spacings (bias 10 V) fabricated on similar material (normalised to account for differing device areas).

UV-photodiode devices that display high levels of discrimination between UV and visible light ($> 10^5 : 1$) and low dark currents (< 1 pA). Such devices may enable high-speed operation to be realised.

16.2.4 Applications

It is apparent that careful device design and material treatment can lead to UV-photodetecting structures that display high performance levels. Photoconductive devices that are sensitive (displaying high gain), have a low dark current and are truly "visible blind" have already been produced. CVD-diamond growth and diamond processing procedures have matured adequately to enable devices routinely to be produced with reliable characteristics. Figure 16.11 shows packaged devices fabricated in the author's laboratories at University College London. The commercial exploitation of diamond devices has already begun, with companies such as Centronic Ltd active in the provision of visible-blind UV photodetectors. Photodiode structures are currently less well developed, but have been demonstrated. Significant challenges remain for the future; for example, improved control over material defects will be required to improve device characteristics further, and to enable the "tailoring" of the actual wavelength at which the device cut-off occurs.

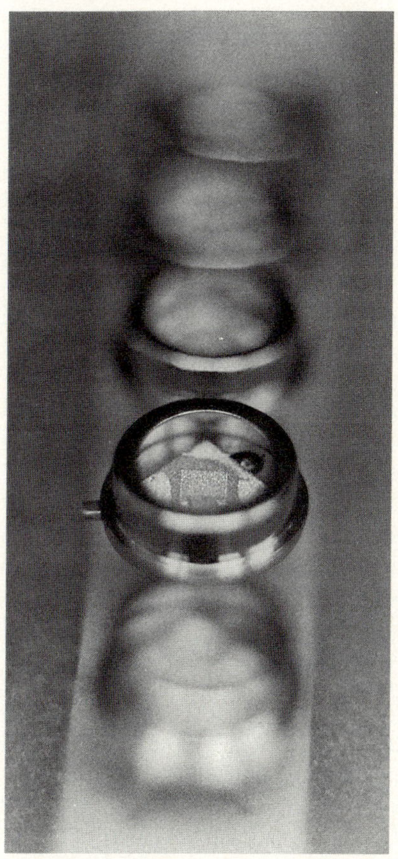

Fig. 16.11: Diamond growth and processing techniques have matured adequately to enable high-performance UV-photoconductive devices routinely to be produced. Pictured here are several devices produced in the author's laboratories at University College London.

The applications for such devices are diverse. Detectors for satellites and terrestrial UV spectroscopy are required. Environmental monitoring is another application area. Water purifiers which operate with UV light require visible-blind UV detectors, as do flame-detection systems within civilian and military sectors. Numerous manufacturing processes involve UV light and consequently require effective detectors. Within the semiconductor industry, the move to 193 nm based lithography that is expected within the next few years offers yet another area of application for diamond devices.

16.3 Particle and High-Energy Radiation Detectors

16.3.1 The Radiation Hardness of Diamond

The need to use high-quality diamond crystals for radiation detection was recognised many years ago, with type II showing the best response and the least tendency to progressive degeneration [16.37]. Irradiation of diamond with 2 MeV electrons produces 0.3 immobile vacancies per electron per cm [16.38–40]; this is around ten times lower than in silicon [16.41]. Many of the defects in silicon are mobile at low temperatures, but they form electrically active defect complexes that leave silicon badly damaged. Early work suggested that neutrons produce around 500 times more vacancies in crystalline diamond than electrons [16.42].

Recently, the damage caused by neutron irradiation of CVD thin-film diamond has been studied in some detail [16.43–45]. Mainwood and co-workers [16.44] have measured the effect of neutron irradiation on the performance of diamond β detectors. Detecting structures were prepared with Ti–Au ohmic contacts on the front and back and utilised a scintillation layer. The β spectrum from a ^{90}Sr source was then monitored. The effect of 1 MeV neutron exposure was a reduction in the device leakage current and an increase in detection efficiency of 30–40%. This surprising result was explained in terms of the production of deep donors and acceptors by the neutron radiation, which compensate pre-existing shallower levels; it was felt that detector performance would be maintained, or improved, for neutron exposures of at least as high as 1.3×10^{15} cm^{-2} [16.43]. The production rate for single neutral vacancies was around 0.5 neutron^{-1} cm^{-1} [16.45]. Bauer and co-workers [16.46] found that polycrystalline CVD-diamond detectors were not significantly degraded by 300 MeV pion irradiation (up to 8×10^{13} cm^{-2}) or 1.2 MeV photon exposure (up to 10 MRad, ^{60}Co gamma source). However, after irradiating the diamond with 10^{15} alpha particles cm^{-2} (~ 1 GRad at 5.5 MeV), the carrier collection distance within the detector was reduced by around 60%.

16.3.2 Alpha- and Beta-Particle Detectors

The direct detection of alpha particles is of considerable interest, as is the detection of alphas that are produced within the material during neutron irradiation. Several groups have investigated the detection of these particles with single-crystal diamond [16.47–54], but these studies have been severely hindered by the highly variable properties of this material. For example, Kozlov and co-workers [16.49] selected type II natural crystals with impurity levels less than 10^{19} cm^{-3} (as judged by carrier lifetime measurements). Contacts were formed on the front and back of samples with thicknesses of 0.1–0.3 mm. The authors argued that since carrier generation will be within the near-surface region of the crystal (5.5 MeV alpha particles will penetrate to a depth of around 15 μm in diamond), the carrier type that drifts towards the back contact will be more heavily "trapped" by defect states, requiring injection of a neutralising carrier from the back contact. To suppress this problem (termed "polarisation") the use of a rectifying–ohmic contact pair was proposed; evaporated Au, Ag and Pt were used as rectifying contacts, while "painted" and annealed Ag, Au, Pt or C or ion implanted and annealed Al or B were used for ohmic contacts. Despite the initial selection criteria, only "good" crystals showed low dark currents (< 1 pA) and a capacitance independent of applied bias and switching frequency (indicating low defect densities). For these structures, the charge collected during irradiation with 5.5 MeV alpha particles (^{241}Am) rose as the applied bias was increased, with a peak value being evident at each bias level. The size of the collected charge at a given bias was compared to that of silicon, with a known ionisation energy of 3.62 eV, giving a diamond alpha ionisation energy of 13.07 eV. At 250 V the full width at half maximum (FWHM) of the peak corresponded to the charge expected from a 120 keV particle (2.2%). It was stressed, that since scattering and trapping of carriers will affect this value, it should be taken as an upper limit (for it assumes that the peak represents 100% collection of the created carriers). A 120 μm thick detector with an applied field of 10 kV cm^{-1} showed a rise time of 1.1 ns for electron traversal and 1.2 ns for holes corresponding to charge-carrier velocities of 1.0×10^7 cms^{-1} and 1.1×10^7 cm s^{-1}, respectively. In a later study, Kozlov and co-workers [16.50] hermetically sealed a similar detector within a Teflon capsule and operated the device within ^{239}Pu nitrate containing solutions; the number of detected pulses correlated well with plutonium concentration within the solution, although no clear peak was evident in the (weak) collected charge size distribution plot.

More recent work by Kaneko and Katagiri [16.54] investigated alpha detection (5.5 MeV, ^{241}Am) using a *synthetic* type IIa single crystal. Simple back and front (sandwich) Pt contacts were evaporated on to the 0.31mm thick sample; no adverse effects attributable to the similarity of the electrical characteristics of the contacts were reported. For an applied bias of 250 V, a sharp peak was obtained (FWHM ~ 45 keV, 0.81%), corresponding to an ionisation energy for the diamond of 13.07 eV, in agreement with Kozlov and co-workers [16.49]. A typical characteristic is reproduced in Fig. 16.12.

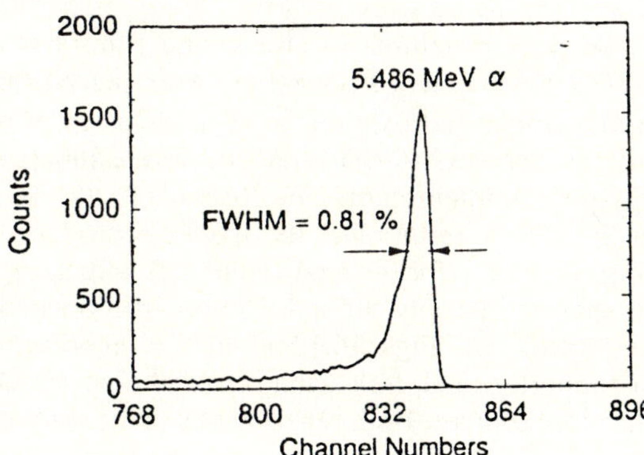

Fig. 16.12: The alpha spectrum from ^{241}Am obtained by a synthetic type IIa diamond detector (sandwich construction) (from [16.54]).

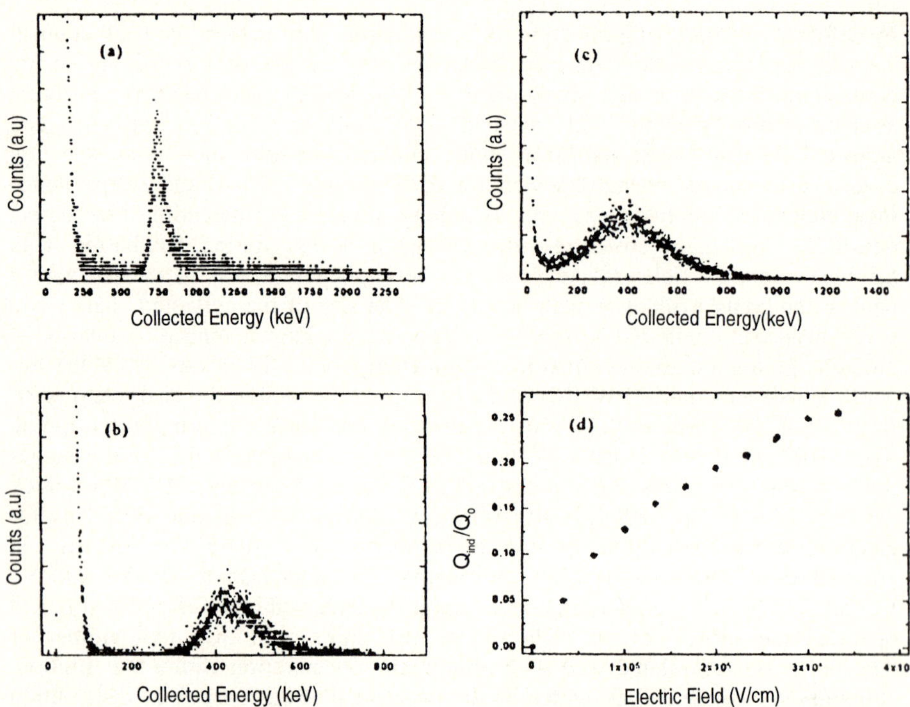

Fig. 16.13: The alpha spectrum from ^{241}Am for sandwich detectors fabricated from **(a)** natural type IIa diamond crystal, 200 μm thick, **(b)** silicon-supported polycrystalline CVD-diamond film, 6–19 μm thick, and **(c)** free-standing polycrystalline CVD-diamond film, 180 μm thick. **(d)** The collection efficiency as a function of bias for device **(c)** (from [16.58]).

CVD thin-film diamond has also been investigated [16.55–59] but, to date, devices have not been reported with efficiencies as high as those fabricated from single-crystal material. The results achieved are also highly variable between differing sources of CVD diamond. Pochet and co-workers at the CEA [16.55, 56] have compared the alpha response of polycrystalline CVD-diamond films from differing sources to a natural type IIa single crystal. Au contacts were placed on the back and front of films of varying origin and exposed to α particles from a ^{241}Am source; the charge collected (as opposed to that created) was estimated using the 13.07 eV ionisation energy for diamond reported by Kozlov [16.49] and an ionisation energy of 3.62 eV for Si [16.60], resulting in the spectra reproduced in Fig. 16.13. The trace shown in Fig. 16.13 a is for the type IIa crystal, indicating that is was of comparatively poor quality, as only around 15% of the created charge is being collected. The traces in Figs. 16.13 b and c represent polycrystalline thin (6–19 μm) Si-supported and free-standing (180 μm) CVD films respectively, although from different sources. Both reveal peaks in the size of the pulses of charge being collected, but are less than 10% efficient; at high fields $(3 \times 10^5 \text{ V cm}^{-1})$, the collection efficiency for the thin-film device rose to around 25% (Fig. 16.13 d). Other free-standing films displayed poorer efficiencies, with no peak in the collected "energy" being apparent. The authors attributed this to inhomogeneity in carrier transport characteristics, caused by grain boundaries and grains of differing crystal orientations and sizes within the film. Radiation hardness tests on the thinnest samples were disappointing, showing a reduction in collection efficiency of a factor of 2 following exposure to 5 MRad of gamma radiation (^{60}Co, 1.25 MeV). However, the thicker structures showed no degradation after 10 MRad exposures.

In a recent collaboration between the author's laboratory and the CEA [16.59], the efficiency of a "sandwich" structure was compared to that of a device with the two electrodes closely spaced on the top surface in an interdigitated pattern *on the same CVD-diamond film*. The electrodes were spaced with a dimension similar to the average grain size within the top of the film (20 µm), to minimise the effect of grain boundaries on carrier transport. Figure 16.14 a shows the charge collected during 5.5 MeV irradiation from a ^{241}Am source for the sandwich structure (left trace) and the planar device (right trace). While neither shows a peak in the size of the collected charge, a significant number of pulses being collected in the planar device represent more than 75% of the charge created by the incident particle. The planar device also only requires low voltages (20 V) to operate, enabling portable devices to be readily envisaged. This is a significant improvement on previously reported CVD-diamond based structures. The planar device showed no sign of degradation following 10 MRad of gamma radiation (^{60}Co). Figure 16.14 b shows a simulation of the field profile expected within the planar structure. While the inhomogeneity of the film over its entire depth may be the reason why a peak was not observed in the sandwich structure, we attributed the lack of a peak in the planar device to the variable field strength in the surface region. Alpha particles arriving at differing positions between the electrodes, or with different trajectories, would experience significantly differing fields, leading to a wide spread of collection efficiencies. The penetration of the field into the near-surface region would appear to offer a route for the production of the first high-performance alpha-detection structures from polycrystalline CVD diamond.

The detection of β particles has been achieved with both single-crystal [16.49, 61, 62] and polycrystalline CVD-diamond structures [16.63]. Han and co-workers [16.63] compared the operation of CVD-diamond devices with "sandwich" configuration electrodes to those with planar contacts separated by 0.5–1.0 mm. The increase in film conductivity in response to 16 MeV electron pulses, with a 25 ps duration, was used as the basis for detection. Both types of structures behaved in a similar manner; the response time for all devices was less than 45 ps, but very low sensitivity was reported (10^{-5}–10^{-6} AW^{-1}). This is not surprising, given the high penetration depth expected for electrons of this energy. The detector response was linear with dose for dose rates up to 10^{13} Rad s^{-1}.

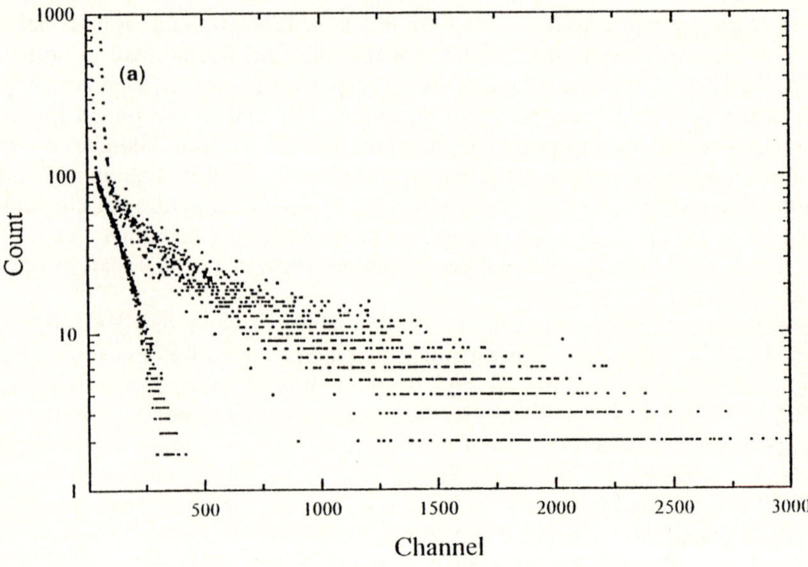

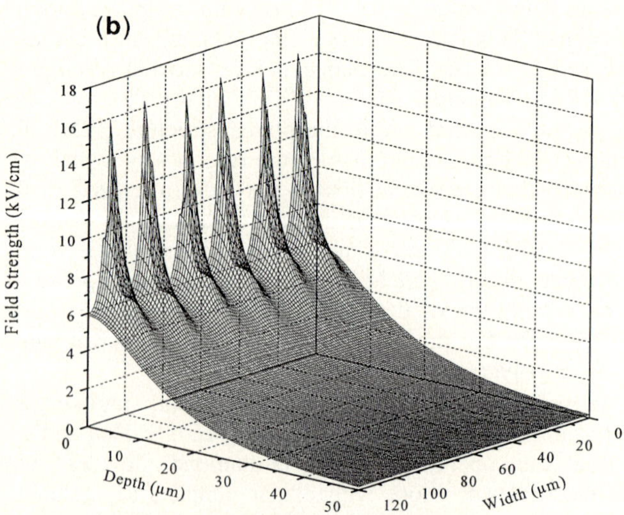

Fig. 16.14: (a) The alpha spectrum from [241]Am for polycrystalline CVD diamond, 100 μm thick, with electrodes in sandwich configuration (left trace) and a planar structure (right trace). **(b)** A simulated profile of the field within the planar structure (20 V applied bias, 40 μm pitch, 20 μm spacings).

16.3.3 Neutron Detectors

Kovalchuck and co-workers [16.64] have reported the use of diamond (type not stated) in the detection of fast neutrons. The nuclear reaction $^{12}C(n, \alpha)\ ^{9}Be$ was promoted by 14.6 MeV neutrons within the volume of the detector (with a thickness of 0.2–0.4 mm); detection of the resultant charged particles led to a multi-peaked spectrum. Some of the collected charge results from the alpha particle and the recoiling Be nucleus and one of the peaks observed was assigned to this reaction. In comparison, a silicon detector ($^{28}Si\ (n,p)\ ^{28}Al$, $^{28}Si\ (n,\alpha)\ ^{25}Mg$) gave a more complex spectrum, leading the authors to conclude that diamond was of significant interest as a neutron detection medium. In a later paper [16.65], the same authors proposed that a diamond crystal is an effective detector for neutrons with energies greater than 5.7 MeV. Pillon et al. [16.66] studied the use of type IIa single-crystal diamond as a 14 MeV neutron detector, again relying upon the $^{12}C(n, \alpha)\ ^{9}Be$ reaction for the creation of the detected charge. Good energy resolution was achieved (4%), but the detector efficiency was 2.5–3.0 times lower than expected; this was attributed to the limited charge collection distance within the crystal, given as around 70 μm. A more detailed discussion of the use of natural-diamond detectors for spectrometry with 154 MeV deuterium–tritium (DT) neutrons has been presented by Krasilnikov et al. [16.67].

More recently, Maqueda and co-workers [16.68] have looked at ways of utilising "typical" as opposed to extra-pure type IIa natural diamond for the detection of neutrons. Of particular interest to these workers was the discrimination of 14 MeV (DT) neutrons from 2.5 MeV deuterium–deuterium (DD) neutrons during fusion "burn" experiments. Interdigitated contacts were placed on one side of a 200 μm thick single crystal and mounted in a proton recoil telescope similar to the silicon diode device designed by Croft et al.[16.69]. In this arrangement, a 2 mm thick polyethylene proton radiator and a 80 μm thick Teflon proton filter are placed in front of the diamond crystal, separated by a 6 mm air gap; the Teflon is thick enough to stop 2.5 MeV neutrons, while the polyethylene has an ideal thickness for the production of the maximum number of protons from a 14 MeV neutron flux. The detector was then exposed to a broad spectrum neutron beam (100 keV to 800 MeV). The authors reported a significant signal from direct neutron absorption in the diamond; at very high energies (> 80 MeV) the observed counts were dominated by direct neutron events. For energies greater than 5 MeV, reasonable sensitivity was reported, with a sensitivity ratio of ~ 6.5 being achieved between DT and DD radiation. While the sensitivity at 14 MeV was similar to Si devices, the authors felt that the device was not suited to applications in which a high level of discrimination between high- and low-energy neutrons was required. Attempts to fabricate similar devices from CVD diamond (type not discussed) were not successful.

Neutrons can be detected by using a "converter" element that produces the charged particle that is then detected within the diamond. The reaction

$$^{10}B + n \text{ ---------> } ^{7}Li + ^{4}He + \text{energy}$$

has been widely used for this purpose. Since boron can be readily incorporated into diamond, this approach may offer a route for the fabrication of high-sensitivity diamond-based neutron detectors [16.58].

16.3.4 X-ray and Gamma-Radiation Detectors

Kozlov and co-workers [16.70] have studied the response of single-crystal diamond detectors (type not specified) to gamma and X-rays. The small absorption coefficient of diamond for photons in the energy range 20–1250 keV made pulse counting difficult and current flow through the detectors was measured instead. A range of devices, fabricated from natural material, were constructed with dimensions of 10–15 mm^2 and thicknesses of 0.2–0.4 mm; it was proposed that complete charge collection would be possible within such structures. Electrode materials were optimised for minimal dark current (silver–gold). The X-ray response (22 keV) was linear with dose up to a dose rate of 500 mRad s^{-1}; beyond this point, the sensitivity of the detectors decreased. Devices with higher dark currents (attributed to more "ohmic"-like contacts) reduced in sensitivity at much higher dose rates. However, for the best device presented, a 300 V bias led to very low detected currents (only up to 5 pA), severely limiting the usefulness of the device. A sensitivity of 1.1×10^{-14} As µRad^{-1} was suggested. A similar sensitivity was found for gamma radiation (661 keV and 1250 keV).

Free-standing polycrystalline-diamond films, grown by CVD techniques, have been investigated by a number of workers [16.55, 56, 58, 71–74]. Beetz and co-workers investigated the use of free-standing films (10 µm thick) that were nitrogen-doped during growth [16.71]. Sandwich structures, with an ohmic and Schottky electrode pair, generated a 30 nA photocurrent when exposed to 30 keV X-rays (Cu target, 50 µA tube current); the dark current was 0.01 pA and the electron collection efficiency was estimated to be at least 25%. The response of polycrystalline CVD-diamond film detectors to 40 keV X-rays has been studied by Foulon et al. [16.72] and compared to structures formed from natural type IIa material. Electrodes were placed on the front and back sides of thin (6 µm) diamond on silicon, on free-standing films (480 µm) and on a natural single crystal (200 µm thick). A sensitivity of 0.8 nA(Gy/h)$^{-1}$ was reported for the natural-diamond device, while the free-standing CVD film gave a value of 1 nA(Gy/h)$^{-1}$; the difference was attributed to the greater thickness of the CVD detector.

16.3.5 Detectors for High-Energy Physics

An international collaboration has been established to investigate the use of diamond as a detector material for the Large Hadron Collider (LHC) being built at CERN (known as the RD42 Diamond Detector Collaboration) [16.75–79]. The detector elements at the innermost radius of the tracking systems of the experiments at the LHC must survive heavy charged particle fluences up to 10^{15} cm^{-2} and fluxes of 10^7 cm^{-2}s^{-1}, and provide position measurements with resolutions of about 10 µm and time resolutions of better than 25 ns [16.79]. Cr–Au layers on each side of the diamond were formed on CVD films; wet etching was then used to define a planar array of top-side electrodes (100 µm pitch, 50 µm strip width) [16.75]. In preliminary experiments, a signal-to-noise ratio of 6 : 1 was achieved with a position resolution of 26 µm. More recent devices, based on higher-quality CVD-diamond substrates, have given a position resolution of 14.3 µm and improved signal strengths; this has been attributed to improved carrier collection distance within the CVD material being used [16.79].

16.3.6 Applications

Particle physicists are clearly interested in the use of diamond radiation detectors, because of the need for a robust replacement for Si in next-generation particle accelerators. There is also potential for diamond alpha- and neutron-radiation detectors to become widely used within the nuclear industry; plutonium emits alpha particles as well as fission-induced neutrons, and transuranian elements such as americium, present in nuclear waste, emit alpha particles. CVD-diamond devices with high sensitivity and long lifetimes appear likely to become available, although no devices based upon thin-film polycrystalline material yet match the performance of the best single-crystal structures. Neutron detectors are also required within fusion reactors, as discussed by Jassby [16.80]. The chemical, physical and thermal robustness of diamond may ultimately mean that only devices based on this material can be used in certain hostile environments. Robust solid-state radiation detectors would also be attractive for environmental monitoring applications. The inherent poor sensitivity of diamond X-ray and gamma-radiation detectors may limit the number of applications that there are for these structures. Spielman and co-workers [16.81] have recently improved the fundamental sensitivity of a diamond photoconductive soft X-ray detector by damaging the diamond crystal with neutron radiation. A rugged device with excellent response times and minimised sensitivity to higher-energy X-rays was produced. For medical photon-beam dosimetry applications, an ideal detector would show a response equivalent to that of human tissue, a property displayed by diamond ($Z = 6$, compared to ~6.7 for tissue). Rustgi and Frye [16.82] have characterised the tissue response ratios of diamond detectors exposed to 6 MeV photon beams, while the calibration of diamond detectors for dosimetry in medical applications has been discussed by Vatnitsky et al. [16.83] and Rustgi [16.84].

16.4 Concluding Remarks

Diamond deposited by CVD techniques on non-diamond substrates is polycrystalline in nature and may contain non-diamond carbon, primarily located at the grain boundaries. Despite this, photodetectors with useful levels of performance can be fabricated, given careful device design and processing; it

16.45 L. Allers and A. Mainwood, Diamond Rel. Mater. **7**, 261 (1998)
16.46 C. Bauer and 46 other authors, Nucl. Instrum. Methods **A367**, 207 (1995)
16.47 F.C. Chapman and F.C. Wright, Proc. Phys. Soc. **A253**, 385 (1959)
16.48 P.J. Dean and J.C. Male, J. Phys. Chem. Solids **25**, 311 (1964)
16.49 S.F. Kozlov, R. Stuck, M. Hage-Ali, and P. Siffert, IEEE Trans. Nucl. Sci. **NS-22**, 160 (1975)
16.50 S.F. Kozlov, E.A. Konorova, M.I. Krapivin, V.A. Nadein, and V.G. Yudina, IEEE Trans. Nucl. Sci. **NS-24**, 242 (1977)
16.51 C. Canali et al., Nucl. Instrum. Methods **160**, 73 (1979)
16.52 P.J. Fallon et al., Appl. Radiat. Isot. **41**, 35 (1990)
16.53 R.J. Keddy and T.L. Nam, Radiat. Phys. Chem. **41**, 767 (1993)
16.54 J. Kaneko and M. Katagiri, Nucl. Instrum. Methods **A383**, 547 (1996)
16.55 F. Foulon, T. Pochet, E. Gheeraert, and A. Deneuville, Mater. Res. Soc. Symp. Proc. **339**, 185 (1994)
16.56 F. Foulon, T. Pochet, E. Gheeraert, and A. Deneuville, IEEE Trans. Nucl. Sci. **NS-41**, 927 (1994)
16.57 C. Manfredotti et al., Nucl. Instrum. Methods **B93**, 516 (1994)
16.58 T. Pochet, A. Brambilla, P. Bergonzo, F. Foulon, C. Jany, and A. Gicquel, Italian Physical Society, Conf. Proc. (Eurodiamond '96) **52**, 111 (1996)
16.59 R.D. McKeag, R.D. Marshall, F. Foulon, P. Bergonzo, C. Jany, and R.B. Jackman, Appl. Phys. Lett. (in press)
16.60 G.F. Knoll, Radiation Detection and Measurement, 2nd edn., Wiley, New York (1989), p. 259
16.61 S.F. Kozlov, V.P. Katkov, and A.J. Krupman, IEEE Trans. Nucl. Sci. **NS-22**, 901 (1975)
16.62 S.F. Kozlov, A.V. Bachurin, S.S. Petrusev, and Y.P. Fedorovsky, IEEE Trans. Nucl. Sci. **NS-24**, 240 (1977)
16.63 S. Han, R.S. Wagner, J. Joseph, M.A. Plano, and M. D. Moyer, Rev. Sci. Instrum. **66**, 5516 (1995)
16.64 V.D. Kovalchuck, V.I. Trotsik, and V.D. Kovallchuck, Nucl. Instrum. Methods **A351**, 590 (1994)
16.65 V.D. Kovalchuck, V.I. Trotsik, and V.D. Kovallchuck, Instr. Exp. Tech. **38**, 14 (1995)
16.66 M. Pillon, M. Angelone, and A.V. Krasilnikov, Nucl. Instrum. Methods **B101**, 473 (1995)
16.67 S.F. Kozlov, A.V. Krasilnikov, and V.M. Bagaev, IEEE Trans. **NS-24**, 235 (1977)
16.68 R.J. Maqueda, C.W. Barnes, S.S. Han, P.A. Staples, and R.S. Wagner, Rev. Sci. Instrum. **68**, 624 (1997)
16.69 S. Croft, D.S. Bond, and N.P. Hawkes, Rev. Sci. Instrum. **64**, 1418 (1993)
16.70 S.F. Kozlov, E.A. Konorova, Y.A. Kuznetsov, Y.A. Salikov, V.I. Redko, V.R. Grinberg, and M.L. Meilman, IEEE Trans. Nucl. Sci. **NS-24**, 235 (1977)
16.71 C.P. Beetz, B. Lincoln, D.R. Winn, K. Segall, M. Vasas, and D. Wall IEEE Trans. Nucl. Sci. **NS-38**, 107 (1991)
16.72 S. Han, R.S. Wagner, and E. Gullikson, Nucl. Instrum. Methods **A380**, 205 (1996)
16.73 F. Foulon, P. Bergonzo, C. Jany, A. Gicquel, and T. Pochet, Nucl. Instrum. Methods **A380**, 42 (1996)
16.74 C. Jany, F. Foulon, P. Bergonzo, A. Brambilla, A. Gicquel, and T. Pochet, Nucl. Instrum. Methods **A380**, 107 (1996)
16.75 C. White, Nucl. Instrum. Methods **A351**, 217 (1994)
16.76 F. Borchelt and 20 co-authors, Nucl. Instrum. Methods **A354**, 318 (1995)
16.77 C. Bauer and 43 co-authors, Nucl. Instrum. Methods **A367**, 202 (1995)
16.78 C. Bauer and 44 co-authors, Nucl. Instrum. Methods **A380**, 183 (1996)
16.79 C. Bauer and 48 co-authors, Nucl. Instrum. Methods **A383**, 64 (1996)

16.80 D.L. Jassby, G. Ascione, H.W. Kugel, A.L. Roquemore, T.W. Barcelo, and A. Kumar, Rev. Sci. Instrum. **68**, 540 (1997)
16.81 R.B. Spielman, L.E. Ruggles, R.E. Pepping, S.P. Breeze, J.S. McGun, and K.W. Struve, Rev. Sci. Instr. **68**, 782 (1997)
16.82 S.N. Rustgi and M.D. Fyre, Med. Phys. **22**, 2117 (1995)
16.83 S. Vatnitsky, D. Miller, J. Siebers, and M. Moyers, Med. Phys. **22**, 469 (1995)
16.84 S.N. Rustgi, Med. Phys. **22**, 567 (1995)

17. Electronic Devices on CVD Diamond

Erhard Kohn and Wolfgang Ebert

Department of Electron Devices and Circuits, University of Ulm,
Albert-Einstein-Allee 45, D-89081 Ulm, Germany
e-mail: ebert@sunrise.e-technik.uni-ulm.de

Springer Series in Materials Processing
Low-Pressure Synthetic Diamond Eds.: B. Dischler and C. Wild
© Springer-Verlag Berlin Heidelberg 1998

17.1 Introduction

At least since the late 1980s, almost all papers dealing with diamond electronics have started with a sentence such as

"diamond is a potential material for high-temperature, high-power and high-speed electronic applications ... ."

This expectation originates from the physical, chemical, and electronic properties as measured for undoped (type IIa) or lightly ($\sim 10^{16}$cm^{-3}) boron doped (type IIb) natural diamond samples (Table 17.1). Due to these properties, diamond is a promising material for three fields of application: high-power electronics, high-temperature electronics, and smart sensors for extreme environment. In this chapter, the high-temperature and high-power applications will be addressed.

Table 17.1: Electronic properties of diamond.

Property	Value
Bandgap	$E_G = 5.4$ eV
Thermal conductivity	$\lambda = 20$ W/(cm·K)
Breakdown field strength	$E_{BR} = 10^7$ V/cm
Relative dielectric permittivity	$\varepsilon_r = 5.5$
Electron saturation velocity	$v_s = 2.7 \times 10^7$ cm/s
Hole saturation velocity	$v_s = 1 \times 10^7$ cm/s
Electron mobility	$\mu_e = 2200$ cm^2/(Vs)
Hole mobility	$\mu_h = 1800$ cm^2/(Vs)

Due to the large bandgap, the intrinsic carrier concentration at 1000°C is comparable to that of silicon at room temperature. Therefore, diamond devices operating at such high temperatures seem feasible. The material itself has been shown to withstand temperatures as high as 2000°C [17.1].

To estimate the power-handling capability of diamond transistors, the figures of merit as defined in Table 17.2 may be used. Using the data listed in Table 17.1, diamond exhibits the highest figures of merit in comparison to all other electronic materials except probably AlN. This is even true for p-channel FETs (Field Effect Transistors), which are realistic devices to date.

Table 17.2: Figures of merit for high-power, high-speed field effect transistors.

Johnson [17.2]	Keyes [17.3]	Baliga [17.4]
$JFM = \dfrac{E_{BR} v_s}{2\pi}$	$KFM = \lambda \sqrt{\dfrac{c v_s}{4\pi\varepsilon}}$	$BFM = \varepsilon\mu E_G^3$

Unfortunately, there have been serious drawbacks which have prevented diamond from becoming a relevant electronic material up to now. Some properties such as the carrier mobility depend strongly on temperature and impurity concentration. In addition, up to now only boron doping has been possible, forming a deep acceptor with an activation energy of $E_A = 0.38$ eV. Donor doping by phosphorus with an activation energy of $E_A = 0.43$ eV has only recently been demonstrated, but not yet implemented in the form of devices [17.5].

In spite of these problems, important technological breakthroughs have been made recently. The improved growth processes, doping techniques, and contact technologies reported here have already enabled Schottky diode operation at 1000°C and allowed to predict an increase in the FET RF power performance to 30 W/mm for 0.1 µm gate length.

17.2 Growth of Active Diamond Layers

The electronic application of diamond imposes several requirements on the quality of CVD-diamond films. These are:

- an almost single-crystalline structure
- a low defect density
- low impurity (nitrogen) incorporation
- a smooth surface
- a well-defined doping concentration and profile

Such "electronic grade" diamond films may be grown on diamond substrates (homoepitaxy), but also on other materials such as silicon and silicon carbide.

For simplicity, the growth on nondiamond substrates will be called "heteroepitaxy," although it may not comply with the exact definition of this term. The commonly used technique for the epitaxial growth of diamond layers for electronic applications is the microwave-plasma chemical vapor deposition technique (MPCVD), which is described in Chap. 2. Therefore, here only special aspects regarding the growth of "electronic-grade" diamond films will be addressed.

17.2.1 Homoepitaxial Growth

Diamond substrates (natural or synthetic) for *homoepitaxy* have two inherent disadvantages: they are very small and expensive. Industrial-scale electronics applications can only be realized by *heteroepitaxial* growth on large-area substrates. The purpose of using homoepitaxial diamond is the demonstration of the ideal case.

The synthetic diamond substrates may be classified into three groups:

- type Ib – insulating by carrier freeze-out, nitrogen-containing (yellow)
- type IIa – insulating, nominally undoped (clear, white)
- type IIb – conducting, boron-containing (deep blue)

Prior to epitaxy, the substrates should be cleaned by a standard RCA process. The CVD growth can be performed within a wide range of parameters (see Chap. 2).

The standard conditions used by the authors are as follows 4 kPa pressure, 950 K substrate temperature, 1.5% CH_4 in H_2, 200 sccm gas flow, and 700 W microwave power in a standard ASTeX 1500 reactor. Typical growth rates are of the order 1 μm/h. For the growth of thicker (> 3 μm) homoepitaxial diamond films, the α-parameter [17.6], which describes the dependence of the growth rate on the crystallographic orientation of the surface, has to be taken into account to avoid twinning and to obtain a smooth surface with a low defect density.

17.2.2 Heteroepitaxial Growth

For large-scale electronics applications, heteroepitaxy (see Chap. 8) is of central importance. The commonly used substrate for heteroepitaxy is [100]-oriented silicon. Three growth steps have to be performed sequentially to obtain a [100]-oriented boron-doped active layer:

1. Bias-enhanced nucleation of diamond seeds by applying a DC [17.7] or AC [17.8] bias voltage between substrate and microwave plasma.

2. Initial outgrowth of a highly oriented, textured buffer layer up to a thickness of 10–20 μm, at which the surface becomes smooth and almost single-crystalline. A temperature rise during growth changes the α-parameter [17.6] from $\alpha \approx 3$ (overgrow of randomly oriented nuclei) to $\alpha \approx 2$ ([100]-oriented growth).

3. Coalescent outgrowth of doped active layers.

Typical growth parameters are shown in Table 17.3. The AC-bias-enhanced nucleation step allows the growth of electronic-grade diamond films homogeneously on areas as large as ~30 cm^2.

Table 17.3: Examples of growth parameters used for diamond heteroepitaxy.

Step	CH_4/H_2 (%)	$T_{substrate}$ (K)	Pressure (kPa)	Particularities
1. Bias-enhanced nucleation	2	1150	2	AC bias (150 V, 10–40 mA)
2. Textured growth	1.5	1020→1070	4	
3. Coalescent outgrowth of the doped active layer	1.5	1100	4	Doping by solid or gas source

17.2.3 Doping Techniques

The concentration, compensation, distribution, and activation of dopants strongly influence the device performance. Usually, electronic devices operate with fully activated doping. Hence shallow n- and p-type dopants are desired. However, to date, the doping of diamond has been limited to boron as acceptor, with a thermal activation energy of $E_A = 0.37$ eV. This high activation energy can be reduced to zero by increasing the doping concentration to above 10^{20}cm^{-3} [17.9]. This effect is due to the formation of a miniband.

The concentration of active acceptors is effectively reduced by nitrogen, which acts as a deep donor (with $E_A \approx 1.7$ eV for substitutional nitrogen and $E_A \approx 4$ eV for agglomerates [17.10]). This compensation leads to an increase in the effective acceptor activation energy and to a decrease in the hole mobility. For the fabrication of high-quality electronic devices, it is therefore necessary to reduce the nitrogen concentration ($< 10^{16}$ cm^{-3}) in the active layer by properly sealing the CVD-system and by using high-purity process gases.

The boron can be incorporated *in situ* during CVD growth by gaseous or solid sources or *ex situ* after epitaxy by ion implantation. Boron implantation has initially been used to prepare conductive layers [17.11] and for localized p$^+$-doping; for example, for the fabrication of ohmic contacts [17.12]. *In situ* doping can be performed using gaseous (e.g. diborane [17.13]), liquid (e.g. trimethylboron [17.14] and trimethylborate [17.15]), or solid (e.g. boron rod or powder [17.16, 17]) precursors.

Figure 17.1 shows a plasma CVD system with a solid boron source, which enables the growth of steep profiles (for profile, see [17.16]) and highly doped p$^+$ layers by changing the immersion time of a boron rod into the plasma from a few seconds to the entire growth period.

Microwave CVD Reactor

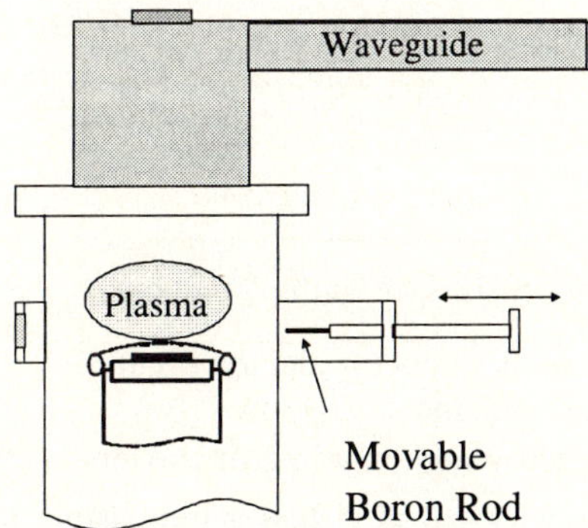

Fig. 17.1: A microwave-plasma CVD reactor with a movable solid boron source for the growth of δ-doped diamond layers.

17.3 Contacts to Diamond

Contacts are an inherent part of any electronic device. Both rectifying and non-rectifying contacts have to meet specific requirements. In general, these are:

- low specific contact resistance for nonrectifying contacts
- low leakage current, high breakdown voltage and high ideality factor for rectifying contacts

On the other hand, the free diamond surface is of equal importance.

17.3.1 The Diamond Surface

The electrical behavior of metal contacts on diamond depends strongly on diamond surface properties such as crystal orientation [17.18], morphology [17.19], and termination [17.20]. Smooth surfaces with a low defect density are generally [100]-oriented.

At room temperature, a hydrogen-terminated (as-grown) diamond surface shows p-type conduction due to a surface Fermi level close to the valence band edge [17.21]. This conductivity can be used for the fabrication of surface channel devices (see Sect. 17.7.2).

The deposition of a metal on to this hydrogen-terminated surface yields a low interface state density and thus a Schottky barrier height (SBH) depending on the electronegativity [17.22] of the metal. For example, Al contacts ($\Phi_B = 0.85$ eV) are rectifying, but Au contacts ($\Phi_B = 0.1$ eV) show ohmic characteristics [17.23]. However, in air the hydrogen termination will convert to oxygen termination at temperatures around 500°C or in an oxygen plasma [17.24]. Its thermal stability is therefore limited [17.25].

The oxygen-terminated surface shows a Fermi-level pinning at $E_V + 1.7$ eV [17.24]. For the formation of high-temperature stable contacts, it is necessary to remove the hydrogen termination of the as-grown diamond surface by an oxygen plasma treatment or by boiling in a strongly oxidizing agent such as a Cr_2O_3 / H_2SO_4 solution.

17.3.2 Nonrectifying Contacts

In principle, there are two ways of obtaining low-resistance ohmic contacts:

- by avoiding energy barriers between the semiconductor and the contact material; for example, by grading the junction, and
- by the formation of thin energy barriers, allowing tunneling of carriers

Due to the wide bandgap of diamond ($E_g = 5.5$ eV), it is virtually impossible to fabricate a metal contact without an energy barrier. Fortunately, the high solubility of boron in diamond allows the incorporation of boron concentrations above 10^{20} cm^{-3}. Due to a neglectable acceptor activation energy at high doping levels, the carrier concentration is also very high. Therefore, tunneling is possible and low resistive ohmic contacts can be fabricated. Such an extremely high doping concentration can be realized by boron implantation [17.26] or p$^+$-epitaxy [17.27]. However, ion implantation poses the problem of damage annealing. This is usually done at high temperature, but may lead in the case of diamond to graphitization [17.28]. Alternatively, a graded interface may be formed by alloying with refractory metals such as Mo, Ta, Ti [17.29-32], or AlSi [17.33]. However, the interdiffusion is difficult to control. Furthermore, ohmic contacts can be realized by destroying the diamond lattice in the surface region using sputtering [17.34], or a laser treatment [17.35].

Boron Implantation

Boron implantation into diamond has already been used in the early stages of diamond device technology to realize conductive channels [17.11]. With this technique, boron surface concentrations of up to 10^{21} cm^{-3} have been obtained [17.35], enabling a specific contact resistance in the 10^{-5} Ωcm^2 range using an alloyed Ti–Au metallization. However, ion implantation raises the problem of damage annealing. This is usually done at high temperatures, but may lead, in the case of diamond, to graphitization [17.36].

P$^+$-Epitaxy

Metal contact on p$^+$-epitaxial areas is the main technique used for the fabrication of ohmic contacts [17.16]. It is a technique compatible with the active layer growth technique and needs no additional equipment. The contact area may be defined either by mesa-etching [17.37] or by selective epitaxy [17.16, 17]. The fabrication steps are illustrated in Fig. 17.2.

The lowest specific contact resistance of $\rho_c = 2\times10^{-7}$ Ωcm^2 has been achieved by combining a high-boron doping concentration with alloying of an Si–Al metallization system [17.38].

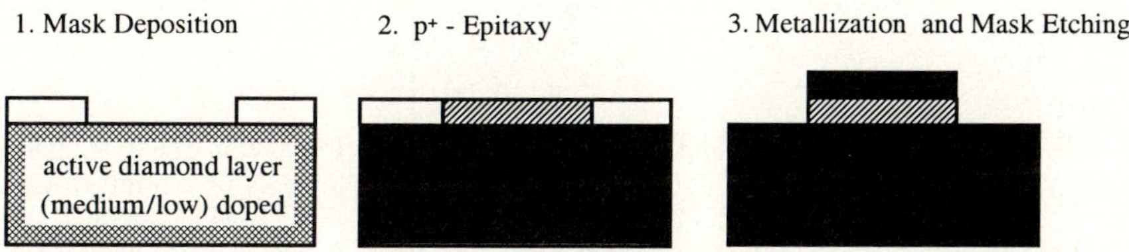

Fig. 17.2: Ohmic contact fabrication by selective epitaxy.

17.3.3 Rectifying Metal Contacts

The main parameters of rectifying metal–semiconductor junctions are the
Schottky barrier height (SBH), Φ_B, and the ideality factor, n. The SBH influences
exponentially the reverse saturation current of the Schottky diode, I_0:

$$I_0 = AA^*T^2 \exp(-q\Phi_B / (kT)) \tag{17.1}$$

where A is the diode area and A^* is the effective Richardson constant.

The ideality factor n describes the slope of the exponential forward I–V
characteristics in the diode equation:

$$I = I_0(\exp(qV / (nkT)) - 1) . \tag{17.2}$$

It indicates whether thermionic emission is the main current transport mechanism
($n \approx 1$), or whether other mechanisms (field emission or tunneling) take part
($n > 1$). Reported SBHs on nonhydrogen-terminated diamond surfaces are in the
range 1.5–2.0 eV [17.39–44] and are not correlated with the metal work function
due to a high interface state density. Au is the material that is mainly used, but the
use of Al [17.45], W [17.46], Cu [17.47], and Ti [17.48] has also been reported.

17.4 Diamond Diodes

17.4.1 p–n Junction Diodes

The first reported diamond diode was a planar p–n diode made by ion
implantation of boron as acceptor and phosphorus as donor. The current obtained
was mainly defect-related [17.49]. A vertical p–n diode was realized by growing a
boron-doped homoepitaxial layer on a synthetic nitrogen-doped diamond [17.50].

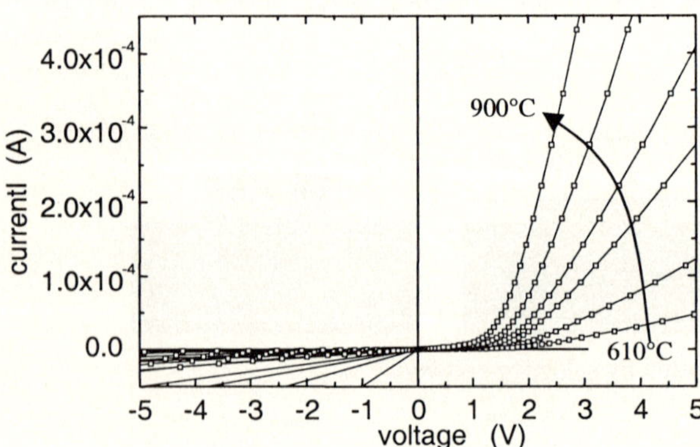

Fig. 17.3: The I–V characteristics of a homoepitaxial p–n diode operated between 610°C and 900°C, after [17.50].

Due to the high activation energy of the nitrogen donors, it could be operated only at temperatures above ~600°C. The *I–V* characteristics of this device are shown in Fig. 17.3.

Alternatively, vertical hetero–junction p–n diode structures are obtained by depositing p-type diamond on n-type Si-substrates [17.51–53]. However, it should be noted that the interface of these diodes is formed by the nucleation layer, and that these structures can only be operated at rather low temperatures due to the small bandgap of Si.

17.4.2 Schottky Diodes

Since a technical diode already needs to operate at room temperature with an acceptable series resistance, special attention has to be paid to the design of the active epitaxial layer configuration (doping–thickness profile). A continuous improvement in device performance has been obtained since the first published report in 1988 [17.54]. In Fig. 17.4, the *I–V* characteristics of a homoepitaxial and heteroepitaxial diode representing the present state of the art are shown. A micrograph of the heteroepitaxial structure is shown in Fig. 17.5.

The analysis of the field distribution in both cases reveals that breakdown is reached at an equal field strength. Thus, it is evident that (up to now) no substantial difference exists between the breakdown strengths of homoepitaxial films on synthetic single-crystal diamond substrates and highly oriented (HOD) heteroepitaxial films on silicon [17.55, 56]. This is very important for the future prospects of diamond devices.

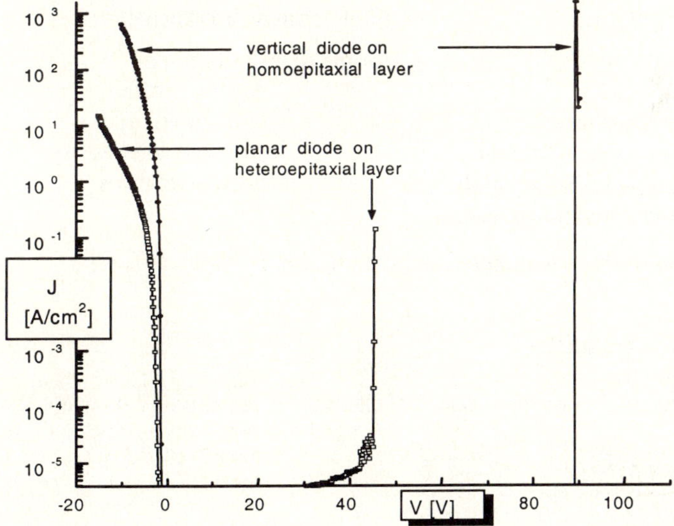

Fig. 17.4: The *I–V* characteristics of Schottky diodes on homoepitaxial diamond and heteroepitaxial diamond on silicon.

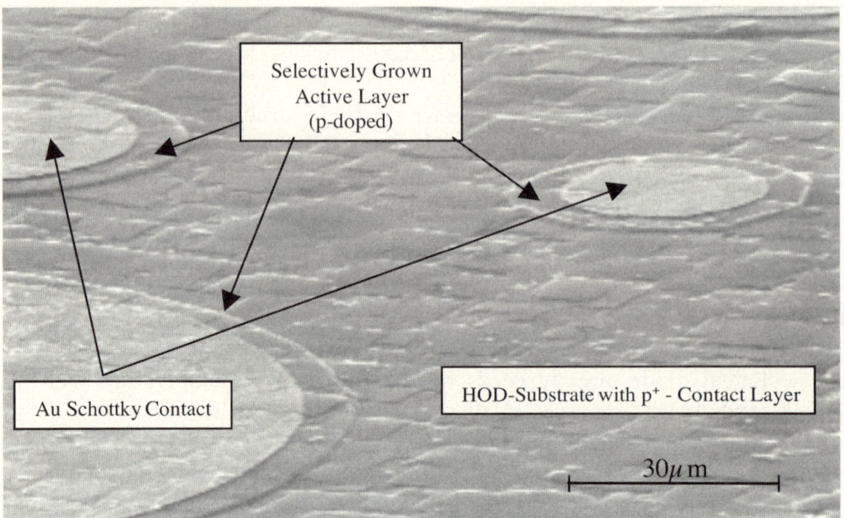

Fig. 17.5: A micrograph of planar Schottky diode structures on HOD film.

17.5 Characterization of Schottky Diodes

There are two widely used basic techniques for the characterization of Schottky diodes, which link the electrical performance to the physical junction properties:

I–V characterization:	*C–V* characterization
for the determination of	for the determination of
- ideality factor	- doping profile
- series resistance	- built-in voltage
- Schottky barrier height	- surface state density
- breakdown field strength	- deep level traps
- thermal activation energies	

A detailed description of these techniques is given in, for example, [17.57].

17.5.1 *I–V* Characteristics

By fitting the forward branch of the diode *I–V* characteristics (Fig. 17.4) using the equation:

$$I = I_0 (\exp(\frac{q(V - I R_S)}{nkT}) - 1)$$
(17.3)

the ideality factor, n, and the series resistance, R_S, of the diode can be extracted.

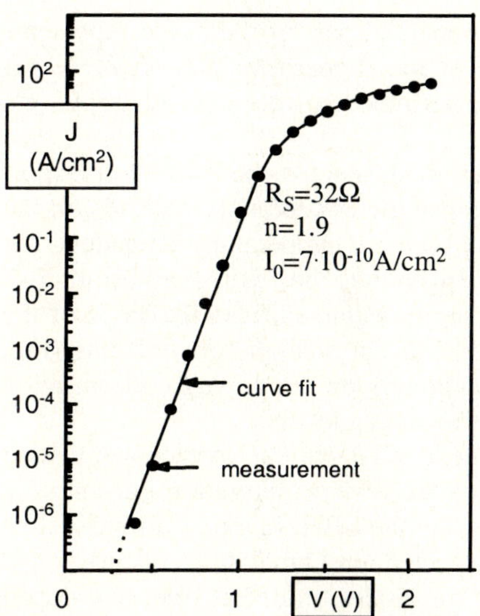

Fig. 17.6: Fitted forward characteristics of a diamond Schottky diode at R.T.

This is shown in Fig. 17.6, where a diode measured at room temperature was fitted. Here the ideality factor is quite high; however, with increasing temperature a decrease down to even below $n = 1.1$ is observed [17.41].

The investigation of the forward characteristics at various temperatures enables the extraction of several important parameters.

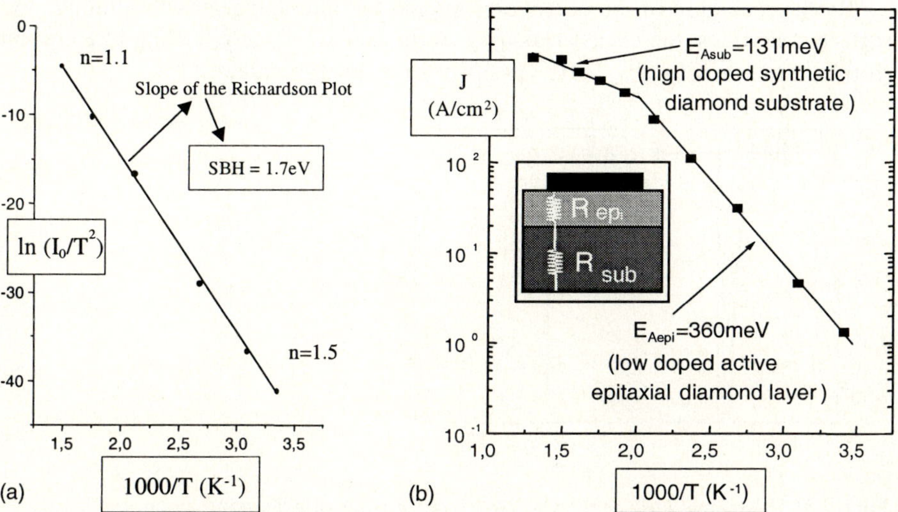

Fig. 17.7: **(a)** Extraction of the Schottky barrier height from the slope of the Richardson plot. **(b)** Extraction of the activation energy from the slope of the Arrhenius plot of the the forward current at high bias.

The current intercept at $V = 0$ of the exponential part of the curve represents the saturation current density I_0. The slope of the Richardson plot (I_0/T^2 versus $1/T$) yields the barrier height, which was determined in this case to be 1.7 eV (Fig. 17.7a).

Furthermore, the activation energy gives evidence for the current transport conditions under forward and reverse bias. Since the conductivity depends on the hole concentration and the hole mobility, and both are temperature-dependent, the exact activation energy of the acceptor may not be extracted from I–V measurements alone. However, an estimate can be obtained from the slope of the Arrhenius plot of the series resistance or the current at high forward bias (Fig. 17.7b). Since the activation energy depends strongly on the doping concentration [17.61], this also allows a rough estimate of the doping levels.

As pointed out before, the SBH of metal contacts on oxygen-terminated surfaces is in the range of 1.5–2.0 eV. Since the reverse saturation current of a Schottky diode depends inverse exponentially on the SBH, reverse currents below the measurement setup resolution limit are expected until breakdown occurs.

This expectation, however, is only valid for diodes on defect-free, low-doped natural diamond [17.58]. For diodes made on CVD-diamond layers, all published I–V characteristics exhibit a strong increase in the reverse current with bias and temperature (Fig. 17.8). The origin of this excess reverse current is not clearly identified yet, but it can be assumed that crystal defects, acting like highly doped ($N_A \approx 10^{19} \mathrm{cm}^{-3}$) small surface area clusters, lower the SBH locally substantially to $\Phi_B = \sim 0.5$ eV or even below [17.59]. Hence, the diode I–V characteristics may be described by a two-area-equivalent circuit, as shown in Fig. 17.9.

By the reduction of the defect density and by optimizing growth, doping, and surface conditions, the excess leakage current can be reduced, leading to a current rectification ratio of $I_{on}/I_{off} \approx 10^9$ [17.60] at room temperature.

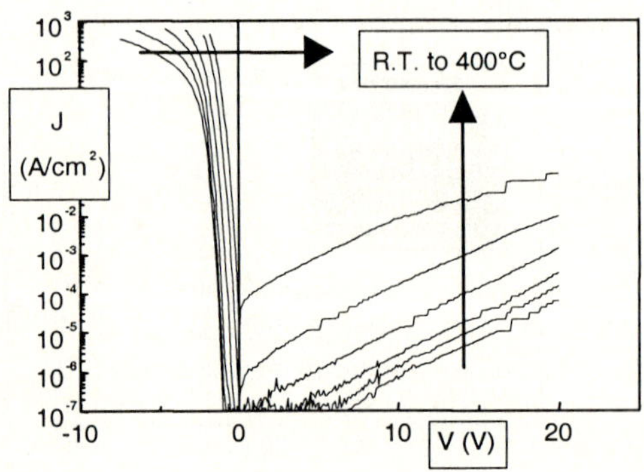

Fig. 17.8: The temperature dependent I–V characteristics of a diamond diode.

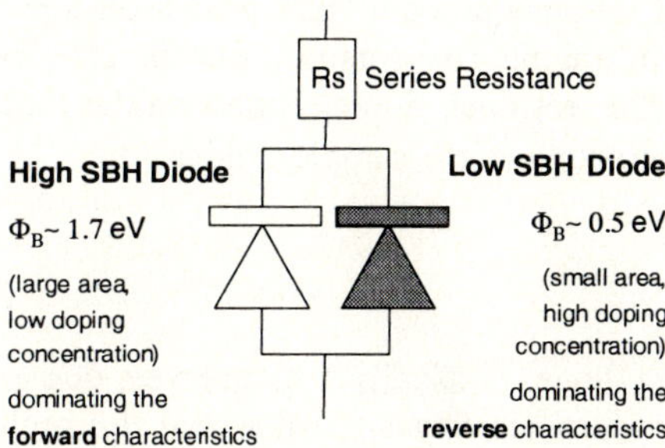

Fig. 17.9: The equivalent circuit of a defect-dominated diode.

17.5.2 *C–V* Characteristics

The capacitance–voltage characterization is a versatile tool to determine the net doping concentration (e.g. $N_A - N_D$) and the built-in voltage (V_B) by plotting $1/C^2$ versus V. The slope of this plot is inversely proportional to the net doping concentration:

$$N_A - N_D = \frac{-2}{q\varepsilon_0\varepsilon_r A^2 \dfrac{\mathrm{d}(1/C^2)}{\mathrm{d}V}} \tag{17.4}$$

where A is the Schottky contact area.

In practice, however, there are additional parameters that have to be taken into account to avoid inaccurate conclusions:

- a variation of the diode area, due to a rough surface
- a fringing capacitance, especially at a small contact area
- the ohmic contact impedance (including imaginary part) [17.62],
- additional *RC* elements in the diode equivalent circuit [17.63–65]
- the bulk resistance of the diamond substrate in comparison to the resistance of the space charge region [17.38]

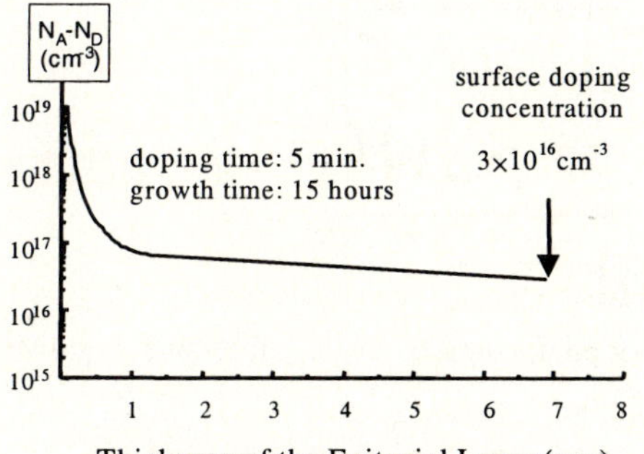

Fig. 17.10: The net doping profile of a homoepitaxial diamond layer [17.53].

Figure 17.10 shows a doping profile of a homoepitaxial layer grown on a p$^+$-diamond substrate. The sample was doped during the first 5 minutes of the growth and then grown for further 15 hours. The residual doping concentration of $N_A \approx 3 \times 10^{16}$ cm^{-3} is caused by boron contamination of the sample-holder.

17.5.3 Breakdown Behavior

At high reverse bias, breakdown in diamond diodes is usually not observed, due to the steady increase of the reverse current up to the forward value and the high thermal stability of the material. However, by improving the quality of the material and the interface, breakdown behavior becomes observable, as illustrated in Fig. 17.4.

A first investigation, on an inhomogeneously doped, successively grown layer [17.53], showed an increase in the breakdown voltage with increasing film thickness (Fig. 17.11). However, the analysis of the depleted part of the profile showed that the thickness itself is not the decisive factor, but that the decreasing surface doping concentration is (Fig. 17.12). From this, the breakdown field strength E_{BR} of the material is estimated.

The extracted value of $E_{BR} = 2 \times 10^6$ V/cm is nearly the same as that measured for thick undoped heteroepitaxial diamond [17.66], but is still one order of magnitude lower than the value measured on natural diamond [17.67]. Therefore, it is still important to evaluate the possibility of CVD-diamond growth with high breakdown strength. For example, an improvement to $E_{BR} = 8 \times 10^6$ V/cm would enable a rectification voltage of $V = 3$ kV for the sample shown in Fig. 17.11, with identical series resistance.

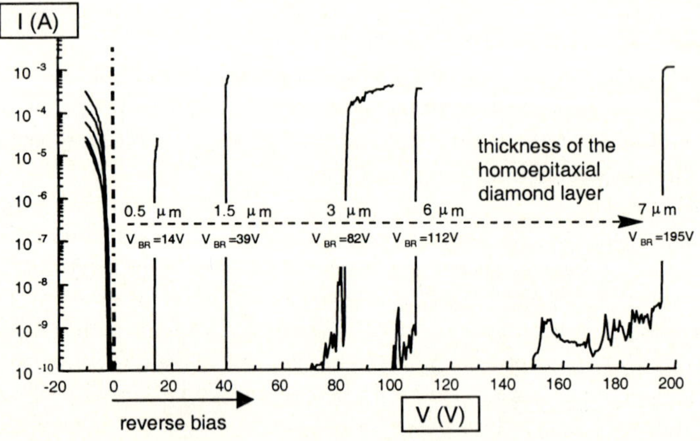

Fig. 17.11: Breakdown voltage of Schottky-point-contacts on a successively grown homoepitaxial diamond layer of Fig. 17.10.

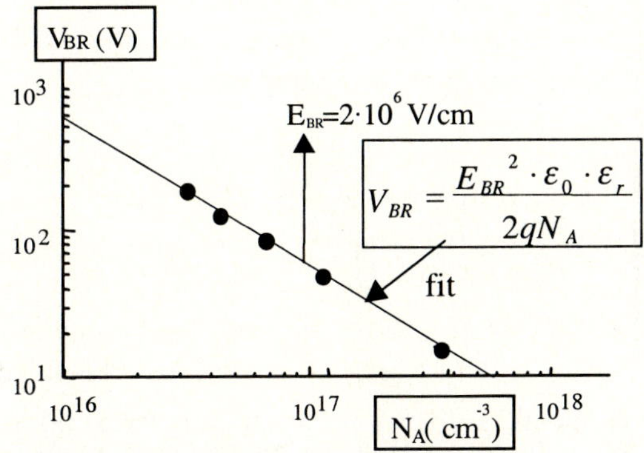

Fig. 17.12: Extraction of the breakdown field strength of doped CVD-diamond.

17.6 High-Temperature Schottky Diodes

17.6.1 High-Temperature Stable Contact Metallization

One of the expected fields of application of diamond electronics is high-temperature operation in the range up to 1000°C. Indeed, early device structures have already reached an operating temperature of 700°C [17.68]. An important prerequisite is the thermal stability of the contact metallization and the diamond metal interface. The temperature stability of the interface of ohmic contacts was found to be not critical, since the contacts are made on highly doped layers and since the alloying with refractory metals is performed at high temperatures [17.69].

In the case of Schottky contacts, however, the formation of carbide interfacial layers leads usually to a reduced SBH and degraded reverse characteristics. On the other hand, noncarbide-forming metals such as Au become soft at elevated temperatures and therefore unreliable at the interface.

Since up to a temperature of ~500°C no SiC formation is observed [17.70], the utilization of highly conductive (amorphous or nano-crystalline) quasi-metallic silicon (qm-Si) as contact material is an interesting possibility. The temperature stability of the silicon/diamond interface may be further improved by thermodynamically stabilizing the Si layer with incorporated oxygen, nitrogen, or carbon atoms. Using such a stabilized qm-Si layer, no SBH changes have been observed up to a temperature of 700°C [17.71]. Due to the high resistivity of the qm-Si layer (as compared to a metal), this layer has to be as thin as possible, and a low-resistance cap layer metal (e.g. Au) has to be applied. Such a cap layer is also necessary for bonding. Furthermore, to avoid chemical reactions between the qm-

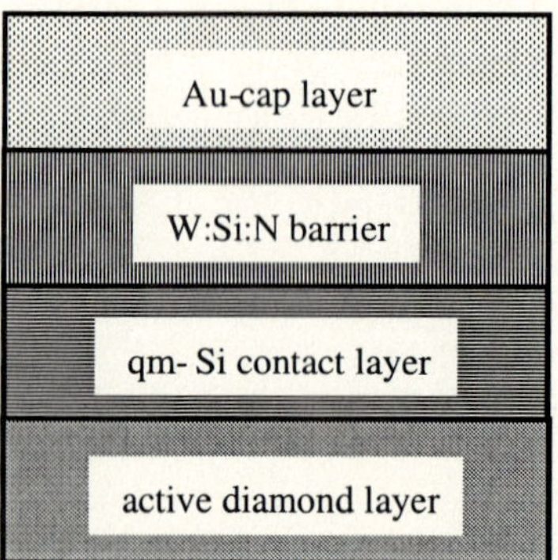

Fig. 17.13: A high-temperature stable multi-layer contact system deposited by sputtering.

Si contact material and the cap layer metal, a diffusion barrier (e.g. W:Si:N) is needed [17.72]. A contact configuration involving a p^+ qm-Si Schottky layer is shown in Fig. 17.13.

17.6.2 High-Temperature Operation

Using the high-temperature stable contact described above, the feasibility of diamond devices operating at temperatures of up to 1000°C has been demonstrated (Figs. 17.14 and 17.15) [17.42]. The contact metallization used in this experiment was similar to that shown in Fig. 17.13, except that the qm-

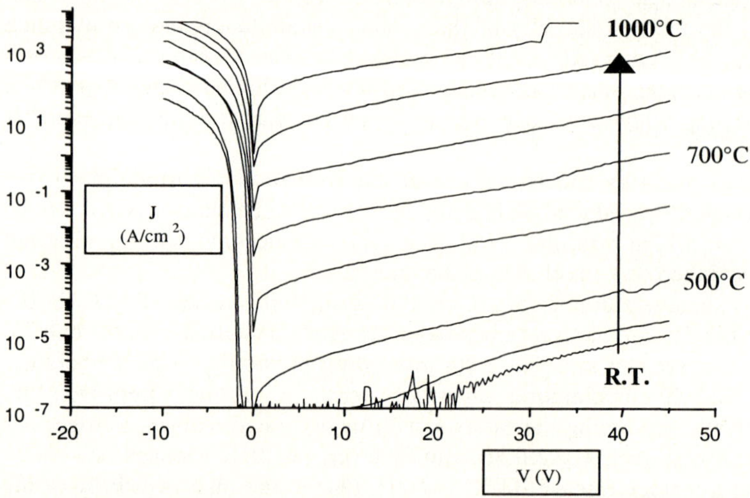

Fig. 17.14: The logarithmic *I*–*V* characteristic of a high-temperature Schottky diode. In this experiment, the diode has been cycled from R.T. up to 1000°C, where the operation failed at 32 V reverse bias. The contact was held at each temperature for 15 min and the bias ramped up to 45 V within 1 min.

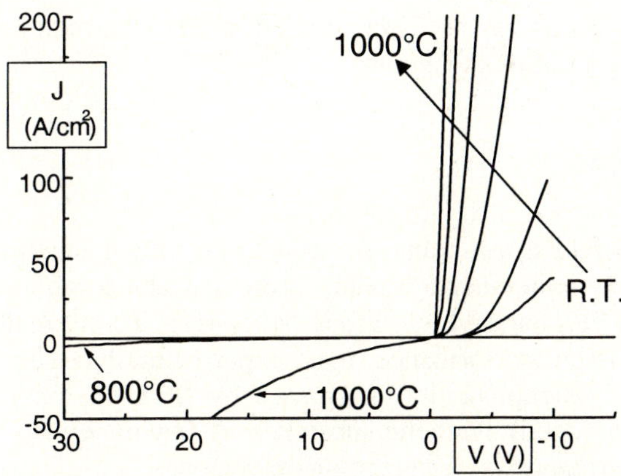

Fig. 17.15: The linear I–V characteristic of a high-temperature Schottky diode.

Si/diamond interface was chemically stabilized and no Au cap layer was used (because of a breakdown of the W:Si:N diffusion barrier at ~800°C).

In the lower part of the characteristics, it can be seen that the reverse leakage current shows an extremely weak thermal activation up to 500°C as compared to earlier results shown in Fig. 17.8. It is thought that this is the combined result of a stabilized interface, possibly passivating defects, a reduced surface doping concentration and a reduced acceptor compensation due to a low-nitrogen background.

In the linear current scale (Fig. 17.15), which is more relevant for the loss determination of a rectifier diode, hardly any reverse leakage current can be observed up to a temperature of 800°C and a reverse voltage of 20 V. However, the forwards resistance was found to vary over two orders of magnitude upon heating to 1000°C.

Above 400°C, the diamond surface has to be shielded against oxidation. This has been achieved by measuring in a vacuum. The structure shown has not been passivated.

17.7 Transistors

Within the framework of the technological key elements described above, transistor structures can be designed and fabricated. Indeed, in the very early stages of diamond technology, all conventional transistor structures have been already attempted.

17.7.1 Bipolar Transistors

In spite of the fact that, up to now, no effective diamond donor doping of technical relevance has become available, two of the three first reported diamond transistors were of bipolar type [17.73-75]. However, the characteristics have lacked far behind theoretical expectations.

Since n-type doping by phosphorus has not been applied to device structures, no results are yet available in respect of this technology.

17.7.2 Field Effect Transistors

Figure 17.16 shows the tree of field effect transistor structures, which emerges when the individual technological components already presented are combined. The variety ranges from junction FETs and MOSFETs to MESFETs. To prove the concepts and obtain the best possible performance, most experiments have been performed on highly insulating homoepitaxial films grown by MPCVD on IIa (undoped) natural or Ib (nitrogen-doped) synthetic substrates. A few experiments have then been transferred to heteroepitaxial films grown on silicon.

Most of the p-channel configurations will be reviewed below. Special emphasis will be given to the requirement for high-power and high-temperature operation. Thus, the key issues are: fully activated doping in a high sheet charge/high aspect ratio structure, high temperature stable and reliable contacts, high breakdown strength, and saturated velocity dominated operation for high current density and speed (which implies sub-μm gate lengths). A high carrier activation is of special importance, since only the activated carriers contribute to the channel current, but the entire net doping concentration has to be modulated by the gate potential. Thus, a partial activation to 10^{-3} will result in only 10^{-3} of the possible output current and transconductance. The channel sheet charge, N_s, (the thickness–doping product), is generally of the order of $10^{12}\,\mathrm{cm}^{-2}$ to $10^{13}\,\mathrm{cm}^{-2}$.

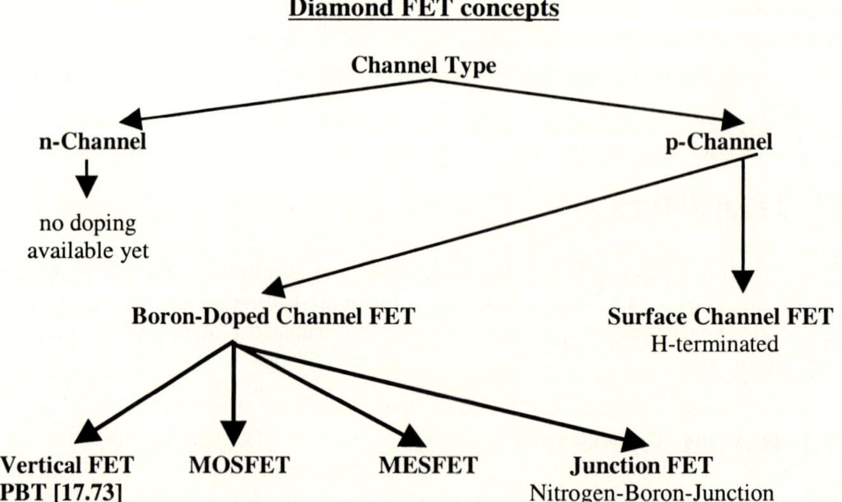

Fig. 17.16: The diamond FET tree.

MOSFET

Initial problems with highly rectifying Schottky barrier contacts on diamond have lead to the development of MOSFET structures with SiO_2 as a gate insulator. The MOS system has been analyzed carefully [17.76] and depletion-mode FET structures have been developed in ion-implanted and epitaxial boron-doped channel material, using standard technologies of mesa-isolation and contact patterning [17.77–79]. Using low-doped channel material, high mobilities have been observed, but also freeze-out at room temperature [17.80]. Therefore, special doping profiles (based on narrow-spike doping) have been employed and the thermal activation could indeed be considerably improved [17.81]. Theoretical estimates yield that output currents above 1 A/mm are feasible [17.82].

These transistor structures have then been implemented into basic logic functions, and a voltage gain could be obtained [17.80, 83] in quasi-static switching. The devices have been operated as a small signal amplifier in the 100 kHz regime [17.80]. The high-temperature operation of these MOSFET devices is attractive due to the increased carrier activation. Operation at elevated temperatures up to 500°C was demonstrated [17.81]. The best results were achieved at 325°C. Here, a maximum channel current of 30 mA/mm and a maximum transconductance of 1.3 mS/mm have been measured at a gate length of 2 μm.

Subsequently, these structures have also been transferred into a diamond-on-Si technology [17.84]. Here, the influence of the surface roughness and grain boundaries is clearly seen in reduced current, transconductance, and voltage ranges, and also increased leakage. However, it could also be clarified that the grain boundaries do not act as internal Schottky barriers preventing current transport along the channel. For a long time, these devices have represented the state of the art.

MESFET

First experiments with MESFETs have reflected the technological difficulties of developing Schottky diodes with good rectification characteristics and low-resistance ohmic contacts [17.85–87]. To reduce carrier freeze-out, high narrow doping profiles are needed. The peak doping concentration needs to be above $10^{19}\,cm^{-3}$ to obtain a noticeable effect in activation energy reduction. At a sheet charge density of $10^{13}\,cm^{-3}$, this implies that the depth of the profile needs to be limited to 10 nm. Such steep profiles may be called spike doping or delta (δ) doping. They require special doping techniques; as, for example, illustrated in Fig. 17.1. Indeed, such a structure has shown improved activation, resulting in a maximum output current of 0.4 mA/mm and a maximum transconductance of 0.12 mS/mm at room temperature [17.87]. Unfortunately, no results on operation at elevated temperatures have been reported.

Combination of the concept of the high-temperature stable contact technology with the δ-doping concept has resulted in a δ-channel FET with a cross-section as

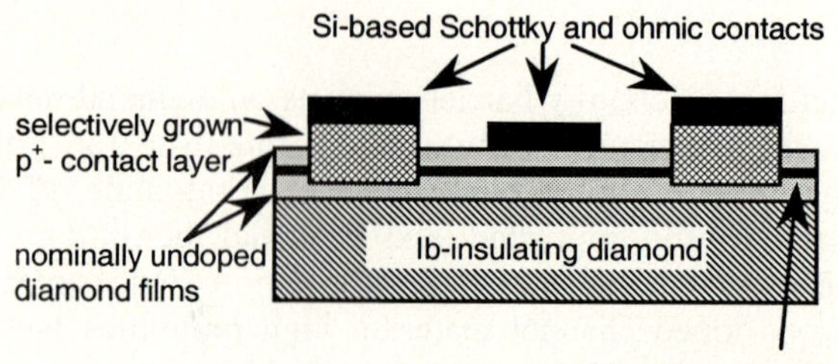

Fig. 17.17: A cross-section of a δ-doped MESFET prepared on a synthetic highly insulating nitrogen doped substrate.

shown in Fig. 17.17. The source and drain contacts were selectively grown as indicated in Fig. 17.2; the metallization, both for the ohmic contacts and the Schottky contact, was prepared using qm-Si [17.88].

The output characteristics are shown in Figs. 17.18a, b, at room temperature and 350°C, respectively. Current activation by more than a factor of 10^2 between these temperatures is observed, consistent with an activation energy of 330 meV and the C–V analysis, indicating a peak doping level of ~5×10^{18} cm^{-3} and an

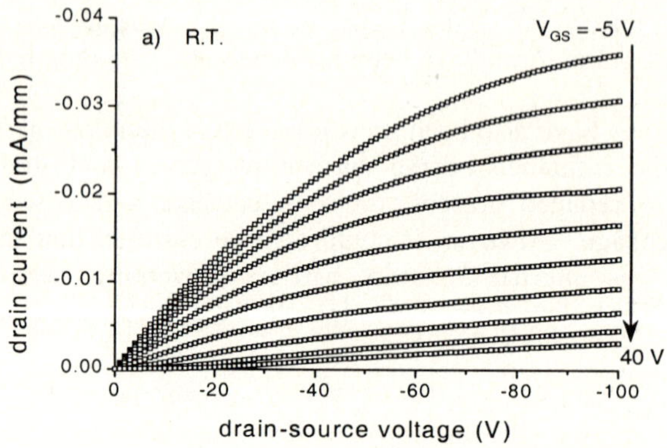

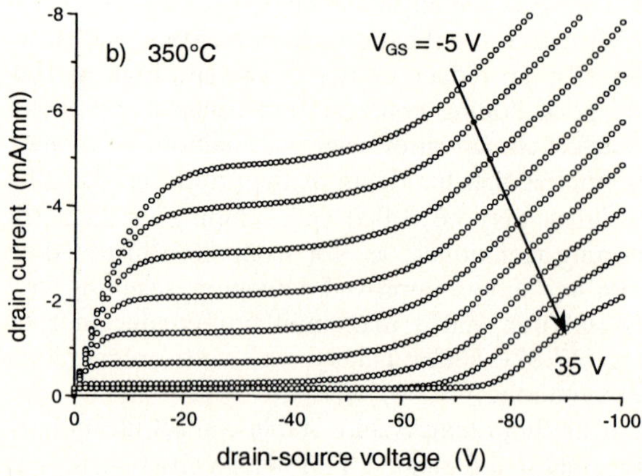

Fig. 17.18: The I–V characteristics of δ-doped FET at **(a)** room temperature and **(b)** 350°C. The gate length was 3.5 µm at a width of 500 µm.

effective pulse width of 15 nm. In this planar structure, freeze-out generates a large series resistance, leading to a high saturation voltage of ~100 V.

At 350°C, the characteristic is essentially improved, the current increases to ~5 mA/mm, and the open-channel saturation voltage is reduced to 25 V. At a high drain bias, above 80 V, the channel cannot be pinched off correctly, possibly due to carrier injection into the highly nitrogen-doped substrate. At 450°C, the gate diode reverse breakdown voltage is reduced slightly and the channel cannot be pinched off completely. Thus, the high-temperature qm-Si contact technology still needs to be further refined, when applied to these FET structures. Nevertheless, a sheet charge density of $8.5\times10^{12}\,\mathrm{cm}^{-2}$ was modulated and a maximum gate field at pinch-off of about 3×10^6 V/cm could be applied.

Junction FET

It has been shown in Sect. 17.4.1 that the nitrogen/boron junction shows the properties of a p/n junction despite the very high activation energy of nitrogen, and may thus be incorporated into an FET design, based on a δ-doped channel. In the structure illustrated in Fig. 17.19, the highly nitrogen doped substrate is used as a control gate from the back. In this case, the full source-to-drain separation represents the electrical gate and channel length simultaneously. No large-channel series resistances are expected. However, the high resistivity of the substrate will cause a high gate series resistance, and the overlapping of the source and drain contacts on to the control gate layer will cause large parasitic capacitances. Nevertheless, such a structure has been operated up to the GHz range in GaAs [17.89].

For diamond, it was found that the δ-channel can indeed be modulated at high current levels at a temperature of 300°C (see Fig. 17.20). At this temperature, full activation is expected according to a channel conduction activation energy of 0.1 eV, as determined by Hall measurements [17.90]. The high gate resistance still causes a low transconductance and a high pinch-off voltage.

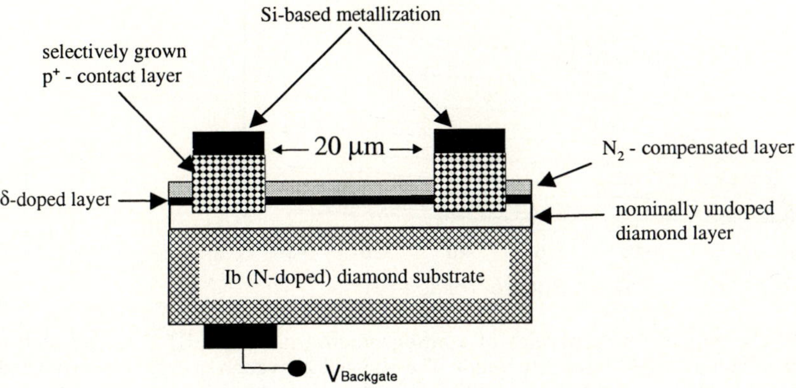

Fig. 17.19: A cross-section of a junction FET using the Ib substrate as gate contact.

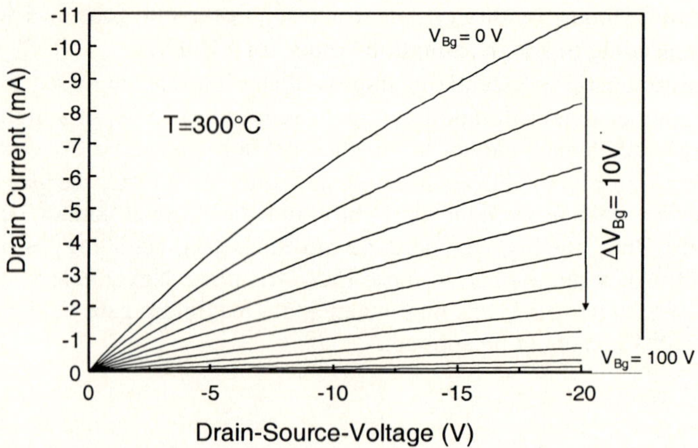

Fig. 17.20: The output characteristics of a back gate junction FET. The source-to-drain separation was 20 μm and the gate width was 200 μm. The estimated channel sheet charge density was 8×10^{12} cm^{-2}, with an effective activation energy of 0.1 eV. The δ-channel to substrate separation was ~0.12 μm.

In an alternative structure, the nitrogen-doped gate control layer was placed on top of the δ-doped channel by selective epitaxy, thus reducing the gate resistance considerably (Fig. 17.21). Indeed, a high output current could be obtained at 200°C. This was possible due to a reduced activation energy of the δ-doped channel of 0.14 eV, resulting in 65% activation [17.90]. Scaling this result to a 1 μm gate length, a maximum output current of 300 mA/mm is expected (see also Fig. 17.26). These preliminary results allow the conclusion that, with δ-doped

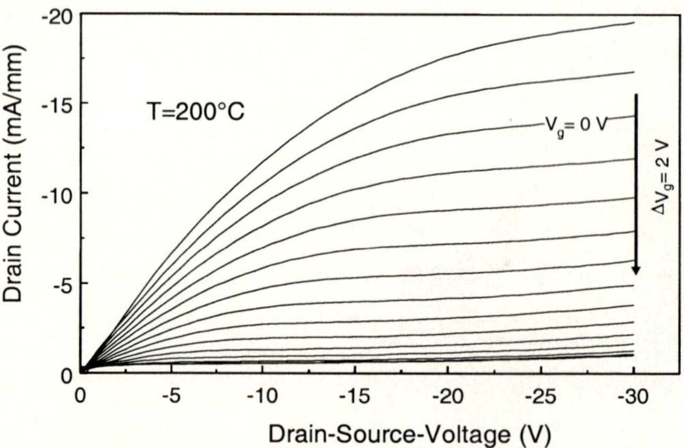

Fig. 17.21: The output cha-racteristics of a nitrogen/boron junction FET structure with a 20 μm gate length and a 200 μm gate width. The channel activation energy was estimated at 0.14 meV and the sheet charge at 8×10^{12} cm^{-3}. The channel gate-layer separation was approximately 0.1 μm. The substrate was not connected.

channel structures, full activation can be reached at moderate temperatures, and that high current levels comparable with these of other wide-gap semiconductor materials can be obtained.

Surface Channel FET

As pointed out in Sect. 17.3.1, the termination of the diamond surface with hydrogen produces a room temperature activated hole conduction at the surface. The sheet charge of this surface channel is of the order of 10^{13} cm^{-2} [17.21]. Thus attempts have been made to utilize this surface layer as the FET channel. Pioneered by Kawarada [17.91–93], high-performance surface p-channel FET structures have indeed been realized. As mentioned previously, depending on the contact material used, ohmic as well as rectifying characteristics can be obtained. Through the specific choice of the gate contact material either enhancement mode (device pinched-off at zero gate bias) or depletion mode (positive gate bias needed for pinch-off) FET characteristics are possible. Implementing these concepts, logic functions have been realized in depletion/enhancement mode technology [17.94]. For these high-performance devices, a maximum channel current of 20 mA/mm and a maximum transconductance of 16.4 mS/mm have been reported for a gate length of 3 µm.

Not addressed in this study [17.94] was the breakdown voltage that can be obtained with this technology. Experiments guided in this direction have indeed led to surprisingly high voltages, which can be tolerated by this surface channel

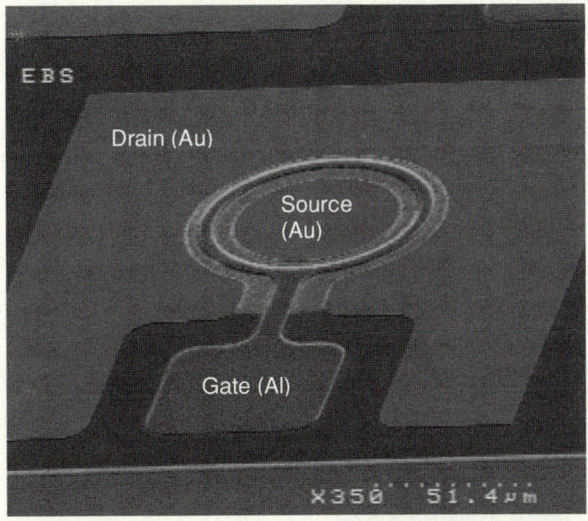

Fig. 17.22: The micrograph shows a surface channel FET structure with circular gate. The hydrogen-terminated surface conductive channel is represented by the bright areas. The insulating parasitic areas are dark. Isolation was obtained by oxygen plasma treatment. For the ohmic source and drain contacts, Au was used. The gate contact material was Al, which results in enhancement mode FET characteristics.

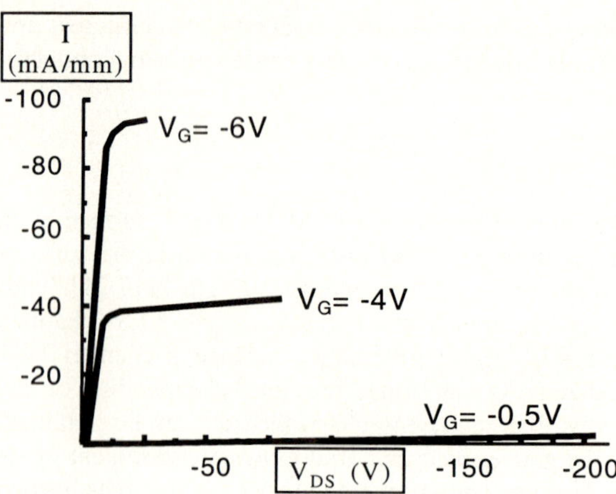

Fig. 17.23: The output characteristics of a surface channel FET prepared on a single-crystal substrate (L_G=3 µm).

diode. The discussion on the physical nature of the surface charge and the exact location at or near the surface is still ongoing, and the reader is referred to the literature [17.92, 95]. In the following, the results of these experiments will be used to estimate the power-handling capability of diamond FETs [17.96]. An SEM micrograph of such a surface device is shown in Fig. 17.22. Figure 17.23 shows the output characteristics of a device with a gate length of 3 µm. The maximum current is 90 mA/mm at a gate bias of $V_G = -6$ V. This is several times the diode barrier height (SBH = 0.8 V) and shows that the characteristics are still dominated by a large series resistance due to a high channel sheet resistance and the relaxed geometry of the circular structure. The maximum transconductance is 25 mS/mm. Breakdown at pinch-off is above $V_{DS} = -200$ V, which was the limit of the source unit.

Using an optimum load line, a maximum RF output power of 2.3 W/mm can be obtained. Assuming a breakdown voltage of 200 V, which will not degrade while reducing the gate length further, this would result in a scaling as shown later

Fig.17.24: A micrograph of a surface channel FET in diamond-on-Si technology [17.96].

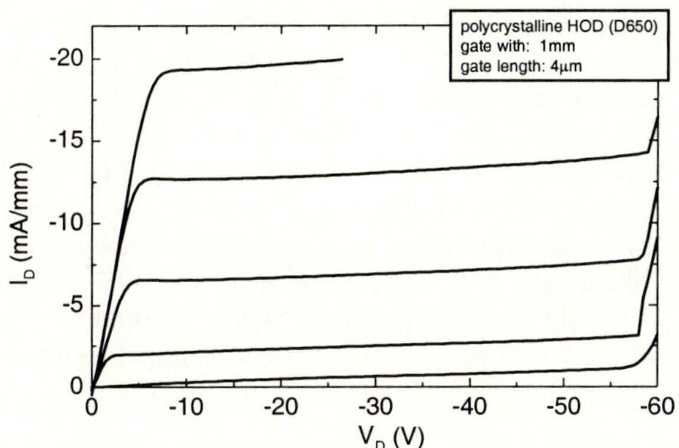

Fig. 17.25: The output characteristics of the surface channel FET shown in Fig. 17.24 [17.96].

in Fig. 17.26. At a gate length of 1 μm, a power-handling capability of more than 6 W/mm can be predicted. For an ultimate gate length of 0.1 μm, where the device starts to operate in a velocity-dominated mode, a maximum open-channel current of more than 1 A/mm may be expected, and a power density as high as 30 W/mm can be projected.

The hydrogen surface termination induced channel is believed to be several tens of nanometers in depth [17.94], and may follow perfectly any surface roughness. Therefore, the experiments may be transferable to heteroepitaxial

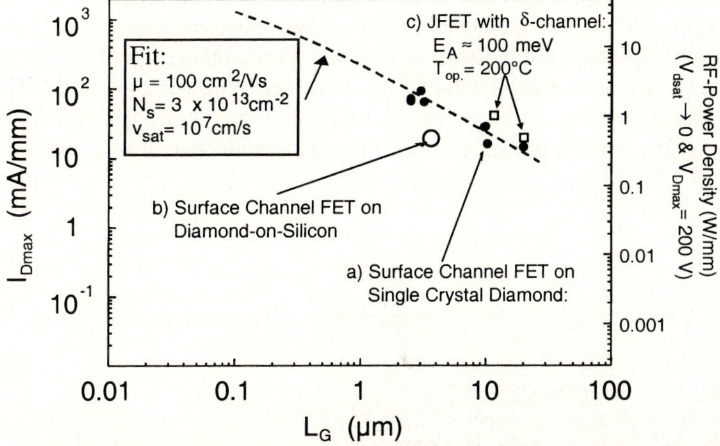

Fig. 17.26: The scaling of diamond FET results with gate length: (*a*) homoepitaxial surface channel FETs on diamond substrate; (*b*) heteroepitaxial surface channel FETs on silicon; and (*c*) homoepitaxial junction FETs. The set of parameters used in the fit is also listed. This fit should be considered as a guide-line, since the individual devices will have deviating parameters in N_s and μ. Shown to the left are the maximum output current densities and to the right the maximum RF power density, assuming that a 200 V maximum drain bias can be sustained and the saturation voltage can be neglected.

layers on silicon. Indeed, first results have been recently obtained [17.97, 98] and Fig. 17.24 shows a micrograph of such a structure. It can be seen that the surface roughness of this highly oriented diamond film is less than the contact metallization thickness (~0.2 µm) and that the gate contact will run across a large number of grain boundaries. Nevertheless, the FET characteristics in Fig. 17.25 show high current levels, good saturation of the output current, and a high breakdown voltage. However, in comparison to the monocrystalline counterparts, the current and breakdown voltage are both reduced to roughly 30%. Thus, presently, the power-handling capability of this device technology would be one order of magnitude below that of the homoepitaxial device on a diamond substrate (see also Fig. 17.26).

17.8 Conclusions

In this chapter, the status of diamond as an electronics material has been reviewed. Important advancements in all fields – materials technology, device processing, and device performance – have been described. They already allow us to demonstrate or extrapolate to an extraordinary device performance based on the extraordinary material properties. High-temperature operation up to 1000°C of a metal/diamond contact could be demonstrated. On the basis of the fabricated FET structures, an extremely high power-handling density of 30 W/mm can be predicted. Together with the recent advances in diamond technology for sensor and actuator applications, the material may indeed be ideally suited for applications in which extreme temperature and power performance are desired, or in which extremely aggressive environmental conditions inhibit the use of other semiconductor materials. Nevertheless, many technological challenges are still before us. The exciting results presented here should be understood as a proof of concept. They will certainly encourage further studies.

References

17.1   S. Kumar, P. Ravindranathan, H.S. Dewan, and R. Roy, Diamond Rel. Mater. **5**, 1246 (1994)
17.2   E.O. Johnson, RCA Rev. **26**, 163 (1963)
17.3   R.W. Keyes, Proc. IEEE **60**, 225 (1972)
17.4   B.J. Baliga, IEEE Electron. Devices Lett. **10**, 455 (1989)
17.5   S. Koizumi, M. Kamo, Y. Sato, H. Ozaki, and T. Inuzuka, Appl. Phys. Lett. **71**, 1065 (1997)
17.6   C. Wild , R. Kohl, N. Herres, W. Müller-Sebert, and P. Koidl, Diamond Rel. Mater. **3**, 373 (1994)
17.7   X. Jiang, C.P. Klages, R. Zachai, M. Hartweg, and H.J. Füßer, Appl. Phys. Lett. **62**, 3438 (1993)
17.8   S.D. Wolter, T.H. Borst, P. Gluche, W. Ebert, A. Vescan, and E. Kohn, Appl. Phys. Lett. **68**, 3558 (1996)
17.9   T.H. Borst and O.Weis, Phys. Status Solidi A **154**, 423 (1996)

17.10 A.T. Collins, Semicond. Sci. Technol. **4**, 605 (1989)

17.11 C.R. Zeisse, C.A. Hewett, R. Nguyen, J.R. Zeidler, and R.G. Wilson; IEEE Electron. Devices Lett. **12**, 602 (1991)

17.12 K. Das, V. Venkatesan, T.P. Glass, D.G. Thompson, J. Appl. Phys. **76**, 2208 (1994)

17.13 H. Shiomi, Y. Nishibayashi, and N. Fujimori, Jap. J. Appl. Phys. **30**, 1363 (1991)

17.14 J. Cifre, J. Puigdoller, M.C. Polo, and J. Esteve, Diamond Rel. Mater. **3**, 628 (1994)

17.15 R. Erz, W. Dötter, K. Jung, and H. Ehrhardt, Diamond Rel. Mater. **4**, 469 (1995)

17.16 A. Vescan, P. Gluche, W. Ebert, and E. Kohn, Electron. Lett. **32**, 1419 (1996)

17.17 T.H. Borst and O. Weis, Diamond Rel. Mater. **4**, 948 (1995)

17.18 H. Shiomi, H. Nakahata, T. Imai, Y. Nishibayashi, and N. Fujimori, Jpn. J. Appl. Phys. **28**, 758 (1989)

17.19 N. Fujimori, T. Imai, H. Nakahata, H. Shiomi, and Y. Nishibayashi, Mater. Res. Soc. Symp. Proc. **162**, 23 (1990)

17.20 S.A. Grot, G.S. Gildenblatt, C.W. Hatfield, C.R. Wronsky, A.R. Badzian, T. Badzian, and R. Messier, IEEE Electron. Devices Lett. **11**, 100 (1990)

17.21 K. Hayashi, S. Yamanaka, H. Okushi, and K. Kajimura, Appl Phys. Lett. **68**, 376 (1996)

17.22 W. Mönch, Europhys. Lett. **27**, 479 (1994)

17.23 M. Aoki and H. Kawarada, Jpn. J. Appl. Phys. **33**, L 708 (1994)

17.24 J. Shirafuji and T. Sugino, Diamond Rel. Mater. **5**, 706 (1996)

17.25 R.E. Thomas, R.A. Rudder, and R.J. Markunas, J. Vac. Sci. Technol. **A 10**, 2451 (1992)

17.26 G. Braunstein and R. Kalish, J. Appl. Phys. **54**, 2106 (1983)

17.27 N. Fujimori, T. Imai, and A. Doi, Vacuum **36**, 99 (1986)

17.28 K. Das, C.T. Kao, and L.S. Plano, in Proceedings of the 2nd International Conference on the Application of Diamond Film and Related Materials, pp. 335–40, Tokyo (1993)

17.29 K.L. Moazed, J.R. Zeidler, and M.J. Taylor, J. Appl. Phys. **68**, 2246 (1990)

17.30 K.L. Moazed, R. Nguyen, and J.R. Zeidler, IEEE Electron. Devices Lett. **9**, 350 (1988)

17.31 V. Venkatesan and K. Das, IEEE Electron. Devices Lett. **13**, 126 (1992)

17.32 M. Seal, Engelhard Tech. Bull. **10**, 10 (1969)

17.33 M. Werner, O. Dorsch, H.U. Baerwind, A. Ersoy, E. Obermeier, C. Johnston, S. Romani, P.R. Chalker, V. Moore, and I.M. Buckley-Golder, Diamond Rel. Mater. **3**, 983 (1994)

17.34 T. Tachibana and J.T. Glass, J. Appl. Phys. **72**, 5912 (1992)

17.35 M.W. Geis, M. Rothschild, R.R. Kunz, R.L. Aggarwal, K.F. Wall, C.D. Parker, K.A. McIntosh, N.N. Efremow, J.J. Zayhowski, D.J. Ehrlich, and J.E. Butler, Appl. Phys. Lett. **55**, 2295 (1989)

17.36 V. Venkatesan, D.M. Malta, K. Das, J. Appl. Phys. **72**, 1179 (1993)

17.37 A. Vescan, W. Ebert, T.H. Borst, and E. Kohn, Diamond Rel. Mater. **5**, 747 (1996)

17.38 M. Werner, S. Romani, P.R. Chalker, C. Johnston, and I.M. Buckley-Golder, J. Appl. Phys. **79**, 2535 (1996)

17.39 G.H. Glover, Solid-State Electron. **16**, 973 (1973)

17.40 C.A. Mead and T.C. McGill, Phys. Lett. **58A**, 249 (1976)

17.41 M.C. Hicks, C.R. Wronski, S.A. Grot, G.Sh. Gildenblat, A.R. Badzian, T. Badzian, and R. Messier, J. Appl. Phys. **65**, 2139 (1989)

17.42 W. Ebert, A. Vescan, T.H. Borst, and E. Kohn, IEEE Electron. Devices Lett. **15**, 289 (1994)

17.43 A. Vescan, I. Daumiller, P. Gluche, E. Ebert, and E. Kohn, IEEE Electron. Devices Lett. **18**, (1997)

17.44 F.J. Himpsel, P. Heimann, and D.E. Eastman, Solid State Commun. **36**, 631 (1980)

17.45 M.L. Hartsell, H.A. Wynands, and B.A. Fox, Appl. Phys. Lett. **65**, 430 (1994)

17.46   M.W. Geis, D.D. Rathman, D.J. Ehrlich, R.A. Murphy, and W.T. Lindley, IEEE Electronic. Devices Lett. **8**, 341 (1987)

17.47   P.K. Baumann, T.P. Humphreys, R.J. Nemanich, K. Ishibashi, N.R. Parikh, L.M. Porter, and R.F. Davis, Diamond Rel. Mater. **3**, 883 (1994)

17.48   J. van der Weide and R.J. Nemanich, J. Vac. Sci. Technol. **B10**, 1940 (1992)

17.49   V.S. Vavilow, M.A. Gukasyan, M.I. Gusewa, E.A. Konorowa, and V.F. Sergienko, Sov. Phys. Dokl. **10**, 856 (1972)

17.50   T.H. Borst, S. Strobel, and O. Weis, Appl. Phys. Lett. **67**, 2651 (1995)

17.51   C.L. Ellison, R.M. Cohen, and J.T. Hoggins, Mater. Res. Soc. Symp. **162** (1990)

17.52   H. Kiyota, M. Yoneda, H. Izumiya, H. Okushi, K. Okano, T. Kurosu, and M. Iida, Jpn. J. App. Phys. **31**, L388 (1992)

17.53   T. Phetchakul, H. Kimura, Y. Akiba, T. Kurosu, and M. Iida, Jpn. J. Appl. Phys. **35**, 4247 (1996)

17.54   S.H. Gildenblat, S.A. Grot, C.R. Wronski, A.R. Badzian, T. Badzian, and R. Messier, Appl. Phys. Lett. **53**, 586 (1988)

17.55   W. Ebert, A. Vescan, P. Gluche, T. Borst, and E. Kohn, Diamond Rel. Mater. **6**, 329 (1997)

17.56   P.Gluche, S.D. Walter, T.H. Borst, W. Ebert, A. Vescan, and E. Kohn, IEEE Electron. Devices Lett. **17**, 270 (1996)

17.57   E.H. Rhoderick, R.H. Williams, Metal-Semiconductor Contacts, Clarendon Press, Oxford (1984)

17.58   M.W. Geis, N.N. Efremow, and J.A. von Windheim, Appl. Phys. Lett. **63**, 952 (1993)

17.59   A. Vescan, W. Ebert, T.H. Borst, and E. Kohn, Diamond Rel. Mater. **4**, 661 (1995)

17.60   W. Ebert, A. Vescan, T.H. Borst, and E. Kohn, in International Electron Device Meeting, San Francisco, CA, Technical Digest, pp. 419–22 (1994)

17.61   T.H. Borst and O. Weis, Diamond Rel. Mater. **4**, 948 (1995)

17.62   J.-J. Shiau, A.L. Fahrenbruch, and R.H. Bube, J. Appl. Phys. **61**, 1556 (1987)

17.63   W. Ebert, A. Vescan, and E. Kohn, Diamond Rel. Mater. **3**, 887 (1994)

17.64   V. Venkatesan, K. Das, J.A. von Windheim, and M. Geis, Appl. Phys. Lett. **63**, 1065 (1993)

17.65   J.A. von Windheim, V. Venkatesan, D.M. Malta, and K. Das, J. Electron. Mater. **22**, 391 (1993)

17.66   R. Hessmer, M. Schreck, S. Geier, and B. Stritzker, Diamond Rel. Mater. **3**, 951 (1994)

17.67   P. Liu, R. Yen, and N. Bloembergen, IEEE J. Quantum Electron. **14**, 574 (1978)

17.68   M.W. Geis, D.D. Rathman, D.J. Ehrlich, R.A. Murphy, and W.T. Lindley, IEEE Electron. Devices Lett. **8**, 341 (1987)

17.69   M. Roser, C.A. Hewett, K.L. Moazed, and J.R. Zeidler, J. Electrochem. Soc. **139**, 2001 (1992)

17.70   A. Vescan, W. Ebert, T.H. Borst, M. Pitter, M. Hugenschmidt, R.J. Behm, J. Gerster, R. Sauer, and E. Kohn, Diamond Materials IV, Electrochemical Society Proceedings **95-4**, 324 (1995)

17.71   A. Vescan, I. Daumiller, P. Gluche, W. Ebert, and E. Kohn, Diamond 97, Edinburgh UK, 1997, paper 13.5 (to be published in Diamond Rel. Mater.)

17.72   P. Gluche, A. Vescan, W. Ebert, M. Pitter, E. Kohn, 124th TMS Annual Meeting, Anaheim (USA), Proceedings (1996) pp. 100–103

17.73   J.F. Prins, Appl. Phys. Lett. **41**, 951 (1982)

17.74   M.W. Geis, D.D. Rathman, D.J. Ehrlichman, R.A. Murphy, and W.T. Lindley, IEEE Electron. Devices Lett. **EDL-8**, 341 (1987)

17.75   M.W. Geis, N.N. Efremow, and D.D. Rathman, J. Vac. Sci. Technol. **A6**, 1953 (1988)

17.76   M.W. Geis, J.A. Gregory, and B.B. Pate, IEEE Trans. Electron. Devices **38**, 619 (1991)

17.77 C.R. Zeise, C.A. Hewett, R. Nguyen, J.R. Zeidler, and R.G. Wilson, IEEE Electron. Devices Lett. **12**, 602 (1991)

17.78 K. Kobashi, K. Nishimura, K. Miyata, R. Nakamura, H. Koyama, K. Saito, and D.L. Dreifus, 2nd International Conference on the Application of Diamond Films and Related Materials, Tokyo (1993), pp. 35–42

17.79 G.S. Gildenblat, S.A. Grot, C.W. Hatfield, and A.R. Badzian, IEEE Electron. Devices Lett. **12**, 37 (1991)

17.80 B.A. Fox, M.L. Hartsell, D.M. Malta, H.A. Wynands, C.T. Kao, L.S. Plano, G.L. Tessmer, R.B. Henard, J.S. Holmes, A.J. Tessmer, and D.L. Dreifus, Diamond Rel. Mater. **4**, 622 (1995)

17.81 J.S. Holmes and D.L. Dreifus, in International Electron Device Meeting, San Francisco, CA, Technical Digest, p. 423 (1994)

17.82 K. Mijata, K. Nishimura, and K. Kobashi, IEEE Trans. Electron. Devices **42**, 2010 (1995)

17.83 J.R. Zeidler, C.A. Hewett, R. Nguyen, C.R. Zeisse, and R.G. Wilson, Diamond Rel. Mater. **2**, 1341 (1993)

17.84 A.J. Tessmer, L.S. Plano, and D.L. Dreifus, IEEE Electron. Devices Lett. **14**, 66 (1993)

17.85 H. Shiomi, Y. Nishibashi, and N. Fujimori, Jpn. J. Appl. Phys. **28**, L2153 (1989)

17.86 W. Tsai, M. Delfino, D. Hodul, M. Riaziat, L.Y. Chin, G. Reynolds, and C.B. Cooper, IEEE Electron. Devices Lett. **12**, 157 (1991)

17.87 H. Shiomi, Y. Nishibayashi, N. Toda, and S. Shikata, IEEE Electron. Devices Lett. **16**, 36 (1995)

17.88 A. Vescan, P. Gluche, W. Ebert, and E. Kohn, IEEE Electron. Devices Lett. **18**, 222 (1997)

17.89 C. Vergnolle, R. Funck, and M. Laviron, Solid-State Circuits Conference, Digest of Technical Papers (1975), pp. 66–7

17.90 M. Kunze, Master's Thesis, Department of Electron Devices and Circuits, University of Ulm (1997)

17.91 H. Kawarada, M. Aoki, and M. Ito, Appl. Phys. Lett. **65**, 1563 (1994)

17.92 H. Kawarada, Surf. Sci. Rep. **26**, 205 (1996)

17.93 H. Noda, A. Hokazano, and H. Kawarada, Diamond Rel. Mater. **6**, 865 (1997)

17.94 A. Hokazono, T. Ishikura, K. Nakamura, S. Yamashita, and H. Kawarada, Diamond Rel. Mater. **6**, 339 (1997)

17.95 H. Karawada, M. Aoki, H. Sasaki, and K. Tsugawa, Diamond Rel. Mater. **3**, 961 (1994)

17.96 P. Gluche, A. Aleksov, A. Vescan, W. Ebert, M. Birk, and E. Kohn, 55th Device Research Conference, Fort Collins (CO), paper II.b-6 (1997)

17.97 H.J. Looi, M.D. Whitfield, and R.B. Jackman, Diamond 1997, Edinburgh UK, paper 15.088

17.98 P. Gluche, A. Aleksov, A. Vescan, W. Ebert, and E. Kohn, Diamond 1997, Edinburgh, UK, late news paper

Part IV

Outlook

18. CVD Diamond in the 21st Century

Paul Chalker[1] **and Simon Lande**[2]

[1] University of Liverpool, Department of Materials Science and Engineering,
Liverpool L69 3BX, UK
e-mail: pchalker@liverpool.ac.uk

[2] Magus Research, Rowlandson House,
289-293 Ballards Lane, London N12 8NP, UK

Springer Series in Materials Processing
Low-Pressure Synthetic Diamond Eds.: B. Dischler and C. Wild
© Springer-Verlag Berlin Heidelberg 1998

18.1 Introduction

Diamond has a range of outstanding properties which make it a promising material for a wide variety of applications, as summarised in Table 18.1. It is the hardest known material, it has a high thermal conductivity, and it has a high resistance to chemical attack. It is a wide bandgap electrical semiconductor and is optically transparent from the ultraviolet to the far infrared.

The practical application of these properties has been limited by the scarcity and expense of natural diamond. Over the past 30 years there have been many efforts dedicated to synthesising diamond in the laboratory. More recently, worldwide interest in diamond has increased dramatically, due to the discovery that high-quality polycrystalline diamond films or coatings can be produced by various CVD techniques with properties comparable to those of natural diamond.

Although CVD diamond is already being used for a number of applications, the commercialisation of CVD diamond is still in its infancy. The industry is now focused on developing methods to scale up CVD processes and reduce production costs to the extent that it will be economically viable to exploit diamond films for their true engineering value. The second major area for development is in processing of diamond to manufacture parts and components.

This chapter reviews the current status of CVD-diamond technology for a range of potential applications and the advances that are required to drive the technology into the commercial marketplace of the 21st century.

Table 18.1: Properties of diamond and potential applications.

Property	Tooling	High-power/ temperature electronics	Sensors	Field emission devices	Light-emitting diodes/lasers
Wide bandgap		*	*		*
Extreme mechanical hardness	*		*		
High bulk modulus/low compressibility	*		*		
High thermal conductivity	*	*	*		
Negative electron affinity				*	
Corrosion resistant	*		*		
Short-wavelength emitter					*

18.2 CVD-Diamond Tooling

Hard CVD and PVD (physical vapour deposition) coating materials are already widely used in tooling applications to promote tool lifetime by reducing wear, allowing higher cutting speeds or improving the finish of machined surfaces. The machining (e.g. cutting, milling and tapping) of abrasive materials such as composites, reinforced plastics, non-ferrous alloys and ceramics imposes extreme performance requirements on existing tool technologies. There is a strong commercial motivation to develop tools capable of dealing with these materials, so that their advantageous properties (such as wear resistance, thermal properties, weight and strength) can be exploited in advanced engineering applications. For tooling applications, diamond offers a high hardness, chemical resistance against acid and caustic solutions, and a very high thermal conductivity (five times better than that of copper).

One example is the machining of hyper-eutectic aluminium–silicon alloys in the automotive industry. The addition of silicon improves the casting properties of this class of alloys, but the formation of silicon particles requires exceptional materials to machine and drill holes within them. Existing tools, such as those coated with titanium nitride or titanium carbonitride, which have shown advantages in the cutting of steel, show relatively poor performance in these more severe cutting applications, and the need to use diamond coatings or diamond tools as a replacement is clear [18.1]. The automotive industry is currently one of the largest users of diamond tooling.

A key problem encountered in coupling CVD diamond with conventional tooling is adhesion [18.2, 3]. In the case of thick, free-standing diamond, proprietary brazes can be used to join the cutting blank to a cemented carbide tool, and these are now set to compete commercially with single-point PCD (polycrystalline diamond) tools [18.4]. Thin-film diamond coatings suffer from high thermal residual stresses arising from high CVD-diamond growth temperatures and detrimental chemical reactions between the carbon film and the substrate material. Critical loads of ~10 N for adhesive failure of CVD diamond on untreated tool substrates are typical. One approach to overcoming this issue is the adoption of interlayers: to act as a diffusion barrier between the diamond and substrate; function as a compliant buffer to ameliorate thermal expansion mismatch; or generate an adherent reaction layer during the initial diamond deposition. Silicon-containing interlayers such as Si_3N_4 and SiC have been shown to demonstrate this effect with critical loads of up to 22 N [18.5]. The exact composition and thickness of interlayers used in commercial applications is a closely guarded secret, but the availability of diamond-coated tools suggests that the problem of adhesion has been largely overcome.

There is considerable work being undertaken on the commercialisation of diamond coatings for tools in Japan, the USA and Europe, with a number of companies offering tools with CVD-diamond inserts or coatings. Diamond-coated inserts, tape automated bonding (TAB) tools for the semiconductor industry and drills fabricated from tungsten carbide cobalt cermets are now available. The future of CVD diamond in tooling products rests on the trade-off between performance and price in competition with PCD and coated tool products.

18.3 Optical Components

18.3.1 Optical Coatings

The short C–C bond length of diamond and its crystalline structure give rise to a highly transmissive material over a wide portion of the electromagnetic spectrum, making CVD diamond an attractive prospect for window or optical coating applications.

Infrared detector materials such as germanium, zinc sulphide and zinc selenide can benefit from a protective diamond coating, as they are rather soft or brittle, and susceptible to degradation in humid environments. In the infrared region, an intrinsic absorption feature of diamond is the two-phonon excitation processes between 2700 cm^{-1} and 1800 cm^{-1} (3.7–5.6 μm). If impurities such as nitrogen are present, then the symmetry-forbidden first-order phonon transition becomes allowed and is observed between 1450 cm^{-1} and 700 cm^{-1} (6.9–14.3 μm). Another significant absorption found in the IR spectrum of CVD diamond is due to the C–H stretching modes between 3200 cm^{-1} and 2800 cm^{-1} (3.1–3.6 μm). These absorption features are generally weak and therefore, coupled with its extreme hardness and high thermal conductivity, diamond is an attractive material for high-transmission optical coatings in the 8–12 μm waveband.

Three technical hurdles have to be overcome to deposit CVD diamond on to other detector components. First, the high substrate temperature required, and the reducing environment of the CVD process, severely degrades many optical materials. Second, the thermal stresses and substrate degradation compromise the coating adhesion. Finally, the surface roughness of the diamond coating must not cause too many losses due to scattering [18.6].

One proposed solution to the problem of aggressive diamond growth environments [18.7] has been to grow the diamond on a sacrificial substrate which, once removed, is bonded to a ZnSe or ZnS element using chalcogenide-based glasses. Another approach has been the development of a low-temperature diamond process by the addition of oxygen-based compounds to the CVD feedstock [18.8]. These additions are intended to assist in suppressing the formation of graphitic phases.

More recent attempts to protect the substrate material and promote adhesion have included the use of interlayers optically matched to Ge and combined with micromachined substrates to key mechanically to the coating. This approach has produced adherent diamond coatings which were deposited at 600°C [18.9]. The Ge and ZnS substrates were micromachined using ion-beam milling through a contact mask to produce a "moth eye" structure with features at a pitch of 3 μm spacing. The resulting coated structure showed near-theoretical optical transmission, and the composite optical element had an improved rain impact damage threshold velocity of over 300 m s^{-1}.

The issue of surface roughness has been addressed through the development of CVD-diamond deposition processes with control of surface roughness. This has included the production of nano- or microcrystalline diamond coatings [18.10] and also highly textured crystalline diamond with surface facets parallel to the substrate [18.6].

18.3.2 Diamond Windows

CVD diamond is already used as a window in soft X-ray detectors incorporated into spectrometers for materials analysis. The low atomic number ($Z = 6$) of carbon also means that diamond has a relatively high transmission to soft X-rays. X-ray windows are also required in the development of X-ray lithographic pellicles for the manufacture of ultra-large-scale integrated electronic devices [18.11].

The shorter wavelength of X-rays reduces the diffraction limit of device dimensions that can be projected using ultraviolet light or via electron beam writing. The objective is to produce sub-micron thick pellicles with a projection area of more than 100 mm in diameter and to maintain a surface roughness of less than 20 nm.

CVD diamond has been commercially available as a bulk window material for several years now [18.12]. Transparent "window-grade" material is produced at dimensions of 30 mm diameter and thicknesses of a few tenths of a millimeter, polished on both sides [18.13] (see also Chap. 10).

18.4 Thermal Management

CVD diamond has also been available for several years as a thermal management material for high-end applications. The producers of CVD-diamond thermal management material are now addressing the issues of how to reduce cost (by improving deposition rate, area, uniformity and quality), which will lead to a more widespread adoption of the technology.

In doing so, they are laying the ground for active diamond electronics, which will have to compete on cost and performance with more established semiconductor materials such as silicon and GaAs, as well as with silicon carbide, which is by far the most mature of the emerging wide-bandgap semiconductors. The thermal management grade or bulk material may also find its way into other applications such as surface acoustic wave devices, sensors and radiation detectors (see below).

The advent of CVD diamond has opened up the opportunity for larger heat spreader geometries than were previously economically viable with natural or high-pressure synthetic single-crystal diamond. CVD diamond is, however, a subtly different material, being polycrystalline in nature. The effects of grain boundaries and other defects tend to degrade its thermal conductivity compared with single-crystal diamond, and extensive research is being devoted to characterising the CVD material (see also Chap. 9).

Analysis carried out to determine the predominant factors that influence thermal conductivity has found that particular impurities and lattice defects have the most significant impact, suggesting that diamond films several hundreds of microns thick have concentrations of impurities and/or lattice defects that are significantly higher near the substrate interface than at the free growth surface (see Chap. 9).

Anthony and Banholzer [18.14] have elegantly demonstrated the contribution of isotopes and crystalline defects to the phonon mean free path of diamond. They have achieved this by preparing near isotopically pure samples of ^{12}C and ^{13}C CVD diamond and subsequently using the films as pure carbon sources for the preparation of single crystals using high-pressure/high-temperature processes. The influence of the ^{13}C isotope in diamond of natural isotope abundances is to add phonon scattering centres. By removing the ^{13}C, single crystals and CVD films with k approaching 32 W cm^{-1} K^{-1} have been reported [18.14].

One practical consequence of this observation is that the highest-quality thermal management material requires lapping and polishing to remove the poorer quality interfacial diamond, as well as for planarising the growth side. Because of the extreme hardness of diamond, polishing with abrasive diamond grit tends to be rather slow. Polishing techniques have been developed by various means [18.15] including ion-beam etching, laser planing or ablation, hot-metal diffusion and so on. Improvements in the optical finish of the CVD-diamond surface have been reported recently by reactive ion etching employing a planarising SiO_2 overlayer in an SF_6–O_2 plasma [18.16] and also by using an ECR air plasma [18.17]

CVD diamond can be produced with thermal conductivities in excess of 20 W cm^{-1} K^{-1} and is comparable to IIa diamond but with dimensions suitable for applications in advanced electronic packages such as multi-chip modules [18.18], where the heat output from the increasing density of devices threatens to compromise performance and lifetime. CVD diamond is likely to find its way into a widening range of packaging types as its cost begins to fall and CVD polycrystalline diamond becomes a commodity material.

18.5 Electronic Devices

The active electronic applications for diamond are more demanding in terms of materials and process technologies, however, they offer the greatest scope of applications for CVD diamond not yet realised. For example, doping, structure delineation and contact schemes have been widely demonstrated, and prototype devices are showing potential benefits in sensors, detectors, photonics and cold cathodes.

Expectations for diamond electronics have been largely based upon extrapolations from a few measurements made on either natural or high-pressure synthetic single crystals of diamond. The hole ($1600\,\mathrm{cm^2\,V^{-1}\,s^{-1}}$) and electron ($2200\,\mathrm{cm^2\,V^{-1}\,s^{-1}}$) mobilities of diamond hint at high-frequency applications. Thermal conductivities of up to $20\,\mathrm{W\,cm^{-1}\,K^{-1}}$ point towards power devices capable of dissipating the heat generated by enormous current throughput. The wide bandgap of diamond (5.45 eV) coupled with the relatively high activation energy of the boron acceptor (365 meV) suggests that diamond would have a significant role in high-temperature electronics.

The factors limiting the use of diamond in electronics include the following: The high acceptor activation energy means that carrier saturation is achievable at ~500°C, which has ramifications for obtaining material with sufficiently low sheet resistances for some applications. Then, it is a partially compensated p-type semiconductor. Furthermore, the hole mobility of diamond drops with temperature as a factor of $T^{-2.8}$, yielding a figure of ~125 $\mathrm{cm^2\,V^{-1}\,s^{-1}}$ at 500°C. Coupled with these factors, the thermal conductivity of diamond decreases with temperature.

18.5.1 Negative Electron Affinity and Field Emission

The potential for using diamond as a negative electron (NEA) affinity material was recognised by Himpsel et al. in 1979 [18.19] and makes CVD diamond a candidate material for cold-cathode applications and field emission displays.

There are a number of practical problems associated with using NEA diamond surfaces in devices. A major issue is how to replenish the electrons lost through NEA emission. Like many other wide bandgap semiconductors, diamond tends to favour one doping type, in this case p-doping. The search for an n-dopant for diamond has been tortuous, and candidates still being debated include phosphorous [18.20] and alkali metals such as lithium [18.21]. If a mechanism for injecting electrons back into NEA devices can be identified, then emission sources with high current densities and low turn-on voltages and low noise will be realisable.

Diamond is an attractive candidate for field emission sources due to its robustness. Diamond also has a relatively large secondary electron coefficient, enabling it to be used as a dynode in electron multipliers yielding gains of five orders of magnitude at picoamp primary currents [18.22]. Various emitter geometries have been investigated, including diamond particles deposited on to silicon structures [18.23], tips [18.24] and emitter diodes with integral anodes, amongst others [18.25].

Most of the commercial activities in the field are based in the USA, with the leaders being perhaps SI Diamond Technology, Inc. (SIDT), and its subsidiary, Field Emission Picture Element Technology, Inc. (FEPET). The technology was initially developed by Microelectronics and Computer Corp. (MCC) in Austin, TX. SIDT has recently demonstrated films capable of delivering high current densities at relatively low extraction fields, making them useful for a number of applications. SIDT aims to make such cathodes simple, economical to manufacture and highly reproducible.

FEPET currently focuses on using this technology to make bright and efficient picture elements for use in large outdoor displays such as electronic billboards. Other applications are also being considered.

Other industrial companies working on diamond cold-cathode technology for flat-panel display and applications include Bell Laboratories, Lucent Technologies and Samsung.

18.5.2 Surface Acoustic Wave Devices

Surface acoustic wave filters are used as radio frequency band-pass filters for mobile communications, resonators and delay lines. Conventional SAW devices consist of bulk piezoelectric crystals, such as $LiNbO_3$ and $LiTaO_3$, with phase velocities up to 4.5 km s^{-1}. As the operating frequency is equal to the SAW velocity divided by the wavelength, interdigitated transducers (IDT) made from conventional low-velocity materials require advanced lithography to produce feature sizes capable of operation in the GHz range. This requirement raises reliability and yield issues, as well as limiting device powers. By combining piezoelectric thin films such as AlN or ZnO with high-velocity substrates, high-frequency devices can be achieved with lower-scale interdigitated feature sizes.

Researchers at Sumitomo Electric [18.26] have fabricated surface acoustic wave devices using a ZnO/diamond system with SAW velocities of 10 km s^{-1}. Using 0.75 μm IDTs, filters with operating frequencies up to 3.5 GHz have been made which would have required quarter-micron lithography with conventional bulk-crystal materials. A 6 dB insertion loss was obtained for this filter geometry (see Chap.15).

The target markets for diamond SAW devices are highly competitive and very price-sensitive. Consequently, it is expected to be several years before diamond is adopted for such applications.

18.5.3 Sensors and Detectors

Diamond has a number of attractive material properties for exploitation in sensors, such as its temperature coefficient of resistance, a relatively high piezoresistivity at elevated temperatures, its superlative mechanical properties and its resistance to chemical attack.

Several examples of diamond thermistors have been reported using various methods of boron doping, patterning and electrical contact schemes [18.27–29].

A significant problem for diamond thermistors is their susceptibility to oxidation above 600°C. A range of device passivation schemes for CVD-diamond thermistors to protect the contacts and diamond film from oxidation at elevated temperatures have been investigated [18.27]. Oxidation tests made on silica-coated diamond films showed a beneficial corrosion protection effect at 650°C in laboratory air. A silica passivated thermistor operated up to marginally higher temperatures of 670°C. At this temperature, the passivation cracked, with subsequent corrosion of the contacts and oxidation of the insulating diamond, leading to etching. Passivation remains a significant issue for diamond devices, and is particularly important if diamond is to find applications in high-temperature electronics above 550°C.

CVD diamond is piezoresistive and has a longitudinal component greater than the transverse one at all degrees of doping [18.30], which exhibits linear behaviour. In a piezoresistive pressure transducer, the change in resistance on exposure to pressure is preferably determined by connecting four resistors in a Wheatstone bridge configuration on the diaphragm. Various piezoresistive pressure transducers have been demonstrated [18.31–33] (see also Chap.13).

Diamond is also an attractive detector material for electromagnetic radiation with energies greater than its bandgap (for ultraviolet, X-rays and γ-rays), since it is relatively insensitive to wide ranges of radiation damage, exhibits optical transparency and has high saturated carrier velocities. Also α, β, neutrons, pions and similar high-energy particles can generate free carriers which form the detector current. The saturated carrier velocity of diamond is 2.2 times that of silicon, and this enables diamond detectors to operate at higher speeds and count rates, which is important in burst events.

Various diamond detectors have been reported: for example metal–insulator–metal [18.34] and metal–semiconductor–metal devices [18.35]. Under UV irradiation, these detectors have a low response to visible light and a low dark current (10^{-12} A), which would make them useful for applications in atmospheric monitoring and in astronomical detectors, amongst others. A significant development in CVD-diamond detector technology has been the optimisation of the collection distance, and hence the charge collection efficiency (see also Chap.16).

18.5.4 Junction Devices

A number of research groups have fabricated diamond metal semiconductor field-effect transistors (MESFET) [18.36] and metal insulator field-effect transistors (MISFET) [18.37, 38]. These are mostly proof-of-concept devices which are still several years from commercial application. One such example is a p-type depletion-mode MISFET with an intrinsic diamond gate [18.39]. The device was operated up to 300°C with low leakage currents. The device was also capable of complete channel current pinch-off and modulation. A transconductance of 174 μS mm^{-1} was measured, which is the highest reported for this type of device to date.

Researchers at Kobe Steel's Electronic Materials Center fabricated the first digital logic circuits operable at high temperatures, by combining two diamond FETs into NAND and NOR circuits [18.40]. Synchronous TTL signals were used as circuit inputs and independent level shifting was used to optimise the circuit performance. The logic circuits were tested between 300°C and 400°C in 50°C increments. The Kobe Steel researchers have identified that the metal–oxide–diamond gate structures employed show significant opportunities for improvement. The frequency dependence of the capacitance of the gate structure is attributable to series resistance. The capacitance–voltage response indicates that interface traps are the likely source. The interface trap density was $\sim 10^{12}$ cm^{-2} eV^{-1}, and is probably due to the misregistry of the SiO$_2$/diamond interface.

These examples demonstrate the potential for integrating diamond with other semiconductor materials, which could extend the potential electronic applications of diamond.

18.6 Trends in Growth Technologies and Costs

A wide range of growth techniques have been developed for the deposition of CVD diamond (see Chaps. 2–7). Each technique has its particular merits and drawbacks, and it is beyond the scope of this chapter to discuss all these issues. However, the key factors which are considered to be pivotal to the technological exploitation diamond and its successful commercialisation are as follows:

- reduction of deposition temperatures
- enhancement of uniform deposition areas
- reduction of deposition costs.

18.6.1 Reduction of Deposition Temperatures

During the rapid growth of research in the early 1980s, the majority of published data on growth chemistry was focused on hydrocarbon–hydrogen feedstocks. Typical growth mixtures for diamond growth were ~1% methane in hydrogen, which required substrate temperatures to be in the approximate range between 700°C and 1000°C. At these temperatures, crystalline materials were generally deposited at growth rates of the order of 0.1–1.0 μm h^{-1} . Decreasing the substrate temperature significantly had two main effects: a decrease in diamond deposition rate; and increasing proportions of non-diamond phases, accompanied by a collapse of the microstructure into a microcrystalline "cauliflower" structure or even soot. This transition could be countered by decreasing the methane content or optimising the gas phase atomic hydrogen concentration. Reports of diamond growth as low as 130°C have been made [18.41], but other work from independent laboratories mutually substantiate a lower limit for diamond deposition, between 360°C and 430°C [18.42, 43].

These temperatures are not unusual for CVD processes, but posed a problem for some of the potential applications of the new CVD diamond. For example, one of the early goals was to coat tool steels with a superhard diamond coating, but the lengthy deposition times at high temperatures would lead to tempering and the tool steel would be unfit for use. Another example already cited above was the application of a durable transparent coating on to soft optical detector materials such as zinc sulphide. Once again, the high temperature and reducing environment seriously degraded the substrate material.

Subsequently, various growth mixtures have been explored as a means of reducing growth temperatures without compromising growth rate and film quality. Four main areas of approach have been widely researched: (1) oxygen additions; (2) CO- and CO_2-based mixtures; (3) additions of halogens such as fluorine and chlorine; (4) and additions of noble gases such as neon and argon. The reader is referred to a review by Muranaka et al. [18.44]. The general consensus would appear to be that growth of crystalline diamond can be achieved by CVD at temperatures of around 300°C. The challenge for researchers in this field is to achieve growth rates in excess of a few microns per hour at these low temperatures, while matching the crystalline quality of the best materials available currently (see also Chap.6).

18.6.2 Enhancement of Uniform Deposition Areas

Another major initiative is the scale-up of deposition areas with high degrees of uniformity in terms of deposition rate and material quality, and this is largely being led by equipment manufacturers and some larger CVD-diamond vendors.

In the vanguard of this development are two competing growth technologies: microwave-plasma CVD and arcject or plasma-torch CVD. In the case of microwave-plasma CVD, a significant body of research has been carried out using 2.45 GHz systems based on either the NIRIM-style tube reactor or the bell jar configuration (see Chap. 2). The frequency and power of available magnetron technology limits growth to a few centimetres in diameter (< 10 cm) using predominantly single-mode excited plasmas. More recently, there has been a drive towards 915 MHz frequencies and multiple-mode excited plasmas with power inputs of many tens of kilowatt. It is likely that deposition of high-quality diamond will be achievable over diameters larger than 30 cm using this approach. Likewise, the development of arcjet deposition technologies has benefited from nationally funded research programs. At relatively high pressures and hundreds of kilowatt power inputs, growth rates of ten to several hundred μm h^{-1} are achievable. This can also be combined with steering of the arc or rotating substrates to provide large-area deposition.

It is likely that both of these technologies will be pursued in the 21st century, but the high capital costs associated with the plant will be a significant entry barrier for potential diamond producers. It is also worth mentioning that hot-filament-assisted CVD has also been scaled by using distributed filaments and, in comparison with the microwave and arcjet technologies, is likely to be relatively inexpensive, which may make it a viable technology for the highly cost-competitive tool-coating market.

18.6.3 Reduction of Deposition Costs

The cost of CVD diamond will be a predominant factor in determining its adoption rate and commercial exploitation. Projections for the expected reduction in deposition costs are summarised in Table 18.2.

Various cost models have been applied to a range of growth processes, and it is clear that major contributions to CVD-diamond production costs come from capital depreciation and labour. Depending on the particular growth technology, electricity charges and feedstock gases may also be highly significant. In these respects, CVD-diamond production is no different from any other manufacturing process, and the goal for diamond producers will continue to be: to gain the maximum yield from their production plant.

Table 18.2: The projected decrease in deposition costs of CVD diamond.

Year	1994	1996	1998	2000	2002
Deposition cost reduction ($/carat)	120	15	4.3	2.5	< 1.0

Because of the strategic importance of CVD-diamond technology, government funded programmes for cost reduction have been instigated in Europe, the Far East and the USA. From a US DARPA programme, it has been estimated that the 1991 production cost of a 1 mm thick 1 cm^2 piece of CVD diamond was of the order of $10 000, which had fallen to ~$12 by 1995, and was expected to fall to ~$1 by the year 2000.

Similarly, a European BRITE-EuRam project (Diamond for Applications in Thermal Management, BREU-5121) set a target of 50 ecu (approximately 1 ecu ~ $1) for a 0.5 mm thick, 1 cm^2 piece of diamond with a thermal conductivity of at least 12 $W\,cm^{-1}\,K^{-1}$. Within these figures, it might be conservatively estimated that the cost of CVD diamond will be of the order of a few dollars for a 1 mm thick 1 cm^2 substrate. This raises the question of what *price* the material will be to the consumer and whether enough producers will risk and survive the entry costs to ensure the volume of CVD diamond at such competitive prices.

18.7 Market Perspectives

18.7.1 Enabling versus Enhancing Applications

So far, the market for CVD diamond has materialised more slowly than originally expected. Estimates for the size of the market vary widely depending on definitions, but the general consensus is that in 1993 the CVD-diamond market reached about $35 million.

Most of the current usage is for "lower-end" applications of CVD diamond, where defect-free substrates and sophisticated downstream processing are not required, such as heat spreaders for thermal management, and protective industrial coatings. None of the more advanced applications, such as sensors, active electronics or field emission devices (FEDs), have yet materialised in significant volumes.

A key factor in the relatively slow adoption of CVD diamond is that an application that is truly *enabled,* rather than just *enhanced,* by diamond technology has yet to be identified.

An enabling technology can be defined as one for which there is no realistic alternative, as illustrated in Table 18.3. For example, the gallium arsenide (GaAs) low-noise HEMT (high electron mobility transistor) is the enabling technology for the entire satellite television industry. There is no alternative to the use of a GaAs HEMT at the front end of the satellite TV receiver to perform the downconversion from the satellite signal at 12–14 GHz to 1 GHz.

SiC and GaN are enabling technologies for blue LEDs and, in the longer term, blue lasers, by virtue of their bandgap properties, an inherent physical characteristic. Similarly, the AlGaAs/GaAs solid-state laser diode has enabled the entire compact disc industry.

Table 18.3: Examples of enabling and enhancing technologies.

Material	Device	System enabled/enhanced
Enabling technologies		
GaAs	Low-noise HEMT	Satellite TV receiver
SiC/GaN	Blue LED	Full-colour LED displays
AlGaAs/GaAs	Laser diode	Compact disc
Enhancing technologies		
GaAs	Power FET	Mobile telephone
SiC	Power FET	PCS base stations
Diamond	FED	Flat-panel displays
Diamond	High-temperature sensor	Industrial process control

In contrast, *enhancing* technologies are those that provide an improvement (typically measured by a price–performance trade-off) over an existing technology, but are not essential to enable the system. For example, GaAs power amplifiers are used in many mobile telephones, to provide better performance (higher power added efficiency) than their silicon counterparts. In this application, they compete very fiercely with silicon, in contrast to the satellite TV market, in which silicon is not a competitor. Similarly, SiC compact and efficient power amplifiers are being developed for personal communication systems (PCS) base stations, where they can potentially deliver improved performance, but the technology is not essential to enable these systems, and will need to show significant price–performance advantages over GaAs or silicon alternatives in order to be adopted.

So far, a high-volume enabling technology for CVD-diamond electronics has not been identified. As a performance enhancer, it can potentially target several markets, but much greater financial investment is attracted by enabling technologies, rather than those that just enhance. For example, the scale of investment in gallium nitride *optoelectronics*, which is a critical enabling technology for blue LEDs and the blue laser, is at least two orders of magnitude greater than the research investment in GaN *electronics,* where it will compete with many other technologies.

18.7.2 Market Potential

CVD diamond's short~medium-term future will continue to lie in those applications where defect-free substrates are not essential, such as heat spreaders for thermal management, protective industrial coatings, simple sensors, and as electron-emitting cold cathodes.

In the longer term, the electronics market presents a major opportunity, as shown in Table 18.4. Diamond is targeting about 10% of these markets. As the first applications, diamond sensors and SAWs are likely to emerge, due to their relative simplicity. Most of the processes required to manufacture sensors have been demonstrated and engineering samples are available from a number of companies. Devices are currently being field tested in applications such as geophysical measurements and nuclear power stations. The competition from other technologies is very strong, and over the next few years diamond is likely to be used in low-volume value-added applications, until reliability is proven and costs fall.

Table 18.4: The total available market for diamond electronics in the years 1995 and 2000.

Device	1995	2000
Thermal, UV and pressure sensors	$2.5 billion	$4.1 billion
Power semiconductors	$4.7 billion	$8.2 billion
High-temperature electronics	$140 million	$435 million
Flat-panel displays	$8 billion	$16 billion

For active electronics, although diamond's theoretical figures of merit for high-temperature (and high-power/high-frequency) applications are higher than for any other semiconductor, this sector will almost certainly be dominated by silicon carbide. SiC is much more mature technologically, is receiving much greater investment and has synergies from the drive in optoelectronics. Currently, there are very few major corporate efforts on diamond electronics.

For FEDs, diamond potentially offers lower power, high brightness, high resolution and improved viewing angles. Amongst the many competing technologies are thin-film transistor active matrix liquid crystal displays (TFT AMLCD), other microtip-based FEDs, plasma displays and electroluminescent displays. The initial applications for diamond FEDs will be in small screen, high-performance, monochrome systems in the least price-sensitive markets, such as military and avionics. Larger-volume applications, such as laptop PCs, personal digital assistants (PDAs) or handheld TVs, are highly competitive and unlikely to be within the grasp of diamond for at least ten years, until the viability of scale-up is proven and costs fall dramatically.

18.8 Conclusions

In conclusion, CVD-diamond technology has made rapid progress over the past five years; and this is expected to accelerate still further, but the markets have not materialised as fast as originally expected.

However, a number of high-volume applications could potentially materialise in the 21st century. Many of the technical issues associated with applications such as tooling, optics and thermal management have been resolved at the prototype stage. An active area for future research remains in electronic applications. Much of the on-going work being done worldwide is on addressing issues such as increasing growth rates, decreasing deposition temperatures and improving film quality. CVD diamond is expected gradually to find increasing acceptance in applications for the 21st century.

References

18.1   J. Oakes, X.X. Pan, R. Haubner, and B. Lux, Surf. Coat. Technol. **47**, 600 (1991)
18.2   C.T. Kuo, T.Y. Yen, and T.H. Huang, J. Mater. Res. **5**, 2515 (1990)
18.3   E.J. Oles, A. Inspektor, and C.E. Bauer, Diamond Rel. Mater. **5**, 617 (1996)
18.4   R. Hay, Cutting Tool Eng. **45**, 52 (1993)
18.5   I. Endler, A. Leonhardt, H.-J. Scheibe, and R. Born, Diamond Rel. Mater. **5**, 299 (1996)
18.6   C. Wild, W. Müller-Sebert, T. Eckermann, and P. Koidl, in Applications of Diamond Films and Related Materials, ed. Y. Tzeng, M. Yoshikawa, M. Murakawa, and A. Feldman, Elsevier Amsterdam (1991), p. 197
18.7   W.D. Partlow, R.E. Witkowski, and J.P. McHugh, in Applications of Diamond Films and Related Materials, ed. Y. Tzeng, M. Yoshikawa, M. Murakawa, and A. Feldman, Elsevier Amsterdam (1991), p. 163
18.8   M. Ulczynski, D.K. Reinhard, M. Prystajko, and J. Asmussen, in Applications of Diamond Films and Related Materials, ed. A. Feldman, Y. Tzeng, W.A. Yarbrough, M. Yoshikawa, and M. Murakawa, NIST Special Publication **885**, US Government Printing Office, Washington (1995), p. 573
18.9   A.J. Miller, D.M. Reece, M.D. Hudson, C.J. Brierly, and J.A. Savage, Diamond Rel. Mater. **6**, 386 (1997)
18.10 W.A. Yarborough and R. Roy, Proc. Mater. Res. Soc. EA-**15**, 33 (1988)
18.11 M.F. Ravet and F. Rousseaux, Diamond Rel. Mater. **5**, 812 (1996)
18.12 R.S. Sussmann, Ind. Diamond Rev. **53** (533), 63 (1993)
18.13 R. Sussmann, J.R. Brandon, G.A. Scarsbrook, C.G. Sweeney, T.J. Valentine, A.J. Whitehead, and C.J.H. Wort, Diamond Rel. Mater. **3**, 303 (1994)
18.14 T.R. Athony and W.F. Banholzer, Diamond Rel. Mater. **1**, 717 (1992)

18.15 B. Bhushan, V.V. Subramanian, and B.K. Gupta, Diamond Films Technol. **4**, 71 (1994)

18.16 C. Vivensang, L. Ferlazzo-Manin, M.F. Ravet, G. Turban, F. Rousseaux, and A. Gicquel, Diamond Rel. Mater. **5**, 840 (1996)

18.17 H. Buchkremer-Hermanns, C. Long, and H. Weiss, Diamond Rel. Mater. **5,** 845 (1996)

18.18 K.J. Gray and P.M. Fabis, Diamond Rel. Mater. **6**, 191 (1997)

18.19 F.J. Himpsel, J.A. Knapp, J.A. VanVechten, and D.E. Eastman, Phys. Rev. B **20**, 624 (1979)

18.20 J.F. Prins, Diamond Rel. Mater. **4**, 580 (1995)

18.21 S. Prawer, C. Uzan-Saguy, G. Braunstein, and R. Kalish, Appl. Phys. Lett. **63**, 2502 (1993)

18.22 G.T. Mearini, I.L. Krainsky, and J.A. Dayton, Jr, in Applications of Diamond Films and Related Materials, ed. A. Feldman, Y. Tzeng, W.A. Yarbrough, M. Yoshikawa, and M. Murakawa, NIST Special Publication **885,** US Government Printing Office, Washington (1995), p. 13

18.23 W.P. Kang, J.L. Davidson, Q. Li, J.F. Xu, D.L. Kinser, and D.V. Kerns, in Applications of Diamond Films and Related Materials, ed. A. Feldman, Y. Tzeng, W.A. Yarbrough, M. Yoshikawa, and M. Murakawa, NIST Special Publication **885** (1995), p. 37

18.24 E.I. Givargizov, V.V. Zhirnov, A.N. Stepanova, E.V. Rakova, A.N. Kiselev, and P.S. Plekhanov, Appl. Surf. Sci. **87/88**, 24 (1995)

18.25 D. Hong and M. Aslam, in Applications of Diamond Films and Related Materials, ed. A. Feldman, Y. Tzeng, W.A. Yarbrough, M. Yoshikawa, and M. Murakawa, NIST Special Publication **885** (1995), p. 49

18.26 S. Shikata, H. Nakahata, K. Higaki, S. Fujii, A. Hachigo, H. Kitabayashi, Y. Seki, K. Tenabe, and N. Fujimori, in Applications of Diamond Films and Related Materials, ed. A. Feldman, Y. Tzeng, W.A. Yarbrough, M. Yoshikawa, and M. Murakawa, NIST Special Publication **885** (1995), p. 29

18.27 P.R. Chalker, C. Johnston, J.A.A. Crossley, J.C. Ambrose, C.F. Ayres, R.E. Harper, I.M. Buckley-Golder, and K. Kobashi, Diamond Rel. Mater. **2**, 1100 (1993)

18.28 R. Job, A.V. Denishenko, A.M. Zaitsev, M. Werner, and A.A. Melnikov, Mater. Res. Soc. Symp. Proc. **416**, 249 (1996)

18.29 L.M. Edwards, J.L. Davidson, Diamond Rel. Mater. **2**, 808 (1993)

18.30 O. Dorsch, K. Holzner, M. Werner, E. Obermeier, R.E. Harper, C. Johnston, P.R. Chalker, and I.M. Buckley-Golder, Diamond Rel. Mater. **2**, 1096 (1993)

18.31 J.L. Davidson and W.P. Kang, Mater. Res. Soc. Symp. Proc. **416**, 397 (1996)

18.32 M. Deguchi, N. Hase, M. Kitabatake, H. Kotera, S. Shima, and M. Kitagawa, Diamond Rel. Mater. **6**, 367 (1997)

18.33 P.R. Chalker and C. Johnston, Phys. Status Solidi A **154**, 455 (1996)

18.34 D.R. Kania, M.I. Landstrass, M.A. Plano, L.S. Pan, and S. Han, Diamond Rel. Mater. **2**, 1012 (1993)

18.35 S.C. Binari, M. Marchywka, D.A. Koolbeck, H.B. Dietrich, and D. Moses, Diamond Rel. Mater. **2**, 1020 (1993)

18.36 W. Tsai, M. Delfino, D. Hodul, M. Riazat, L.Y. Ching, G. Reynolds, and C.B. Cooper, IEEE Electron. Devices Lett. **12**, 157 (1991)

18.37 K. Nishimera, K. Kumagai, R. Nakamura, and K. Kobashi, J. Appl. Phys. **76**, 8142 (1994)

18.38 B.A. Fox, M.L. Hartsell, D.M. Malta, H.A. Wynands, C.-T. Kao, L.S. Plano, G.J. Tessmer, R.B. Hernand, J.S. Holmes, A.J. Tessmer, and D.L. Dreifus, Diamond Rel. Mater. **4**, 622 (1995)

18.39 L.Y.S. Pang, S.S.M. Chan, C. Johnston, P.R. Chalker, and R.B. Jackman, Diamond Rel. Mater. **6**, 333 (1997)

18.40 B.A. Fox, M.L. Hartsell, D.M. Malta, H.A. Wynands, C.-T. Kao, L.S. Plano, G.J. Tessmer, R.B. Hernard, J.S. Holmes, A.J. Tessmer, and D.L. Dreifus, Diamond Rel. Mater. **4**, 622 (1992)

18.41 M. Ihara, H. Maeno, K. Miyamoto, and H. Komiyama, Appl. Phys. Lett. **59**, 1473 (1991)

18.42 J. Stiegler, T. Lang, M. Nygård-Furguson, Y. von Kaenel, and E. Blank, Diamond Rel. Mater. **5**, 226 (1996)

18.43 Y. Liou, A. Inspektor, R. Weimer, and R. Messier, Appl. Phys. Lett. **55**, 631 (1989)

18.44 Y. Muranaka, H. Yamashita, and H. Miyadera, Diamond Rel. Mater. **3**, 313 (1994)

Subject-Index

Printing: Mercedesdruck, Berlin
Binding: Buchbinderei Lüderitz & Bauer, Berlin